Pathophysiology and Clinical Applications of Nitric Oxide

Part A

Pathophysiology and Clinical Applications of Nitric Oxide

Part A

Edited by

Gabor M. Rubanyi

Berlex Biosciences
Richmond, California
USA

harwood academic publishers

Australia • Canada • China • France • Germany • India
Japan • Luxembourg • Malaysia • The Netherlands
Russia • Singapore • Switzerland

Amsteldijk 166
1st Floor
1079 LH Amsterdam
The Netherlands

British Library Cataloguing in Publication Data

A catalogue record for this book is available from the British Library.

ISBN: 90-5702-415-2

ISSN: 1384-1270

DEDICATION

During the printing of this book we received the news of the award of the 1998 Nobel Prize in Medicine to Dr Robert Furchgott, Dr Louis Ignarro and Dr Ferid Murad for their pioneering work in the discovery of nitric oxide as a novel mediator of biological functions. In the name of all contributors to this book and numerous other scientists in the field we congratulate Bob, Lou and Ferid on this great personal accomplishment and we would like to dedicate this volume to them. The Nobel Prize also reflects a well-deserved appreciation of the entire field, some important recent progress of which is presented here.

CONTENTS

Preface xi

Contributors xiii

Part A

1 The Beginnings of Research on Nitric Oxide in Biology: A Historical
   Perspective 1
   R.F. Furchgott

I. Generation and Biological Actions of Nitric Oxide

2 Structural Variations on a Theme of Nitric Oxide Production by Three
   Isoforms of Nitric Oxide Synthase 17
   B.S. Masters, P. Martásek and *L.J. Roman*
3 Nitric Oxide Synthase Gene Regulation 39
   K.K. Wu
4 Localization and Subcellular Targeting of Nitric Oxide Synthases 51
   G. García-Cardeña and *W.C. Sessa*
5 Molecular Mechanisms of Peroxynitrite Reactivity 59
   A.J. Gow and H. Ischiropoulos
6 Nitric Oxide, Peroxynitrite and Poly (ADP-Ribose) Synthetase:
   Biochemistry and Pathophysiological Implications 69
   C. Szabó
7 Nitric Oxide and Gene Transcription 99
   J.K. Liao and *P. Libby*
8 Nitric Oxide and the Regulation of Vasoactive Genes 121
   D.V. Faller
9 Cyclooxygenase: An Important Transduction System for the Multifaceted
   Roles of Nitric Oxide 155
   D. Salvemini
10 Regulation of Nitric Oxide Synthase Expression and Activity by
   Hemodynamic Forces 171
   B.E. Sumpio, B.O. Oluwole, X. Wang and M.A. Awolesi
11 Nitric Oxide Synthase Regulation by 17β-Estradiol 195
   K. Kauser and *G.M. Rubanyi*

viii

Contents

12 Mice Deficient in eNOS and nNOS Isoforms   209
   P.L. Huang

Index   I/1

Part B

II. Nitric Oxide and Pathophysiology of Diseases

13 Nitric Oxide and Its Role in Hemostasis   229
   E.M. Battinelli, M.R. Trolliet and *J. Loscalzo*
14 Hypertension   251
   E. Nava and T.F. Lüscher
15 Pulmonary Hypertension   267
   J. Dupuis, D. Langleben and *D.J. Stewart*
16 Atherosclerosis: Role of NO   283
   J.P. Cooke and *P.S. Tsao*
17 Nitric Oxide in Myocardial and Splanchnic Ischaemia/Reperfusion   301
   A.M. Lefer and *R. Scalia*
18 Nitric Oxide and Circulatory Shock   323
   C. Thiemermann
19 Kidney Diseases   335
   P. Gross, H. Dorniok, D. Reimann and *M. Plug*
20 The Role of Nitric Oxide in Preterm Labour and Pre-eclampsia   349
   R.E. Garfield, I. Buhimschi, G.R. Saade and *K. Chwalisz*
21 Multiple Sclerosis   375
   J.E. Merrill
22 Regulation of Nitric Oxide and Inflammatory Mediators in Human Osteoarthritis-Affected Cartilage: Implication for Pharmacological Intervention   397
   A.R. Amin, M.G. Attur and *S.B. Abramson*
23 The Role of Nitric Oxide in Rheumatoid Arthritis   413
   M. Stefanovic-Racic and *C.H. Evans*
24 Gastrointestinal Function in Shock   425
   A.L. Salzman

III. Therapeutic Applications

25 Nitric Oxide Donors   455
   J.A. Hrabie and *L.K. Keefer*
26 Nitric Oxide Inhalation   471
   A.M. Atz and *D.L. Wessel*
27 Nitric Oxide Synthase Inhibitors   505
   J.F. Parkinson, J.J. Devlin and *G.B. Phillips*
28 The Pharmacology of Peroxynitrite-Dependent Neurotoxicity Blockade   523
   J.S. Althaus, G.J. Fici and *P.F. Von Voigtlander*

29 Hemoglobin: Its Role as a Nitric Oxide Scavenger 539
    S.R. Fischer and *D.L. Traber*
30 Therapeutic Implications of Recombinant Endothelial Nitric Oxide
    Synthase Gene Expression in Cerebral and Peripheral Arteries 555
    A.F.Y. Chen, T. O'Brien and *Z.S. Katusic*
31 Gene Therapy Approaches with iNOS 569
    E. Tzeng and *T.R. Billiar*

Index I/1

PREFACE

The identification that nitric oxide (NO) mediates endothelium-dependent vasodilation and that NO is synthesized by nitric oxide synthase (NOS) from L-arginine represented key discoveries of the 1980s. These observations opened the way to unprecedented research activity in the early 1990s which led to new insights into the physiology and pathophysiology of numerous biological and disease processes.

Research on NO has continued at a high pace in the past 5 years and it is by now clear that in addition to the cardiovascular system, NO is implicated in the function and disease of several other organs and systems as well (e.g., inflammatory diseases in the brain, joints, lung and gut; immunological disorders, cancer, diseases of the reproductive organs, etc.). It is also evident that we are entering into a new phase of NO research: several of the basic research observations are turning into novel therapeutic principles, and these are being pursued in clinical trials. Selective inhibitors of nitric oxide synthase isoforms are being discovered, novel NO-donors are being developed and the first reports on gene therapy approaches have been published — just to name a few examples which illustrate the point.

Athough some excellent books have been published in the past about NO research, this work is the first that summarizes the quantum leap from basic sciences to clinical applications of novel therapeutic principles emerging from this decade-long research activity.

The book is divided into two parts. Part A starts with a historical perspective written by Dr Robert Furchgott. His cornerstone observation in 1980 about the essential role of the endothelium in acetylcholine-induced vasorelaxation undoubtedly represented the start of this whole field. The rest of this volume is dedicated to the theme Generation and Biological Actions of Nitric Oxide. Leading experts give state-of-the-art overviews of various aspects of this theme, including the description of NOS function and regulation, the biological actions of NO, and the functional consequences of NOS gene knock-out in mice.

Part B has two sections. The first of these describes the known role of NO (its deficiency and/or excess) in the pathophysiology of diseases, such as those of the cardiovascular, kidney, reproductive, nervous and skeletal systems. The second section of Part B summarizes the recent progress achieved with therapeutic applications of NO. This section describes, among other topics, the discovery and therapeutic application of new NO donors, the therapeutic use of NO inhalation, selective inhibitors of NOS isoforms, and gene therapy approaches with both the constitutive and inducible forms of NOS.

Based on the scope, the excellent contributions by the leading experts, and the very efficient and professional work of the publisher, I believe this book on the pathophysiology and clinical application of nitric oxide is a one-of-a-kind work which will be of interest and benefit to both the experts working in the field and interested professionals of a wide variety of disciplines.

CONTRIBUTORS

Abramson, Steven B
Department of Rheumatology and
 Medicine
Hospital for Joint Diseases
Rm. 1401, 301 East 17th Street
New York, NY 10003
USA

Althaus, John S
CNS Diseases Research Unit 7251
Pharmacia and Upjohn Inc.
301 Henrietta Street
Kalamazoo, MI 49001
USA

Amin, Ashok R
Department of Rheumatology and
 Medicine
Hospital for Joint Diseases
Rm. 1600, 301 East 17th Street
New York, NY 10003
USA

Attur, Mukundan G
Department of Rheumatology and
 Medicine
Hospital for Joint Diseases
Rm. 1600, 301 East 17th Street
New York, NY 10003
USA

Atz, Andrew M
Cardiac Intensive Care Unit
Children's Hospital, Farley 653
300 Longwood Avenue
Boston, MA 02115
USA

Awolesi, Mark A
Department of Surgery (Vascular)
Yale University School of Medicine
333 Cedar Street, FMB 137
New Haven, CT 06510
USA

Battinelli, Elisabeth M
Department of Medicine and
 Biochemistry
Whitaker Cardiovascular Institute
Boston University School of Medicine
80 E Concord Street, W-507
Boston, MA 02118-2394
USA

Billiar, Timothy R
Department of Surgery
University of Pittsburgh
479 Scaife Hall
Pittsburgh, PA 15261
USA

Buhimschi, Irina
Department of Obstetrics and Gynecology
University of Texas Medical Branch
Medical Research Building, Rm. 11.104
301 University Boulevard
Galveston, TX 77555-1062
USA

Chen, Alex FY
Department of Anesthesiology and
    Pharmacology
Mayo Clinic
St Mary's Hospital
200 First Street SW
Rochester, MN 55905
USA

Chwalisz, Kristof
Research Laboratories
Schering AG
Berlin
Germany

Cooke, John P
Division of Cardiovascular Medicine
Stanford University School of Medicine
300 Pasteur Drive, CVRC
Stanford, CA 94305-5406
USA

Devlin, James J
Department of Immunology and
    Pharmaceuticals Discovery
Berlex Biosciences
15049 San Pablo Avenue
Richmond, CA 94804-0099
USA

Dorniok, Heike
Department of Medicine
Universitätsklinikum CG Carus
Fetscherstraße 76
D-01307 Dresden
Germany

Dupuis, Jocelyn
Terrence Donnelly Heart Centre
St Michael's Hospital
30 Bond Street, Rm. 712-B
Toronto, Ontario
Canada M5B 1W8

Evans, Christopher H
Room C-313
Presbyterian University Hospital
200 Lothrop Street
Pittsburgh, PA 15261
USA

Faller, Douglas V
Cancer Research Center, Rm. K701
Boston University Medical Campus
80 East Concord Street
Boston, MA 02118
USA

Fici, Gregory J
CNS Diseases Research Unit 7251
Pharmacia and Upjohn Inc.
301 Henrietta Street
Kalamazoo, MI 49001
USA

Fischer, Stefanie R
Investigational Intensive Care Unit
The University of Texas Medical Branch
610 Texas Avenue
Galveston, TX 77555-0833
USA

Furchgott, Robert F
Department of Pharmacology
SUNY Health Science Center at
    Brooklyn
450 Clarkson Avenue, PO Box 29
Brooklyn, NY 11203
USA

García-Cardeña, Guillermo
Department of Pharmacology
Yale University School of Medicine
Boyer Center for Molecular Medicine
295 Congress Avenue
New Haven, CT 06536-0812
USA

Garfield, Robert E
Department of Obstetrics and Gynecology
University of Texas Medical Branch
Medical Research Building, Rm. 11.104
301 University Boulevard
Galveston, TX 77555-1062
USA

Gow, Andrew J
University of Pennsylvania Medical
  Center
1 John Morgan Building
3620 Hamilton Walk
Philadelphia, PA 19104-6068
USA

Gross, Peter
Department of Medicine
Universitätsklinikum CG Carus
Fetscherstraße 76
D-01307 Dresden
Germany

Hrabie, Joseph A
SAIC Frederick
National Cancer Institute – Frederick
  Cancer Research and Development
  Center
Bldg. 469-2, PO Box B
Frederick, MD 21702
USA

Huang, Paul L
Cardiovascular Research Center
Massachusetts General Hospital – East
149 East 13th Street, 4th Floor
Charlestown, MA 02129-2060
USA

Ischiropoulos, Harry
University of Pennsylvania Medical Center
1 John Morgan Building
3620 Hamilton Walk
Philadelphia, PA 19104-6068
USA

Katusic, Zvonimir S
Departments of Anesthesiology and
  Pharmacology
Mayo Clinic
St Mary's Hospital
200 First Street SW
Rochester, MN 55905
USA

Kauser, Katalin
Department of Cardiovascular Research
Berlex Biosciences
15049 San Pablo Avenue
Richmond, CA 94804-0099
USA

Keefer, Larry K
Chemistry Section
Laboratory of Comparative Carcinogenesis
National Cancer Institute – Federick
  Cancer Research and Development
  Center
Building 538
Frederick, MD 21702-1202
USA

Langleben, David
Jewish General Hospital
3755 Cote St Catherines Rd
Montreal, Quebec
Canada H3T 1E2

Lefer, Allan M
Department of Physiology
Jefferson Medical College
Thomas Jefferson University
1020 Locust Street
Philadelphia, PA 19107-6799
USA

Liao, James K.
Atherosclerosis and Vascular Medicine Unit
Brigham and Womens Hospital and
  Harvard Medical School
221 Longwood Avenue, LMRC-316
Boston, MA 02115-5817
USA

Libby, Peter
Atherosclerosis and Vascular Medicine Unit
Brigham and Womens Hospital and
  Harvard Medical School
221 Longwood Avenue, LMRC-316
Boston, MA 02115-5817
USA

Loscalzo, Joseph
Department of Medicine and Biochemistry
Whitaker Cardiovascular Institute
Boston University School of Medicine
80 E Concord Street, W-507
Boston, MA 02118-2394
USA

Lüscher, Thomas F
Division of Cardiology
University Hospital Bern
Römistrasse 100, CH 8091
Zürich
Switzerland

Martasek, Pavel
The Robert A Welch Foundation
Department of Biochemistry
The University of Texas Health Science
  Center at San Antonio
7703 Floyd Curl Drive
San Antonio, TX 78284-7760
USA

Masters, Bettie Sue
The Robert A Welch Foundation
Department of Biochemistry
The University of Texas Health Science
  Center at San Antonio
7703 Floyd Curl Drive
San Antonio, TX 78284-7760
USA

Merrill, Jean E
Hoechs Marion Roussel Inc.
Route 202-206
PO Box 6800 Bridgewater
NJ 08807-0800
USA

Nava, Eduardo
Department of Physiology
Univeristy of Murcia School of Medicine
Murcia
Spain

O'Brien, Timothy
Divisions of Endocrinology and
  Metabolism
Mayo Clinic
St Mary's Hospital
299 First Street SW
Rochester, MN 55905
USA

Oluwole, Babalola O
Department of Surgery (Vascular)
Yale University School of Medicine
333 Cedar Street, FMB 137
New Haven, CT 06510
USA

Parkinson, John F
Department of Immunology and
  Pharmaceuticals Discovery
Berlex Biosciences
15049 San Pablo Avenue
Richmond, CA 94804-0099
USA

Phillips, Gary B
Department of Immunology and
  Pharmaceuticals Discovery
Berlex Biosciences
15049 San Pablo Avenue
Richmond, CA 94804-0099
USA

Plug, Maria
Department of Medicine
Universitätsklinikum CG Carus
Fetscherstraße 76
D-01307 Dresden
Germany

Reimann, Doreen
Department of Medicine
Universitätsklinikum CG Carus
Fetscherstraße 76
D-01307 Dresden
Germany

Roman, Linda J
The Robert A Welch Foundation
Department of Biochemistry
The University of Texas Health Science
    Center at San Antonio
7703 Floyd Curl Drive
San Antonio, TX 78284-7760
USA

Rubanyi, Gabor M
Berlex Biosciences
15049 San Pablo Avenue
Richmond, CA 94804-0099
USA

Saade, George R
Department of Obstetrics and Gynecology
University of Texas Medical Branch
Medical Research Building, Rm. 11.104
301 University Boulevard
Galveston, TX 77555-1062
USA

Salvemini Daniela
Searle Research and Development
c/o Monsanto Company
800 N. Lindbergh Boulevard
St Louis, MO 63167
USA

Salzman, Andrew L
Division of Critical Care Medicine
Children's Hospital Medical Center
3333 Burnet Avenue
Cincinatti, OH 45229-3039
USA

Scalia, Rosario
Department of Physiology
Jefferson Medical College
Thomas Jefferson University
1020 Locust Street
Philadelphia, PA 19107-6799
USA

Sessa, William C
Department of Pharmacology
Yale University School of Medicine
Boyer Center for Molecular Medicine
295 Congress Avenue
New Haven, CT 06536-0812
USA

Stefanovic-Racic, Maja
Room C-313
Presbyterian University Hospital
200 Lothrop Street
Pittsburgh, PA 15261
USA

Stewart, Duncan J
Terrence Donnelly Heart Centre
St Michael's Hospital
30 Bond Street, Rm. 712-B
Toronto, Ontario
Canada M5B 1W8

Sumpio, Bauer E
Department of Surgery (Vascular)
Yale University School of Medicine
333 Cedar Street, FMB 137
New Haven, CT 06510
USA

Szabó, Csaba
Division of Critical Care Medicine
Children's Hospital Medical Center
3333 Burnet Avenue
Cincinatti, OH 45229-3039
USA

Thiemermann, Christoph
The William Harvey Research Institute
St Bartholomew's Hospital and the
  Royal London School of Medicine
  and Dentistry
Charterhouse Square
London, EC1M 6BQ
UK

Traber, Daniel L
Investigational Intensive Care Unit
The University of Texas Medical Branch
610 Texas Avenue
Galveston, TX 77555-0833
USA

Trolliet, Maria R
Department of Medicine and Biochemistry
Whitaker Cardiovascular Institute
Boston University School of Medicine
80 E Concord Street, W-507
Boston, MA 02118-2394
USA

Tsao, Philip S
Division of Cardiovascular Medicine
Stanford University School of Medicine
300 Pasteur Drive, CVRC
Stanford, CA 94305-5406
USA

Tzeng, Edith
Department of Surgery
University of Pittsburgh
497 Scaife Hall
Pittsburgh, PA 15261
USA

Von Voigtlander, Philip F
CNS Diseases Research Unit 7251
Pharmacia and Upjohn Inc.
301 Henrietta Street
Kalamazoo, MI 49001
USA

Wang, Xiujie
Department of Surgery (Vascular)
Yale University School of Medicine
333 Cedar Street, FMB 137
New Haven, CT 06510
USA

Wessel, David L
Cardiac Intensive Care Unit
Children's Hospital, Farley 653
300 Longwood Avenue
Boston, MA 02115
USA

Wu, Kenneth K
Division of Hematology
Department of Internal Medicine
University of Texas at Houston
  Medipcal School
6431 Fannin, MSB 5.016
Houston, TX 77030
USA

1 The Beginnings of Research on Nitric Oxide in Biology: a Historical Perspective

Robert F. Furchgott

Department of Pharmacology, Box 29, State University of New York,
Health Science Center at Brooklyn, 450 Clarkson Ave., Brooklyn, NY 11203, USA
Tel: (718) 270-1355; Fax: (718) 270-2241

The discovery in 1980 of the release of endothelium-derived relaxing factor (EDRF) by acetylcholine from blood vessels *in vitro* stimulated much research. Many agents and also shear stress were found to activate release of EDRF from endothelial cells. Also, EDRF was found to stimulate guanylate cyclase in the vascular smooth muscle, generating cyclic GMP which mediated relaxation. However, the chemical identity of the labile EDRF remained unclear until 1986, when many similarities of its properties and those of NO as relaxing agents were recognized. By 1987, there was convincing evidence that EDRF is NO, and by 1988, NO was shown to be derived from the guanidino nitrogen of L-arginine by NO synthase. Inhibitors of this enzyme revealed that endothelium-derived NO contributed to regulation of vascular tone *in vivo* as well as *in vitro*. In addition, cross-stimulation between research on EDRF and research in other areas was helpful for advances in both. In each of these other areas (platelet function, cytotoxicity of activated macrophages, peripheral NANC nerve transmission, and activation of guanylate cyclase by stimulation of glutamate receptors in the CNS) the involvement of NO and NO synthase isozymes was documented by 1989.

Key words: EDRF, cyclic GMP, perfusion-bioassays, hemoglobin, nitric oxide synthase.

INTRODUCTION

In the recounting in this chapter of some of the research that led to the beginning of the still expanding field of nitric oxide (NO) in biology, it will be evident that the perspective of the author stems mainly from his involvement in research on vascular pharmacology and physiology, especially on the significance and nature of the endothelium-derived relaxing factor (EDRF). Thus, most of this chapter will be concerned with studies on EDRF that eventually led to its identification as NO. It will also include early findings (1988–1989) on the enzymology of nitric oxide synthase of endothelial cells, and on the use of inhibitors of the enzyme for evaluating the physiologic significance of NO. Finally, there will be some brief discussion concerning findings in research on other than vascular systems — e.g., platelets, peripheral and central nervous systems, and activated macrophages — that allowed new insights into the findings on vascular systems. [For a very comprehensive and masterful review of the field of nitric oxide in biology up to 1991, see Moncada *et al.* (1991).]

THE DISCOVERY OF ENDOTHELIUM-DERIVED RELAXING FACTOR

In 1980 Furchgott and Zawadzki reported that relaxation of isolated preparations of rabbit thoracic aorta in response to acetylcholine (ACh) and other muscarinic agonists depended

1

on the presence of endothelial cells. If these cells were removed from the intimal surface of the preparation either by gentle rubbing or by incubation with collagenase, relaxation in response to muscarinic agonists was lost. How an accidental finding as a result of a research assistant's error led to the discovery of endothelium-dependent relaxation has been previously described in some detail (Furchgott, 1993, 1996). It will suffice here to point out that the first kind of preparation developed and used for many years in the author's laboratory for *in vitro* studies on vasoactive agents, namely the helical (spiral) strip of the rabbit thoracic aorta (Furchgott and Bhadrakom, 1953), had never exhibited relaxation in response to ACh because the endothelial cells of the strip were unintentionally rubbed off in the standard method of preparation. Once it was recognized that the relaxation observed in rings of the aorta in response to ACh depended on the presence of endothelial cells, it was a simple matter of modifying the procedure for cutting strips so as to avoid rubbing of the intimal surface and thus allow endothelium-dependent relaxation of strips as well as rings (Furchgott and Zawadzki, 1980).

In the early organ chamber experiments, it was found that in addition to the thoracic aorta, a number of other arteries from the rabbit and various arteries from other species of laboratory animals exhibited relaxation in response to ACh only if endothelial cells were present. The relaxation was not altered by cyclooxygenase inhibitors. An attractive hypothesis to explain the obligatory role of endothelial cells was that ACh acting on a muscarinic receptor of these cells stimulates them to release a non-prostanoid substance that diffuses to and activates relaxation of the vascular smooth muscle cells. Direct evidence for this hypothesis was obtained with a so-called "sandwich mount" procedure in which a transverse strip of aorta which had been denuded of endothelial cells and therefore failed to relax in response to ACh, was remounted for recording with its endothelium-free intimal surface layered against the endothelium-containing intimal surface of a longitudinal strip of aorta. The transverse strip now relaxed in response to ACh, thus demonstrating that ACh stimulated the release of a diffusible relaxing substance (or substances) from the endothelial cells of the longitudinal strip (Furchgott and Zawadzki, 1980). The relaxing substance was later referred to as the endothelium-derived relaxing factor or EDRF (Cherry *et al.*, 1982).

EARLY STUDIES ON AGENTS INDUCING AND INHIBITING ENDOTHELIUM-DEPENDENT RELAXATION

Within a few years after the discovery that ACh-induced relaxation of isolated arteries was endothelium-dependent, a number of other endothelium-dependent relaxing agents were identified in organ chamber experiments using rings or strips of blood vessels from a variety of species, including man. Among these were the calcium ionophore A23187, ATP and ADP, substance P, bradykinin, histamine, thrombin, serotonin and vasopressin. [For references to first reports on endothelium-dependent relaxation by these and other agents, see Furchgott and Vanhoutte (1989) and Furchgott (1990)]. The early finding that the calcium ionophore A23187 was a very potent endothelium-dependent relaxing agent on all vessels tested suggested a central role for calcium in the release, or synthesis and release, of EDRF by endothelium-dependent relaxing agents (Furchgott, 1981).

It was soon recognized in this early work that, depending on the agent, endothelium-dependent relaxation might be limited to certain species, or even to specific arteries from a given species. For example, whereas ACh relaxed by this mechanism practically all systemic arteries from all mammalian species tested, histamine, another potent vasodilator, produced endothelium-dependent relaxation of rat (van de Voorde and Leusen, 1983) but not of rabbit arteries (Furchgott, 1983). Another example was bradykinin, which gave endothelium-dependent relaxation of isolated canine and human arteries but not of rabbit or cat arteries (Cherry *et al.*, 1982). As examples of vessel specificity, it was found that noradrenaline activated endothelium-dependent relaxation in isolated canine coronary arteries by acting on α_2-adrenoceptors of the endothelial cells, but failed to activate this kind of relaxation in canine systemic arteries (Cocks and Angus, 1983), and that serotonin activated endothelium-dependent relaxation in canine coronary arteries but not in canine systemic arteries (Cocks and Angus, 1983; Cohen *et al.*, 1983).

Among the conditions and agents which were found to inhibit endothelium-dependent relaxation of arteries in the early organ chamber experiments were anoxia and some agents that were recognized as inhibitors in the metabolic pathway of arachidonic acid (AA) from phospholipids through lipoxygenase (Furchgott and Zawadzki, 1980; Furchgott *et al.*, 1981). Because it had already been reported that in certain smooth muscles there was a positive relationship between an increase in cyclic guanosine monophosphate (cGMP) and relaxation (Katsuki and Murad, 1977; Böhme *et al.*, 1978; Murad *et al.*, 1978, 1979) and that guanylate cyclase (G-cyclase) was markedly stimulated by hydroperoxides of AA (Hidaka and Asano, 1977; Goldberg *et al.*, 1978), the present author speculated that (1) EDRF is either a short-lived hydroperoxide or free radical arising as an intermediate product in the oxidation of liberated AA by the lipoxygenase pathway, and (2) that EDRF stimulates the G-cyclase of the vascular smooth muscle, causing an increase in cGMP which then somehow activates relaxation (Furchgott *et al.*, 1981). As it turned out, this early speculation about the nature of EDRF was wrong, but that about the role of cGMP proved to be right.

CYCLIC GMP AS A MEDIATOR FOR EDRF AND NITROVASODILATORS

The speculation that EDRF released by ACh would stimulate an increase in cGMP was soon proven correct in a number of independent studies — e.g., Rapoport and Murad (1983) using rat aorta, Diamond and Chu (1983) and Furchgott *et al.* (1984) using rabbit aorta, Holzmann (1982) using bovine coronary artery and Ignarro *et al.*, (1984) using bovine pulmonary artery. All of these investigators showed that a rise in cGMP accompanied the ACh-induced relaxation of the blood vessel, consistent with the proposal that the increase in cGMP in the muscle has a causal role in the relaxation. Endothelium-dependent relaxation of rat aorta by A23187 and by histamine (Rapoport and Murad, 1983) and of rabbit aorta by A23187 (Furchgott *et al.*, 1984) was also accompanied by a rise in cGMP.

Early in 1986, two laboratories reported that EDRF released from endothelial cells of blood vessels can react directly with soluble G-cyclase to stimulate synthesis of cGMP (Forstermann *et al.*, 1986; Ignarro *et al.*, 1986).

It is well to stress at this point that Murad and coworkers (see Murad *et al.*, 1978 and 1979 for reviews) had previously found that nitric oxide (NO) is a very potent stimulus of G-cyclase, and had proposed that a number of agents (e.g. glyceryl trinitrate, nitroprusside, azide) which they and others had found to increase cGMP and produce relaxation of certain non-vascular smooth muscle preparations, do so by generating NO as the

4   *Robert F. Furchgott*

proximal activator of G-cyclase. Also, Ignarro and his coworkers (Gruetter *et al.*, 1979) had found that the relaxation of a vascular preparation (bovine coronary artery) by NO as well as by nitroprusside was associated with an increase in cGMP. However, despite these early findings with NO, none of the investigators who several years later showed that endothelium-dependent relaxation of arteries in organ chamber experiments was also correlated with an increase in cGMP (see above) hypothesized that EDRF might be the free radical NO. Certain key findings had yet to be made to foster this novel proposal.

FURTHER STUDIES ON EDRF, INCLUDING THOSE WITH PERFUSION-BIOASSAY PROCEDURES

Hemoglobin and Methylene Blue as Inhibitors

Two agents that became widely used as inhibitors of endothelium-dependent relaxation were hemoglobin (Hb) and methylene blue (MB). Hb had been found by Murad *et al.* (1978) to inhibit guanylate cyclase in cellular extracts, and by Bowman *et al.* (1982) to inhibit the relaxation of bovine retractor penis muscle in response either to nonadrenergic, noncholinergic (NANC) nerve stimulation or to an acid-activated extract of that tissue, referred to as the bovine retractor penis inhibitory factor (BRPIF). Gruetter *et al.* (1981) had reported that methemoglobin (metHb) interferes with NO-induced increases in vascular cGMP. In the present author's laboratory, work with Hb began in 1983 when William Martin, who had studied the properties of BRPIF while a graduate student (Gillespie *et al.*, 1981; Bowman *et al.*, 1981), arrived as a postdoctoral fellow. It was soon found that Hb was an extremely potent and fast acting inhibitor of endothelium-dependent relaxation and of the associated rise in cGMP induced by ACh, A23187 and other agents on rabbit and rat aorta (Furchgott *et al.*, 1984; Martin *et al.*, 1985a, 1986). Myoglobin was as effective as Hb, but metHb only inhibited very weakly (Martin *et al.*, 1985b). The degree of potentiation by Hb of contractions of arterial rings by vasoconstrictors such as phenyle-phrine also provided a measure of the relaxing activity of basally released EDRF (Martin *et al.*, 1986).

It was originally proposed that Hb inhibits relaxation and the rise in cGMP by binding EDRF as it diffuses from the endothelial cells, thus preventing it from reaching and activating the G-cyclase in the muscle cells (Martin *et al.*, 1985a). However, after the identification of EDRF as NO, it became apparent that the scavenging of EDRF by Hb under the usual aerobic conditions was the result of the very rapid reaction of HbO_2 with NO to form NO_3^- and metHb, as first reported by Doyle and Hoekstra (1981).

MB had first been shown by Gruetter *et al.* (1981) to inhibit stimulation of G-cyclase by nitrovasodilators in cell-free systems and to inhibit both the relaxation and increase in cGMP elicited by these agents in vascular smooth muscle. Subsequently, MB was found to inhibit both the endothelium-dependent relaxation and accompanying increase in cGMP produced by ACh in bovine arteries (Holzman, 1982; Ignarro *et al.*, 1984) and by ACh and A23187 in rabbit aorta (Furchgott *et al.*, 1984; Martin *et al.*, 1985a). Holzman (1982) and Ignarro *et al.* (1984) concluded that the inhibition of relaxation was the result of the inhibition of G-cyclase by the MB. On the other hand, the present author and his coworkers, although agreeing that such enzyme inhibition accounted for the sustained inhibition by

MB developed during prolonged exposure of the rabbit aorta, proposed that the extremely fast inhibition of endothelium-dependent relaxation immediately after addition of MB to an organ chamber resulted from some more direct inactivation of EDRF by the MB (Furchgott *et al.*, 1984; Martin *et al.*, 1985a). It now appears that this more direct inactivation is due to superoxide anions, generated by reactions of the MB in the tissue (also see Wolin *et al.*, 1990).

Results with Perfusion-Bioassay Procedures

For more direct studies on the properties of EDRF than were possible in organ chamber experiments, perfusion-bioassay procedures were developed. In such procedures, endothelial cells are upstream in the perfusion (or superfusion) cascade and an endothelium-denuded arterial preparation (usually a ring or strip) whose contractile state is monitored, is downstream to bioassay EDRF in the perfusion fluid. In some procedures, the endothelial cells upstream were native cells on the luminal surface of a perfused artery (e.g. Forstermann *et al.*, 1984; Griffith *et al.*, 1984; Rubanyi *et al.*, 1985); in other procedures, they were cultured cells (usually from bovine or porcine aortas) on the surface of microcarrier beads contained in a perfused column (e.g. Cocks *et al.*, 1985; Gryglewski *et al.*, 1986). Perfusion-bioassay procedures allowed one to determine clearly whether an inhibitor of endothelium-dependent relaxation interfered with the synthesis/release of EDRF or inactivated the EDRF after its release. In addition, by altering the transit time of the perfusate between the endothelial cells and the bioassay preparation, it was possible to study the rate of decay of EDRF. Surprisingly, estimated half-life values for EDRF ranged from as low a 4–6 s to as high as 50 s depending on the biological preparations and the experimental conditions. The likely explanation of these discrepant values came with the important finding that superoxide dismutase (SOD) markedly stabilizes EDRF released from either endothelial cells on arteries (Rubanyi and Vanhoutte, 1986) or cultured endothelial cells on microcarrier beads (Gryglewski *et al.*, 1986), clearly showing that the superoxide anion (O_2^-) very rapidly inactivates EDRF. Thus, the marked differences in half-life of EDRF reported from different laboratories could largely be attributed to differences in concentrations of O_2^- in the perfusion fluid leaving the endothelial cells under the different experimental conditions used. Subsequently, Moncada *et al.* (1986) were able to show with their perfusion-bioassay procedure that inhibition of endothelium-dependent relaxation by certain reducing agents (e.g. hydroquinone, phenidone and pyrogallol) originally demonstrated in organ chamber experiments, was principally the result of inactivation of EDRF by O_2^- formed in a single-electron oxidation of the reducing agent by O_2.

Development of perfusion-bioassay procedures also led to the important finding that increases in shear stress on endothelial cells of the perfused artery with increases in rate of flow of the perfusion fluid resulted in increases in the rate of release of EDRF (e.g. Holtz *et al.*, 1984; Rubanyi *et al.*, 1986). Increase in flow through arteries both *in situ* (Holtz *et al.*, 1984; Kaiser and Sparks, 1986; Pohl *et al.*, 1986) and *in vitro* produced vasodilation that could not be blocked by inhibitors of cyclooxygenase but was abolished by removal of the endothelium. These findings concerning increased EDRF release with increased shear stress provided an explanation of the well recognized dilation of large arteries *in vivo* when flow through them is augmented as a result of decreased arteriolar resistance downstream.

THE IDENTIFICATION OF EDRF AS NITRIC OXIDE

The Original Proposals

At an international symposium on mechanisms of vasodilatation in July, 1986, two investigators independently proposed in their presentations that EDRF is NO (Ignarro *et al.*, 1988; Furchgott, 1988). Ignarro noted the similar characteristics of EDRF and NO in stimulating G-cyclase, and also presented results of a study on the ability of Hb, metHb and MB to inhibit relaxations of isolated bovine pulmonary artery produced by EDRF (released by ACh) and by known concentrations of NO (solutions of the gas) in organ chamber experiments (Ignarro *et al.*, 1988). He found such a remarkable similarity between the results with ACh-released EDRF and with added NO to lead him to suggest that EDRF is either NO or some closely related unstable radical species.

At the 1986 symposium, the present author reported results from experiments comparing the characteristics of relaxation in rings of rabbit aorta produced by ACh-released EDRF and by added NO, and found the characteristics so similar that he too proposed that EDRF is NO (Furchgott, 1988). (In his experiments reported at the symposium, he had used acidified sodium nitrite solutions as a source of NO, but soon after obtained similar results with solutions of NO gas.) In addition to comparing the effects of Hb on relaxation by ACh and by NO, he also studied the effects of superoxide generators and SOD on the transient relaxations produced by NO in organ chamber experiments, and found their inhibitory and stabilizing effects, respectively, to resemble those that had been observed in the case of EDRF in perfusion-bioassay experiments on canine coronary arteries (Rubanyi and Vanhoutte, 1986) and on cultured porcine aortic endothelial cells (Gryglewski *et al.*, 1986; Moncada *et al.*, 1986). He proposed that the very fast reaction between the two free radicals, NO and O_2^-, generated biologically inert NO_3^- (Furchgott, 1988). Later, it was established that the immediate product is short-lived, reactive peroxynitrite (Beckman *et al.*, 1990).

The Convincing Evidence

Following the 1986 symposium referred to above, several laboratories utilized perfusion-bioassay procedures to compare more accurately the biological and chemical characteristics of EDRF and NO, Ignarro *et al.* (1987) using bovine pulmonary vessels, Palmer *et al.* (1987) using cultured porcine aortic endothelial cells, and Khan and Furchgott (1987) using rabbit aorta as the source of EDRF, all found EDRF released upstream and NO infused upstream to have similar rates of decay, similar susceptibility to inhibitors like Hb and superoxide generators, similar stabilization by SOD, etc. In addition, Ignarro *et al.* (1987) showed spectroscopically that the product of the reaction of released EDRF and Hb was similar to that of the reaction between NO and Hb, and Palmer *et al.* (1987) showed that the amount of NO released by bradykinin (as determined by a chemiluminescence assay) could quantitatively account for the relaxation of the bioassay strip by the EDRF released by the peptide. Subsequently, Kelm *et al.* (1988), using a difference spectrophotometric method for measuring NO based on the very rapid reaction between NO and oxyHb to yield metHb, also concluded that NO could account quantitatively for the relaxing activity of EDRF released by bradykinin or ATP from cultured endothelial cells.

Despite the evidence for the identity of EDRF and NO, some results obtained in various laboratories in the late 1980s did not appear to be consistent with this conclusion. [For a review of these apparent or real discrepancies, see Furchgott *et al.* (1990a,b).] Nevertheless, as of the time of this writing, it has become generally accepted that EDRF is either NO or some labile adduct that readily releases NO. This does not preclude a contribution to endothelium-dependent relaxation in some vessels from the release of a non-nitric oxide endothelium-derived hyperpolarizing factor (EDHF) along with EDRF. [For a review, see Garland *et al.*, (1995).]

THE SYNTHESIS OF NO IN ENDOTHELIAL CELLS

L-Arginine as the Substrate for NO Synthase

Within two years of the first proposals that EDRF is NO, the laboratory of Salvador Moncada reported that L-arginine is the substrate for the enzymatic synthesis of NO in cultured endothelial cells (Palmer *et al.*, 1988a). D-arginine is not a substrate. With the use of ^{15}N-labelled L-arginine and mass spectrometry, it was demonstrated that the endothelial enzyme synthesized NO from the terminal guanidino nitrogen atom(s) of L-arginine. Moreover, it was found that the synthesis of NO by endothelial cells in culture could be inhibited in an enantiomerically specific manner by N^G-monomethyl-L-arginine (L-NMMA) (Palmer *et al.*, 1988b), an arginine analogue that Hibbs *et al.* (1987a) had found to be an inhibitor of the generation of inorganic nitrite and citrulline from L-arginine in murine macrophages (see below). Independently, Schmidt *et al.* (1988) also reported results showing that L-arginine is the physiological precursor of NO in endothelial cells.

Using endothelial cell homogenates, Palmer and Moncada (1989) found that the formation of citrulline from L-arginine by NO synthase was NADPH-dependent and was inhibited by L-NMMA. These early results were consistent with NO and citrulline being coproducts of the same enzymatic reaction. In addition, it was soon found with endothelial cell cytosol that the NO synthase is a Ca^{2+}-dependent enzyme (Moncada and Palmer 1990; Mayer *et al.*, 1990) and that the Ca^{2+}-dependent stimulation is mediated by calmodulin (Busse and Mulsch, 1990).

Inhibitors of NO Synthase

L-NMMA, the first arginine analogue used as a competitive inhibitor of NO synthase, proved to be an extremely useful tool in studies on the biological significance of the L-arginine:NO pathway in the cardiovascular system. This agent attenuated endothelium-dependent relaxation of arterial preparations *in vitro* in response to agents like ACh, substance P and A23187 (Rees *et al.*, 1989a). It also inhibited the release of NO from cultured endothelial cells and from endothelial cells of segments of rabbit aorta in perfusion-bioassay experiments (Palmer *et al.*, 1988b; Rees *et al.*, 1989a). These effects were reversed in large part by the addition of high concentrations of L-arginine, but not D-arginine. In anesthetized rabbits, intravenous L-NMMA, but not D-NMMA, produced increases in blood pressure that could be reversed by intravenous L-arginine (Rees *et al.*, 1989b). Similar results with L-NMMA were obtained with anesthetized guinea pigs (Aisaka *et al.*,

1989). In man, infusion of L-NMMA into the brachial artery produced local arteriolar vasoconstriction, which could be reversed by infusion of L-arginine (Vallance *et al.*, 1989). Results such as these with L-NMMA, and with other inhibitors of NO synthase led to the conclusion that there is a physiologic, nitric oxide-dependent vasodilator tone that has a major role in regulating blood flow and blood pressure.

SOME EXAMPLES OF CROSS-STIMULATION BETWEEN RESEARCH ON EDRF AND RESEARCH IN OTHER AREAS

Platelets

Mellion *et al.* (1981) first reported that NO and nitroprusside inhibited the aggregation of platelets, and that this action was closely associated with a stimulation of G-cyclase. Prior to the proposal in 1986 that EDRF is NO (see above), Azuma *et al.* (1986) found that EDRF inhibited platelet aggregation. Shortly after the proposal was made, Radomski *et al.* (1987a, 1987b) showed that the properties of EDRF and NO as inhibitors of platelet aggregation were similar, and differed from those of prostacyclin; and that NO also inhibited platelet adhesion to vascular endothelium (Radomski *et al.*, 1987c). Subsequently, it was shown that NO synthase is present in human platelets, and is involved in the regulation of aggregation (Radomski *et al.*, 1990).

Activated Macrophages

In Tannenbaum's laboratory, endogenous NO_3^- synthesis had been observed in laboratory animals as well as humans (Green *et al.*, 1981; Wagner *et al.*, 1983). In addition, the blood levels and urinary excretion of NO_3^- were found to increase in cases of infection in humans, and the urinary output of NO_3^- in rats was markedly elevated following treatment of the animals with bacterial lipopolysaccharide (LPS). These and other findings suggested that immunostimulation was responsible for the increased NO_3^- synthesis, and led to experiments by Stuehr and Marletta (1985) in which they showed that activated peritoneal macrophages from LPS-treated mice exhibited an increased production of both NO_2^- and NO_3^- *in vitro*. These workers went on to show that murine peritoneal macrophages in culture and cells in a macrophage cell line could be induced *in vitro* to synthesize NO_2^- and NO_3^- by treatment with LPS and/or interferon-γ (Stuehr and Marletta, 1985, 1987). Using a macrophage cell line, they next were able to identify L-arginine as the substrate for synthesis of NO_2^-, NO_3^- and citrulline, and to show with the use of ^{15}N-labelled L-arginine that both NO_2^- and NO_3^- were derived from the terminal guanidino nitrogen atom(s) of that amino acid (Iyengar *et al.*, 1987).

Independently, and slightly prior to the publication of Iyengar *et al.* (1987), Hibbs and coworkers published two important papers in which they showed that the cytotoxicity of activated macrophages against tumor cells in co-culture was dependent on the presence of L-arginine in the medium (Hibbs *et al.*, 1987a), and that L-arginine was also the substrate from which citrulline and NO_2^- were synthesized by the activated macrophages (Hibbs *et al.*, 1987b). Moreover, these investigators found that the arginine derivative L-NMMA inhibited both the synthesis of these products and the expression of cytotoxicity (Hibbs *et al.*, 1987a,b).

These findings on activated macrophages attracted the interest of some investigators working on EDRF, since EDRF had already been tentatively identified as NO, and it seemed possible that NO could be the intermediate synthesized from L-arginine which gave rise to NO_2^- and NO_3^- in the macrophage. As noted in an earlier section of this chapter, Moncado's laboratory, taking advantage of the perfusion-bioassay procedure with cultured endothelial cells, was able to demonstrate with [^{15}N]L-arginine that the NO (EDRF) was synthesized from the guanidino nitrogen of the amino acid (Palmer *et al.*, 1988a), and that L-NMMA was a competitive inhibitor of the release of NO from the cultured cells (Palmer *et al.*, 1988b). Shortly thereafter, three groups working with activated macrophages demonstrated that NO was the precursor of the NO_2^- and NO_3^- formed from L-arginine by these cells (Marletta *et al.*, 1988; Hibbs *et al.*, 1988; Stuehr *et al.*, 1989).

It was soon evident that the induction of an NO synthase in cells exposed to LPS and/ or cytokines was not limited to macrophages. Between 1989 and 1990, induction was demonstrated in a variety of cells in culture as well as cells and tissues *in vivo*, including vascular smooth muscle. Cell-free preparations of induced NO synthase, unlike the constitutive synthases from endothelial cells and brain (see below), were not dependent on added Ca^{2+} for activity. [For an early review, see Moncado *et al.* (1991).]

NANC Nerves

As noted earlier in this chapter, Hb had been found to be a potent inhibitor of the relaxation of the bovine retractor penis muscle in response to NANC nerve stimulation and in response to an acid-activatable inhibitory factor (BRPIF) in extracts prepared from that muscle (Bowman *et al.* 1982). In addition, shortly after the first reports that cGMP mediates relaxation of vascular smooth muscle in response to EDRF, Bowman and Drummond (1984) found that this cyclic nucleotide also mediates relaxation of the retractor penis muscle in response to NANC nerve stimulation. When William Martin, who had carried out his thesis work on the BRPIF (Gillespie *et al.*, 1981; Bowman *et al.*, 1981), worked with the present author on EDRF, the latter learned more about the properties of this smooth muscle inhibitory material. At a later date, he (the present author) was struck by the markedly similar properties of acidified $NaNO_2$ and the acidified BRPIF in producing relaxation of smooth muscle preparations in organ bath experiments (dependence on pH, transiency of responses, inhibition by Hb, etc.), and suggested that the acid-activatable component of BRPIF is inorganic NO_2^-, which is protonated under acid conditions (pK_a, 3.2) to give HNO_2 which undergoes reversible dismutation to liberate some NO (Furchgott, 1988). The results with acidified nitrite solutions, which he presented at the 1986 symposium on mechanisms of vasodilatation, constituted a crucial part of the evidence for proposing that EDRF is NO. In addition, it led him to propose that the "relaxing factor released on stimulation of noncholinergic, nonadrenergic nerves . . . . may be the same substance as EDRF" (Furchgott, 1988).

Martin *et al.* (1988) subsequently reported evidence that the inhibitory factor extracted from the bovine retractor penis is nitrite, which yields a "stabilized" nitric oxide derivative on acid-activation, and it is this derivative that produces the relaxation. Also, using the rat anococcygeus muscle for *in vitro* studies on NANC nerve stimulation, Gillespie *et al.* (1989) and Li and Rand (1989) found that the relaxation in response to stimulation was blocked by the NOS inhibitor L-NMMA and that L-arginine could reverse this effect of the L-NMMA. These were among the first of many reports which demonstrated the ability of L-NMMA and other NOS inhibitors to interfere with NANC nerve stimulation of smooth muscle relaxation, and which supported the view that the transmitter from NANC nerves is NO or some adduct of NO.

Central Nervous System

Ferendelli *et al.* (1974) had shown that glutamate caused an accumulation of cGMP in rat cerebellar slices, and Miki *et al.* (1977) had shown that NO stimulated soluble G-cyclase in homogenates of mouse brain. Deguchi (1977) found that there was an endogenous activating factor for G-cyclase in the synaptosomal soluble fraction of rat brain, and Deguchi and Yoshioka (1982) subsequently identified this factor as L-arginine. Being aware of these findings and having made the discovery that L-arginine is the substrate for the synthesis of NO (EDRF) in endothelial cells (Palmer *et al.*, 1988a), Moncado and his coworkers began investigating the existence of an L-arginine:NO pathway in brain. Using rat synaptosomal cytosol, they clearly showed that this pathway did exist and that the NO synthase in brain, like that in endothelial cells, was dependent on Ca^{2+} and on NADPH (Knowles *et al.*, 1989).

While the investigation by Knowles *et al.* (1989) was in progress, the very interesting paper of Garthwaite *et al.* (1988) was published, in which it was shown that the stimulation of rat cerebellar cells in suspension with the glutamate receptor agonist N-methyl-D-aspartate (NMDA) not only induced an elevation in cGMP levels but also resulted in release of a very labile EDRF-like relaxant of vascular smooth muscle. The stimulation of an increase in cGMP by NMDA was enhanced by L-arginine and inhibited by L-NMMA in a competitive manner (Garthwaite *et al.*, 1989; Bredt and Snyder, 1989), clearly indicating that the increase in cGMP was mediated by NO generated by an NO synthase. By 1990, Bredt and Snyder, using extracts from rat cerebellum, achieved the first purification of an isozyme of NO synthase.

ACKNOWLEDGEMENT

Research of the author referred to in this chapter was supported by U.S.P.H.S. grant HL21860.

REFERENCES

Aisaka, K., Gross, S.S., Griffith, O.W. and Levi, R. (1989) N^G-methyl-arginine, an inhibitor of endothelium-derived nitric oxide synthesis. Is a potent pressor agent in the guinea pig: Does nitric oxide regulate blood pressure *in vivo*? *Biochem. Biophys. Res. Commun.*, **160**, 881–886.

Azuma, H., Ishikawa, M. and Sekizaki, S. (1986) Endothelium-dependent inhibition of platelet aggregation. *Br. J. Pharmacol.*, **88**, 411–415.

Beckman, J.S., Beckman, T.W., Chen, J., Marshall, P.A. and Freeman, B.A. (1990) Apparent hydroxyl radical production by peroxynitrite: implications for endothelial injury from nitric oxide and superoxide. *Proc. Natl. Acad. Sci. U.S.A.*, **87**, 1620–1624.

Bowman, A. and Drummond, A.H. (1984) Cyclic GMP mediates neurogenic relaxation in the bovine retractor penis muscle. *Br. J. Pharmacol.*, **81**, 665–674.

Bowman, A., Gillespie, J.S. and Martin, W. (1981) Actions on the cardiovascular system of an inhibitory material extracted from the bovine retractor penis. *Br. J. Pharmacol.*, **72**, 365–372.

Bowman, A., Gillespie, J.S. and Pollock, D. (1982) Oxyhaemoglobin blocks nonadrenergic inhibition in the bovine retractor penis muscle. *Eur. J. Pharmacol.*, **85**, 221–224.

Bredt, D.S. and Snyder, S.H. (1989) Nitric oxide mediates glutamate-linked enhancement of cGMP levels in the cerebellum. *Proc. Natl. Acad. Sci. U.S.A.*, **86**, 9030–9033.

Bredt, D.S. and Snyder, S.H. (1990) Isolation of nitric oxide synthetase, a calmodulin-requiring enzyme. *Proc. Natl. Acad. Sci. U.S.A.*, **87**, 682–685.

Busse, R. and Mulsch, A. (1990) Calcium-dependent nitric oxide synthesis in endothelial cytosol is mediated by calmodulin. *FEBS Letts.*, **265**, 133–136.

Cherry, P.D., Furchgott, R.F., Zawadzki, J.V. and Jothianandan, D. (1982) The role of endothelial cells in the relaxation of isolated arteries by bradykinin. *Proc. Nat. Acad. Sci.*, **79**, 2106–2110.

Cocks, T.M. and Angus, J.A. (1983) Endothelium-dependent relaxation of coronary arteries by noradrenaline and serotonin. *Nature*, **305**, 627–630.

Cocks, T.M., Angus, J.A., Campbell, J.H. and Campbell, G.R. (1985) Release and properties of endothelium-derived relaxing factor (EDRF) from endothelial cells in culture. *J. Cell Physiol.*, **123**, 310–320.

Cohen, R.A., Shepherd, J.T. and Vanhoutte, P.M. (1983) 5-Hydroxytryptamine can mediate endothelium-dependent relaxation of coronary arteries. *Am. J. Physiol.*, **245**, H1077–1080.

Deguchi, T. (1977) Endogenous activating factor for guanylate cyclase in synaptosomal-soluble fraction of rat brain. *J. Biol. Chem.*, **252**, 7617–7619.

Deguchi, T. and Yoshioka, M. (1982) L-arginine identified as an endogenous activator for soluble guanylate cyclase from neuroblastoma cells. *J. Biol. Chem.*, **257**, 10147–10152.

Diamond, J. and Chu, E.B. (1983) Possible role for cyclic GMP in endothelium-dependent relaxation of rabbit aorta by acetylcholine. Comparison with nitroglycerin. *Res. Comm. Chem. Pathol. Pharmacol.*, **41**, 369–381.

Doyle, M.P. and Hoekstra, J.W. (1981) Oxidation of nitrogen oxides by bound dioxygen in hemoproteins. *J. Inorg. Biochem.*, **14**, 351–358.

Ferrendelli, J.A., Chang, M.M. and Kinscherf, D.A. (1974) Elevation of cyclic GMP levels in central nervous system by excitatory and inhibitory amino acids. *J. Neurochem.*, **22**, 535–540.

Forstermann, U., Trogisch, G. and Busse, R. (1984) Species-dependent differences in the value of endothelium-derived vascular relaxing factor. *Eur. J. Pharmacol.*, **106**, 639–643.

Forstermann, U., Mulsch, A., Bohme, E. and Busse, R. (1986) Stimulation of soluble guanylate cyclase by an acetylcholine-induced endothelium-derived factor from rabbit and canine arteries. *Circ. Res.*, **58**, 531–538.

Furchgott, R.F. (1981) The requirement for endothelial cells in the relaxation of arteries by acetylcholine and some other vasodilators. *Trends Pharmacol. Sci.*, **2**, 173–176.

Furchgott, R.F. (1983) Role of endothelium in the responses of vascular smooth muscle. *Circ. Res.*, **53**, 557–573.

Furchgott, R.F. (1988) Studies on relaxation of rabbit aorta by sodium nitrite: the basis for the proposal that the acid-activatable inhibitory factor from retractor penis is inorganic nitrite and the endothelium-derived relaxing factor is nitric oxide. In: *Vasodilatation: Vascular Smooth Muscle, Peptides and Endothelium*, edited by P.M. Vanhoutte, pp. 401–414. New York: Raven Press.

Furchgott, R.F. (1990) The 1989 Ulf von Euler Lecture: Studies on endothelium-dependent vasodilation and the endothelium-derived relaxing factor. *Acta Physiol. Scand.*, **139**, 257–270.

Furchgott, R.F. (1993) The discovery of endothelium-dependent relaxation. *Circulation*, **87**(Suppl. V), V3–V8.

Furchgott, R.F. (1996) The discovery of endothelium-derived relaxing factor and its importance in the identification of nitric oxide. *J.A.M.A.*, **276**, 1186–1188.

Furchgott, R.F. and Bhadrakom, S. (1953) Reactions of strips of rabbit aorta to epinephrine, isopropylarterenol, sodium nitrite and other drugs. *J. Pharmacol. Exp. Therap.*, **108**, 129–143.

Furchgott, R.F. and Vanhoutte, P.M. (1989) Endothelium-derived relaxing and contracting factors. *FASEB J.*, **3**, 2007–2018.

Furchgott, R.F. and Zawadzki, J.V. (1980) The obligatory role of endothelial cells in the relaxation of arterial smooth muscle by acetylcholine. *Nature*, **288**, 373–376.

Furchgott, R.F., Zawadzki, J.V. and Cherry, P.D. (1981) Role of endothelium in the vasodilator response to acetylcholine. In: *Vasodilatation*, edited by P. Vanhoutte and I. Leusen, pp. 49–66. New York: Raven Press.

Furchgott, R.F., Cherry, P.D., Zawadzki, J.V. and Jothianandan, D. (1984) Endothelial cells as mediators of vasodilation of arteries. *J. Cardiovasc. Pharmacol.*, **6** (Suppl. 2), S336–S344.

Furchgott, R.F., Jothianandan, D. and Freay, A.D. (1990b) Endothelium-derived relaxing factor: some old and new findings. In *Nitric Oxide from L-Arginine: A Bioregulatory System*, edited by S. Moncada and E.A. Higgs, pp. 5–17. Amsterdam: Elsevier.

Furchgott, R.F., Khan, M.T. and Jothianandan, D. (1990a) Comparison of properties of nitric oxide and endothelium-derived relaxing factor: Some cautionary findings. In *Endothelium-derived Relaxing Factors*, edited by G.M. Rubanyi and P.M. Vanhoutte, pp. 8–21. Basel: Karger.

Garland, C.J., Plane, F., Kemp, B.K. and Cocks, T.M. (1995) Endothelium-dependent hyperpolarization: a role in the control of vascular tone. *Trends Pharmacol. Sci.*, **16**, 23–30.

Garthwaite, J., Charles, S.L. and Chess-Williams, R. (1988) Endothelium-derived relaxing factor release on activation of NMDA receptors suggests role as intercellular messenger in the brain. *Nature*, **336**, 385–388.

Garthwaite, J., Garthwaite, G., Palmer, R.M.J. and Moncada, S. (1989) NMDA receptor activation induces nitric oxide synthesis from arginine in rat brain slices. *Eur. J. Pharmacol.*, **172**, 413–416.

Gillespie, J.S., Hunter, J.C. and Martin, W. (1981) Some physical and chemical properties of the smooth muscle inhibitory factor in extracts of the bovine retractor penis muscle. *J. Physiol.*, **315**, 111–125.

Gillespie, J.S., Liu, X. and Martin, W. (1989) The effect of L-arginine and N^G-monomethly-L-arginine on the response of the rat anococcygeus to NANC nerve stimulation. *Br. J. Pharmacol.*, **98**, 1080–1082.

Green, L.C., Wagner, D.A., Ruiz de Luzuriaga, K., Istan, N., Young, V.R. and Tannenbaum, S.R. (1981) Nitrate biosynthesis in man. *Proc. Natl. Acad. Sci. U.S.A.*, **78**, 7764–7768.

Griffith, T.M., Edwards, D.H., Newby, A.C., Lewis, M.J., Newby, A.C. and Henderson, A.H. (1984) The nature of the endothelium-derived vascular relaxant factor. *Nature*, **305**, 645–647.

Gruetter, C.A., Barry, B.K., McNamara, D.B., Gruetter, D.Y., Kadowitz, P.J. and Ignarro, L.J. (1979) Relaxation of bovine coronary artery and activation of coronary arterial guanylate cyclase by nitric oxide, nitroprusside and carcinogenic nitrosoamine. *J. Cyclic Nucleotide Res.*, **5**, 211–214.

Gruetter, C.A., Gruetter, D.Y., Lyon, J.E., Kadowitz, P.J. and Ignarro, E.F. (1981) Relationship between cyclic guanosine 3':5'-monophosphate formation and relaxation of coronary arterial smooth muscle by glyceryl trinitrate, nitroprusside, nitrite and nitric oxide: Effects of methylene blue and methemoglobin. *J. Pharmacol. Exp. Ther.*, **219**, 181–186.

Gryglewski, P.J., Palmer, R.M. and Moncada, S.A. (1986) Superoxide anion is involved in the breakdown of endothelium-derived relaxing factor. *Nature*, **320**, 454–456.

Hibbs, J.B. Jr., Taintor, R.R. and Vavrin, Z. (1987a) Macrophage cytotoxicity: role for L-arginine deiminase and imino nitrogen oxidation to nitrite. *Science*, **235**, 473–476.

Hibbs, J.B. Jr., Vavrin, Z. and Taintor, R.R. (1987b) L-arginine is required for expression of the activated macrophage effector mechanism causing selective metabolic inhibition in target cells. *J. Immunol.*, **138**, 550–565.

Hibbs, J.B. Jr., Taintor, R.R., Vavrin, Z. and Rachlin, E.M. (1988) Nitric oxide: a cytotoxic activated macrophage effector molecule. *Biochem. Biophys. Res. Commun.*, **157**, 87–94.

Holtz, J., Förstermann, U., Pohl, U., Giesler, M. and Bassenge, E. (1984) Flow-dependent, endothelium-mediated dilation of epicardial coronary arteries in conscious dogs: effects of cyclooxygenase inhibition. *J. Cardiovasc. Pharmacol.*, **6**, 1161–1169.

Holzmann, S. (1982) Endothelium-induced relaxation by acetylcholine associated with larger rises in cyclic GMP in coronary arterial strips. *J. Cyclic Nucleotide Res.*, **8**, 409–419.

Ignarro, L.J., Burke, T.M., Wood, K.S., Wolin, M.S. and Kadowitz, P.J. (1984) Association between cyclic GMP accumulation and acetylcholine-elicited relaxation of bovine intrapulmonary artery. *J. Pharmacol. Exp. Ther.*, **228**, 682–690.

Ignarro, L.J., Harbison, R.G., Wood, K.S. and Kadowitz, P.J. (1986) Activation of purified soluble guanylate cyclase by endothelium-derived relaxing factor from intrapulmonary artery and vein: stimulation by acetylcholine, bradykinin and arachidonic acid. *J. Pharmacol. Exp. Ther.*, **237**, 893–900.

Ignarro, L.J., Buga, G.M., Wood, K.S., Byrns, R.E. and Chaudhuri, G. (1987) Endothelium-derived relaxing factor produced and released from artery and vein is nitric oxide. *Proc. Natl. Acad. Sci. U.S.A.*, **84**, 9265–9269.

Ignarro, L.J., Byrns, R.E. and Wood, K.S. (1988) Biochemical and pharmacological properties of endothelium-derived relaxing factor and its similarity to nitric oxide radical. In: *Vasodilatation: Vascular Smooth Muscle, Peptides and Endothelium*, edited by P.M. Vanhoutte, pp. 427–435. New York: Raven Press.

Iyengar, R., Stuehr, D.J. and Marletta, M.A. (1987) Macrophage synthesis of nitrite, Nitrate and N-nitrosamines: Precursors and role of the respiratory burst. *Proc. Natl. Acad. Sci. USA.*, **84**, 6369–6373.

Kaiser, L. and Sparks, H.V., Jr. (1986) Mediation of flow-dependent arterial dilation by endothelial cells. *Circulatory Shock*, **18**, 109–114.

Kelm, M., Feelisch, M., Spahr, R., Piper, H.-M., Noack, E. and Schrader, J. (1988) Quantitative and kinetic characterization of nitric oxide and EDRF released from cultured endothelial cells. *Biochem. Biophys. Res. Commun.*, **154**, 236–244.

Khan, M.T. and Furchgott, R.F. (1987) Additional evidence that endothelium-derived relaxing factor is nitric oxide. In: *Pharmacology*, edited by M.J. Rand and C. Raper, pp. 341–344. Amsterdam: Elsevier.

Knowles, R.G., Palacios, M., Palmer, R.M.J. and Moncada, S. (1989) Formation of nitric oxide from L-arginine in the central nervous system: A transduction mechanism for stimulation of the soluble guanylate cyclase. *Proc. Natl. Acad. Sci. U.S.A.*, **86**, 5159–5162.

Li, C.G. and Rand, M.J. (1989) Evidence for a role of nitric oxide in the neurotransmitter system mediating relaxation of the rat anococcygeus muscle. *Clin. Exp. Pharmacol. Physiol.*, pp. 933–938.

Marletta, M.A., Yoon, P.S., Iyengar, R., Leaf, C.D. and Wishnok, J.S. (1988) Macrophage oxidation of L-arginine to nitrite and nitrate: Nitric oxide is an intermediate. *Biochemistry*, **27**, 8706–8711.

Martin, W., Villani, G.M., Jothianandan, D. and Furchgott, R.F. (1985a) Selective blockade of endothelium-dependent and glyceryl trinitrate-induced relaxation by hemoglobin and by methylene blue in the rabbit aorta. *J. Pharmacol. Exp. Ther.*, **232**, 708–716.

Martin, W., Villani, G.M., Jothianandan, D. and Furchgott, R.F. (1985b) Blockade of endothelium-dependent and glyceryl trinitrate-induced relaxation of rabbit aorta by certain ferrous hemoproteins. *J. Pharmacol. Exp. Ther.*, **233**, 679–685.

Martin, W., Furchgott, R.F., Villani, G.M. and Jothianandan, D. (1986) Depression of contractile responses in rat aorta by spontaneously released endothelium-derived relaxing factor (EDRF). *J. Pharmacol. Exp. Ther.*, **237**, 529–538.

Martin, W., Smith, J.A., Lewis, M.J. and Henderson, A.H. (1988) Evidence that inhibitory factor extracted from bovine retractor penis is nitrite, whose acid-activated derivative is stabilized nitric oxide. *Br. J. Pharmacol.*, **93**, 579–586.

Mayer, B., Schmidt, K., Humbert, R. and Bohme, E. (1989) Biosynthesis of endothelium-derived relaxing factor: a cytosolic enzyme in porcine aortic endothelial cells Ca^{2+}-dependently converts L-arginine into an activator of soluble guanylyl cyclase. *Biochem. Biophys. Res. Commun.*, **164**, 678–685.

Mellion, B.T., Ignarro, L.J., Ohlstein, E.H., Pontecorvo, E.G., Hyman, A.L. and Kadowitz, P.J. (1981) Evidence for the inhibitory role of guanosine 3',5'-monophosphate in ADP-induced human platelet aggregation in the presence of nitric oxide and related vasodilators. *Blood*, **57**, 946–955.

Miki, N., Kawabe, Y. and Kuriyama, K. (1977) Activation of cerebral guanylate cyclase by nitric oxide. *Biochem. Biophys. Res. Commun.*, **75**, 851–856.

Moncada, S. and Palmer, R.M.J. (1990) The L-arginine: nitric oxide pathway in the vessel wall. In: *Nitric Oxide from L-Arginine: A Bioregulatory System*, edited by S. Moncada and E.A. Higgs, pp. 19–33. Amsterdam: Elsevier.

Moncada, S., Palmer, R.M.J. and Gryglewski, R.J. (1986) Mechanism of action of some inhibitors of endothelium-derived relaxing factor. *Proc. Natl. Acad. Sci. U.S.A.*, **83**, 9164–9168.

Moncada, S., Palmer, R.M.J. and Higgs, E.A. (1991) Nitric oxide: physiology, pathophysiology and pharmacology. *Pharmacol. Rev.*, **43**, 109–142.

Murad, F., Mittal, C.K., Arnold, W.P., Katsuki, S., Kimura, H. (1978) Guanylate cyclase activation by azide, nitro compounds, nitric oxide and hydroxyl radical and inhibition by hemoglobin and myoglobin. *Adv. Cyclic Nucleotide Res.*, **9**, 145–158.

Murad, F., Arnold, W.P., Mittal, C.K. and Braughler, J.M. (1979) Properties and regulation of guanylate cyclase and some proposed functions of cyclic GMP. *Adv. Cyclic Nucleotide Res.*, **11**, 175–204.

Palmer, R.M.J. and Moncada, S. (1989) A novel citrulline-forming enzyme implicated in the formation of nitric oxide by vascular endothelial cells. *Biochem. Biophys. Res. Commun.*, **158**, 348–352.

Palmer, R.M.J., Ferrige, A.G. and Moncada, S. (1987) Nitric oxide release accounts for the biological activity of endothelium-derived relaxing factor. *Nature (Lond.)*, **327**, 524–526.

Palmer, R.M.J., Ashton, D.S. and Moncada, S. (1988a) Vascular endothelial cells synthesize nitric oxide from L-arginine. *Nature*, **333**, 664–666.

Palmer, R.M.J., Rees, D.D., Ashton, D.S. and Moncada, S. (1988b) L-Arginine is the physiological precursor for the formation of nitric oxide in endothelium-dependent relaxation. *Biochem. Biophys. Res. Commun.*, **153**, 1251–1256.

Pohl, U., Holtz, J., Busse, R. and Bassenge, E. (1986) Crucial role of endothelium in the vasodilator response to increased flow *in vivo*. *Hypertension*, **8**, 37–44.

                                 Robert F. Furchgott

Radomski, M., Palmer, R.M.J. and Moncada, S. (1987a) Comparative pharmacology of endothelium-derived relaxing factor, nitric oxide and prostacyclin in platelets. *Br. J. Pharmacol.*, **92**, 181–187.

Radomski, M.W., Palmer, R.M.J. and Moncada, S. (1987b) The anti-aggregating properties of vascular endothelium: interactions between prostacyclin and nitric oxide. *Br. J. Pharmacol.*, **92**, 639–646.

Radomski, M.W., Palmer, R.M.J. and Moncada, S. (1987c) The role of nitric oxide and cGMP in platelet adhesion to vascular endothelium. *Biochem. Biophys. Res. Commun.*, **148**, 1482–1489.

Radomski, M.W., Palmer, R.M.J. and Moncada, S. (1990) An L-arginine: nitric oxide pathway present in human platelets regulates aggregation. *Proc. Natl. Acad. Sci. U.S.A.*, **87**, 5193–5197.

Rapoport, R.M. and Murad, F. (1983) Agonist-induced endothelium-dependent relaxations in rat thoraic aorta may be mediated through cGMP. *Cir. Res.*, **52**, 352–357.

Rees, D.D., Palmer, R.M.J., Hodson, H.F. and Moncada, S. (1989a) A specific inhibitor of nitric oxide formation from L-arginine attenuates endothelium-dependent relaxation. *Br. J. Pharmacol.*, **96**, 418–424.

Rees, D.D., Palmer, R.M.J. and Moncada, S. (1989b) Role of endothelium-derived nitric oxide in the regulation of blood pressure. *Proc. Natl. Acad. Sci. U.S.A.*, **86**, 3375–3378.

Rubanyi, G.M. and Vanhouette, P.M. (1986) Superoxide antions and hyperoxia inactive endothelium-derived relaxing factor. *Am. J. Physiol.*, **250**, H822–H827.

Rubanyi, G.M., Lorenz, R.R. and Vanhoutte, P.M. (1985) Bioassay of endothelium-derived relaxing factor. *Am. J. Physiol.*, **215**, H1077–H1080.

Rubanyi, G.M., Romero, J.C. and Vanhoutte, P.M. (1986) Flow-induced release of endothelium-derived relaxing factor. *Am. J. Physiol.*, **250**, H1145–H1149.

Schmidt, H.H.H.W., Nau, H., Wittfoht, W., Gerlach, J., Prescher, K.-E., Klein, M.M., Niroomand, F. and Bohme, E. (1988) Arginine is a physiological precursor of endothelium-derived nitric oxide. *Eur. J. Pharmacol.*, **154**, 213–216.

Stuehr, D.J. and Marletta, M.A. (1985) Mammalian nitrate biosynthesis: mouse macrophages produce nitrite and nitrate in response to Escherichia coli lipolysaccharide. *Proc. Natl. Acad. Sci. U.S.A.*, **82**, 7738–7742.

Stuehr, D.J. and Marletta, M.A. (1987) Synthesis of nitrite and nitrate in murine macrophage cell lines. *Cancer Res.*, **47**, 5590–5594.

Stuehr, D., Gross, S., Sakuma, I., Levi, R. and Nathan, C. (1989) Activated murine macrophages secrete a metabolite of arginine with the bioactivity of endothelium-derived relaxing factor and the chemical reactivity of nitric oxide. *J. Exp. Med.*, **169**, 1011–1020.

Vallance, P., Collier, J. and Moncada, S. (1989) Effects of endothelium-derived nitric oxide on peripheral arteriolar tone in man. *Lancet*, **ii**, 997–1000.

Van de Voorde, J. and Leusen, I. (1983) Role of endothelium in the vasodilator responses of rat thoriac aorta to histamine. *Eur. J. Pharmacol.*, **87**, 113–120.

Wagner, D.A., Young, V.R. and Tannenbaum, S.R. (1983) Mammalian nitrate biosynthesis: Incorporation of $^{15}NH_3$ into nitrate is enhanced by endotoxin treatment. *Proc. Natl. Acad. Sci. U.S.A.*, **80**, 4518–4521.

Wolin, M.S., Cherry, P.D., Rodenburg, J.M., Messina, E.J. and Kaley, G. (1990) Methylene blue inhibits vasodilation of skeletal muscle arterioles to acetylcholine and nitric oxide via the extracellular generation of superoxide anion. *J. Pharmacol. Exp. Ther.*, **254**, 872–876.

Generation and Biological Actions
of Nitric Oxide

2. Structural Variations on a Theme of Nitric Oxide Production by Three Isoforms of Nitric Oxide Synthase

Bettie Sue Masters, Pavel Martásek and Linda J. Roman

The Robert A. Welch Foundation, Department of Biochemistry, The University of Texas, Health Science Centre at San Antonio, San Antonio TX 78284-7760, USA

The following is an attempt to summarize the state-of-the-art in structure-function relationships of the isoforms of nitric oxide synthase. The literal explosion of information in the scientific literature, which encompasses a number of disciplines, has made it impossible to be comprehensive, but we have attempted to discuss various structural properties of these complex, prosthetic group-containing, cofactor-requiring enzymes which enable them to perform their individualized functions in the cells of various tissues. We will not encompass mechanistic aspects of these enzymes, as that subject will be addressed elsewhere, but, instead, this review will discuss the many experimental approaches that have been applied thus far to determining the structural properties of these interesting enzymes. We have found it fascinating that three genetically determined isoforms of an enzyme, which utilizes the same basic mechanism of catalysis in each case, are capable of being adapted for unique cellular functions by clever structural modifications at the transcriptional and post-transcriptional levels.

Key words: nitric oxide, nitric oxide synthase, structure/function, L-arginine, superoxide anion.

INTRODUCTORY REMARKS

Interest in the biology and chemistry of nitric oxide (NO) formation was germinated by the finding by Furchgott and Zawadski in 1980 that endothelial cells play an obligatory role in acetylcholine-mediated relaxation of vascular smooth muscle cells. The mediator of this action, endothelium derived relaxing factor (EDRF), was later shown to be nitric oxide (Palmer *et al.*, 1987; Ignarro *et al.*, 1987; Furchgott, 1988), synthesized from L-arginine (Palmer *et al.*, 1988; Sakuma *et al.*, 1988; Schmidt *et al.*, 1988). Tannenbaum and colleagues (Green *et al.*, 1981a,b) also showed that nitrate and nitrite, breakdown products of NO from ^{15}N-labelled L-arginine, were excreted in the urine of rats and man. Subsequently, Stuehr and Marletta (1985) showed that *E. coli*-derived lipopolysaccharide stimulated the production of these nitrogen oxides (referred to as NO_x) in macrophages *in vitro*. The explosion of this new field led to the discovery of a family of three nitric oxide synthases, their respective purification from different sources, cloning and characterization. These enzymes, designated NOSI (nNOS), NOSII (iNOS), and NOSIII (eNOS), were first purified from rat brain (Bredt and Snyder, 1990), murine macrophages (Xie *et al.*, 1992), and membrane fractions of cultured bovine aortic endothelial cells (Pollock *et al.*, 1991), respectively.

For all isoforms of NOS, the same catalytic mechanism has been invoked in which NO and L-citrulline are formed in a complex stepwise reaction with N^{ω}-hydroxy-L-arginine as the obligatory intermediate. The reaction is believed to involve a five-electron oxidation of the terminal guanidino-nitrogen of L-arginine with molecular oxygen as co-substrate (Marletta, 1993).

It was recognized that deeper understanding of the structural features of the various isoforms of NOS would be required to determine the common and unique aspects of these enzymes that dictate their localization and regulation. In addressing the distinct tissue-specific roles and demands for NO production, the complicated architecture of nitric oxide synthases which subserve these functions with the multistep levels of control by oligomerization status, cofactor availability, intracellular localization, and protein-protein interaction as well as post-translational modifications will be examined. A focussed review of structure/ function relationships among the nitric oxide synthases has recently appeared (Stuehr, 1997).

As originally proposed by Anfinsen in his Nobel Prize-winning studies (Anfinsen, 1973), the peptide architecture of an enzyme is determined by its primary sequence. This architecture then serves to provide scaffolding for the appropriate geometry and dynamics for recognition and catalysis. In addition, this superstructure provides a secluded environment for the active site, preventing unwanted interactions with solvent and other reactive or disruptive agents (Greaves and Rotello, 1997)

General Features of Nitric Oxide Synthase Isoforms

All three isoforms are bidomain, dimeric enzymes. The heme domain, containing a cysteine thiolate-liganded heme similar to that found in cytochromes P-450, comprises approximately the first half of the protein and is connected by a calmodulin-binding sequence (Bredt and Snyder, 1990) to the reductase domain, which has high sequence homology to NADPH-cytochrome P-450 reductase (Bredt *et al.*, 1991; Stuehr *et al.*, 1991). All isoforms in their dimeric, physiological form contain the following prosthetic groups: heme (iron protoporphyrin IX; McMillan *et al.*, 1992; White and Marletta, 1992; Stuehr and Ikeda-Saito, 1992; Klatt *et al.*, 1992) and both flavin mononucleotide (FMN) and flavin adenine dinucleotide (FAD) [Bredt *et al.*, 1991; Stuehr *et al.*, 1991; Mayer *et al.*, 1991; Hevel *et al.*, 1991]. They are also able to bind BH4 (Tayeh and Marletta, 1989; Kwon *et al.*, 1989; Mayer *et al.*, 1990), NADPH, and L-arginine, the latter two as cosubstrate and substrate, respectively (Bredt and Snyder, 1990). nNOS and eNOS, and most probably also iNOS, are able to bind and interact in their physiological milieux with other proteins.

While the enzymology of NOS will be addressed elsewhere, in this chapter we will focus mainly on the structure-function aspects of the three isoforms of nitric oxide synthase, emphasizing the uniqueness of the modularity, regulatory aspects, and location of each enzyme which make each suitable for its biological function. We will concentrate initially on the constitutive isoforms, NOSI and NOSIII, followed by a discussion of the inducible isoform, NOSII.

NOSI. NEURONAL NITRIC OXIDE SYNTHASE (nNOS)

Bidomain Structure

In pursuit of the structural properties of nNOS, our laboratory (Sheta *et al.*, 1994) applied the technique of limited proteolysis first used in dissecting the *Bacillus megaterium* cytochrome P450 (BM-3), which also contains a flavoprotein domain and a heme domain connected by a short polypeptide (Figure 2-1). Unlike the BM-3 cytochrome P450, however, all of the NOS isoforms contain a calmodulin-binding site between the flavin- and

Neuronal NOS

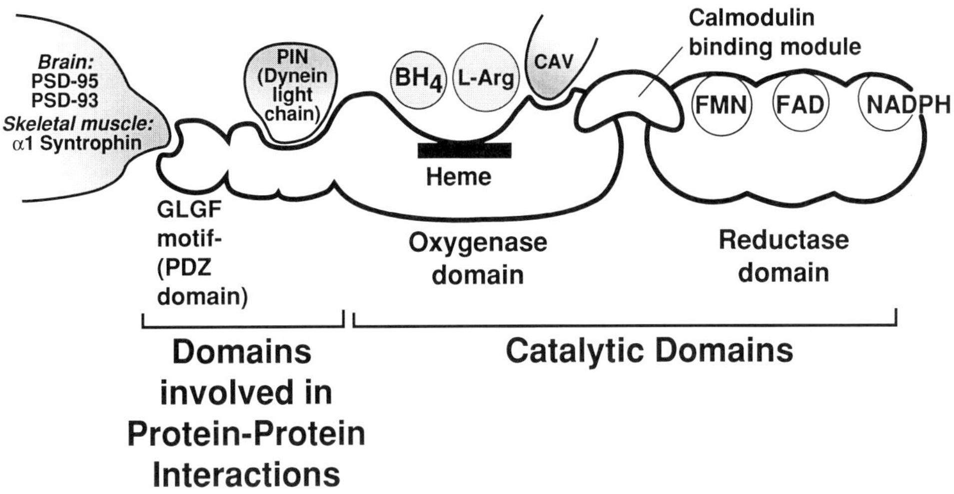

Figure 2-1. Structural representation of the neuronal nitric oxide synthase. Abbreviations are: BH4, tetrahydrobiopterin; L-Arg, L-arginine; FMN, flavin mononucleotide; FAD, flavin adenine dinucleotide; CAV, caveolin; PSD-, post synaptic density proteins; PIN, protein inhibiting nNOS; PDZ, GLGF-containing domain.

heme-binding domains. Sheta *et al.* (1994) subjected neuronal nitric oxide synthase to limited proteolysis by using an immobilized trypsin to cleave the enzyme in the presence and absence of calmodulin. We were able to show that nNOS could be cleaved into two distinct domains, approximately 89 and 79 kDA corresponding to the NH_2-terminus (heme-binding) and the COOH-terminus (flavin-binding) domains. Furthermore, the heme domain exhibited the characteristic absorption maximum at ~ 400 nm as isolated, as well as the low and high spin spectral shifts upon the addition of imidazole and L-arginine, respectively, and the flavin domain presented a typical flavoprotein spectrum and catalyzed the reduction of cytochrome *c*, 2,6-dichlorophenolindophenol, and ferricyanide.

The Sheta *et al.* (1994) experiments also showed that in the absence or presence of calmodulin, nNOS was cleaved at Arg 220, roughly corresponding to the N-terminal segment that differentiates nNOS from the other two isoforms. Increased intensity of the band corresponding to the larger fragment (~130 kDa) in the presence of calmodulin strongly suggested that exposure of this residue was enhanced upon calmodulin binding. Cleavage at Ala 728 in the absence of calmodulin was also obtained, agreeing with previous findings in other calmodulin-binding enzymes that calmodulin-binding sites are generally located in a long exposed surface loop (Hubbard and Klee, 1989). This site had been predicted by Bredt *et al.* (1991) to be in "a basic amphipathic α helix calmodulin-binding sequence at position 725–745".

It was also suggested by Sheta *et al.* (1994) that, due to the high degree of amino acid sequence homology among the three NOS isoforms, such a bidomain structure may be seen with the other NOS isoforms. This has been shown to be the case by Ghosh and Stuehr (1995) for inducible macrophage NOS and by Chen *et al.* (1996) for endothelial NOS.

Oligomeric Status of nNOS; Heme, BH₄ and L-arginine Binding

Ghosh and Stuehr (1995) showed that the trypsinolyzed oxygenase domain of the inducible isoform (NOSII) was sufficient for dimerization, leading to a model in which the heme-binding domains interact head-to-head with the flavin-binding domains extending freely. This structural concept is consistent with the report by Klatt *et al.* (1995), using light scattering, that dimeric nNOS has a prolate ellipsoid or rod-like shape with an approximate axial ratio of 20/1. These investigators also reported the tetrahydrobiopterin (BH4)- and L-arginine-induced conversion of nNOS into a stable homodimer which survives in 2% SDS and 5% 2-mercaptoethanol (Klatt *et al.*, 1995). Using circular dichroism, sedimentation velocity and dynamic light scattering measurements, these authors concluded that the addition of L-arginine and BH4 to nNOS had no effect on any of these parameters. A later study by Klatt *et al.* (1996), with a heme-deficient preparation of rat neuronal NOS-baculovirus-infected insect Sf9 cells, showed that heme is required for dimerization of nNOS and that this heme-deficient enzyme bound neither BH4 nor N^G-nitro-L-arginine (L-NNA). This result was in spite of the fact that most of the structural elements of nNOS were preserved, as indicated by circular dichroism. Nevertheless, this heme-deficient enzyme maintained its full capacity to reduce cytochrome *c* due to its full complement of FAD and FMN.

Data obtained with neuronal NOS (Sheta, Hansen, Demeler, Schwartz, and Masters, unpublished observations; Masters, 1994) in the Beckman XL-A analytical ultracentrifuge have indicated that nNOS can exist in monomeric form when diluted to lower enzyme concentrations. The monomer-dimer equilibrium tends toward increasing monomer concentrations at decreasing enzyme concentration. This, however, may not be indicative of the *in vivo* situation since the association of the nNOS and eNOS isoforms with cellular structures, such as the skeletal muscle dystrophin-glycoprotein complex and endothelial cell caveolae, respectively, could promote an extremely high intracellular concentration of these enzymes, thus inducing the formation of dimers.

It is useful to determine the hydrodynamic behavior of these molecules under a variety of conditions, not the least of which would involve the inclusion of BH4 and/or L-arginine, as compared to preparations which have been depleted of substrate and/or BH4. Experiments with wild type nNOS expressed in *E. coli* (lacking enzymatic machinery for synthesis of BH4) and a C331A mutant (Martasek *et al.*, 1998), which has a decreased binding affinity for both L-arginine and BH4, have shown that both form dimers in the absence of these compounds. This is in contrast to the findings of Klatt *et al.* (1996) who advocated a role for BH4 in the dimerization of nNOS. Differences in experimental results may be explained by the partial or profound lack of heme in the Klatt *et al.* (1996) subunit preparations, which in our experiments appears to be the determining factor in dimerization. As will be seen in later discussions of the endothelial and inducible isoforms of NOS, the binding of L-arginine and/or BH4 has different effects on the dimerization and structure/function aspects of these enzymes.

Spectral Characterization

The substrate-perturbed optical difference spectra with the tissue-purified (McMillan and Masters, 1993) and *E. coli*-expressed constructs of neuronal NOS (McMillan and Masters, 1995; Roman *et al.*, 1995) demonstrated that L-arginine produced a high-spin shift as opposed to the low-spin spectral shift attributable to imidazole, a typical nitrogenous ligand.

These interactions not only permitted the determination of binding constants for ligands of the NOS isoforms but were further refined through examination by electron paramagnetic resonance (EPR) spectroscopy. In the first such study (Salerno *et al.*, 1995), our laboratory, with those of Griffith and Salerno, characterized the interaction of L-arginine and L-thiocitrulline with the heme domain of nNOS. These studies corroborated the optical spectral data showing that purely high spin shifts were obtained upon the addition of L-arginine and, interestingly, showed both the high spin and low spin species associated with the binding of L-thiocitrulline to this isoform (Frey *et al.*, 1994). Based upon these early studies, we were able to apply EPR spectroscopy in binding experiments with various arginine-based ligands and to determine differences in binding modes among the three isoforms (Salerno *et al.*, 1996a,b, 1997a). For example, it was shown for nNOS holoenzyme that the binding of substrate (L-arginine), intermediate (N-hydroxy-L-arginine), product (L-citrulline), and L-arginine-based inhibitors perturbs the spin state equilibrium as detected by EPR to produce a "spectroscopic fingerprint" for each ligand (Salerno, 1996b). These perturbations have been tabulated for nNOS (Salerno, 1996b) and eNOS (Salerno, 1997a), revealing interesting differences in degree and type of perturbation and indicating different modes of binding depending upon substituents on the arginine side chain. Interestingly, L-arginine produces an EPR spectral shift toward high spin species, but N^{ω}-methyl-L-arginine (L-NMA) produces a low spin spectral shift. By comparison, eNOS shows very little perturbation with L-arginine but, coupled with BH4, gives a dramatic high spin shift. On the other hand, as with nNOS, L-NMA produces a dramatic low spin shift with eNOS, but the complex is of lower rhombicity than with either nNOS or iNOS, reflecting that L-NMA is a relatively poor inhibitor of eNOS. Galli *et al.* (1996) showed the proximity of the flavin and heme iron by EPR saturation recovery studies but did not show the high spin shifts due to L-arginine observed by Salerno *et al.* (1996).

To gain further insight into substrate binding modes, our laboratory with Tierney *et al.* (1998) used electron nuclear double resonance (ENDOR) spectroscopic techniques. A combination of EPR and nuclear magnetic resonance (NMR) spectroscopy, ENDOR has permitted the localization of the guanidino nitrogen of L-arginine in the active sites of all three isoforms. Using ^{15}N-L-arginine, this distance was determined to be within ~ 4 Å of the heme iron in all three of the isoforms, information that is presently unavailable from the recently reported crystal structure of the iNOS heme domain (Crane *et al.*, 1997). These studies indicate that subtle differences in the binding modalities observed in EPR upon binding of L-arginine to each of the NOS isoforms is not reflected in changes in the distance of the guanidino nitrogen from heme iron but is due to changes in rhombicity relating to the iron orbitals.

Tissue Localization

The localization of nNOS in brain and other tissues has been of great interest due to its unusual behavior during purification from tissue sources. It has been observed by a number of investigators that the enzyme behaves as a partially particulate as well as a soluble (cytosolic) protein. This behavior can now be explained by the association of nNOS in brain with neuronal postsynaptic density proteins PSD-95 or PSD-93 (Brenman *et al.*, 1996) or its alternatively spliced form μNOS (Silvagno *et al.*, 1996) in skeletal muscle (Brenman *et al.*, 1995, 1996; Chang *et al.*, 1996) with the dystrophin-glycoprotein complex at force transmitting sites, myotendinous junctions, and costameres in the sarcolemma. The previously mentioned 220+ N-terminal residues unique to neuronal NOS (Figure 2-1) targets the enzyme to these structures through the so-called PDZ (or GLGF motif-containing) domain by interaction with α1-syntrophin in skeletal muscle and to the second PDZ/GLGF motif of PSD-95 in brain neurons. These protein-protein interactions do not appear to affect enzymatic activity as measured by NO production. The integral component of the dystrophin-glycoprotein complex in skeletal muscle is caveolin-3 (Song *et al.*, 1996), which binds to nNOS and inhibits its activity (García-Cardeña *et al.*, 1997; Venema *et al.*, 1997b). Another nNOS-inhibiting protein, PIN (protein inhibiting nNOS; Figure 2-1) has been reported by Snyder's group (Jaffrey and Snyder, 1996) and shown to be identical to the light chain of dynein. The physiological significance of this inhibitory protein is unknown.

Recently, an alternatively spliced, truncated form of nNOS has been found in human testicular tissue (Wang *et al.*, 1997). Interestingly, this form is truncated at the N-terminus (Δ336) resulting in an enzyme missing those sequences putatively involved in dimerization of nNOS. If the PIN protein were involved in regulation of dimerization of the full-length neuronal NOS in testes by binding to residues 163–245, then this tissue-specific, truncated NOS would not be expected to be so regulated. This is particularly important since Jaffrey and Snyder (1996) have reported that the PIN transcript is most abundant in testes.

Superoxide Production

Zweier's laboratory (Xia *et al.*, 1996) demonstrated by EPR spin-trapping the possible *in vivo* significance of superoxide production by nNOS-transfected human kidney 293 cells depleted of L-arginine. Ca^{+2} ionophore stimulation was cytotoxic to the cells but this toxicity was largely diminished by superoxide dismutase and L-NAME, suggesting the formation of peroxynitrite.

Considerable controversy still surrounds the conditions for and sources and amounts of superoxide anion produced by the various isoforms of NOS (Stuehr, 1997). Earlier, Mayer's group (Mayer *et al.*, 1991; Heinzel *et al.*, 1992) demonstrated that hydrogen peroxide (H_2O_2) was produced by nNOS. Pou *et al.* (1992) determined by EPR spectroscopy that nNOS produced superoxide anion. More recently, Sheta *et al.* (1994) showed superoxide dismutase-inhibitable cytochrome *c* reduction by wild type nNOS, and Richards *et al.* (1996) demonstrated an O_2 requirement for reduction of cytochrome *c* by a heme-less mutant of intact nNOS. Furthermore, the involvement of the flavoprotein domain of nNOS in superoxide formation has been unequivocally established by Richards *et al.* (1996), Gacchui *et al.* (1996), and Miller *et al.* (1997) using various constructs of nNOS.

NOSIII. ENDOTHELIAL NITRIC OXIDE SYNTHASE (ENOS)

The enzyme responsible for endothelial nitric oxide production was first purified from membrane fractions of cultured bovine aortic endothelial cells (Pollock *et al.* 1991) and was the last cloned from the NOS family (therefore designated as NOS III) (Janssens *et al.*, 1992; Marsden *et al.*, 1992; Nishida *et al.*, 1992; Sessa *et al.*, 1992). Endothelial NOS (eNOS) was later found to be present in cardiac myocytes (Balligand *et al.*, 1993), brain hippocampus (Dinerman *et al.*, 1994), pulmonary (Shaul *et al.*, 1994) and renal (Tracey *et al.*, 1994) epithelium as well as in blood platelets (Sase and Michel, 1995). The

Endothelial NOS

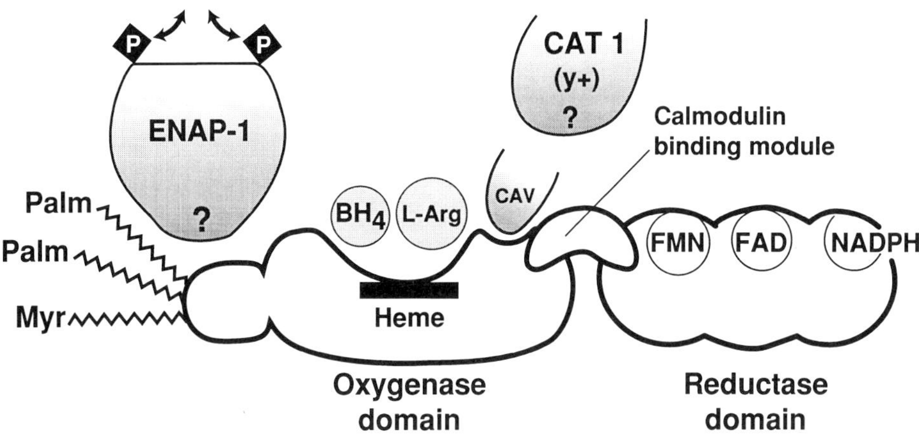

Figure 2-2. Structural representation of the endothelial nitric oxide synthase. Abbreviations are: BH4, tetrahydrobiopterin; L-Arg, L-arginine; FMN, flavin mononucleotide; FAD, flavin adenine dinucleotide; CAV, caveolin; P, putative phosphorylation sites; ENAP-1, eNOS activating protein; CAT 1, cationic amino acid transporter 1; Palm, palmitate; Myr, myristate.

presence of eNOS in such diverse cell types may underline differential regulation (For review, see Sase and Michel, 1997).

eNOS undergoes posttranslational modifications, being myristoylated at Gly-2 and palmitoylated on Cys-15 and Cys-26 at the N-terminal end of the protein (Figure 2-2) (Busconi and Michel, 1993; Liu and Sessa, 1994; Pollock *et al.*, 1992; Robinson and Michel, 1995). Because this N-terminal part of the eNOS does not share any similarity with either nNOS or iNOS, these posttranslational modifications are a unique feature of eNOS, targeting it to the plasma membrane.

Oligomeric Status of eNOS; Heme, BH4, and L-arginine Binding

Recent comparison of the interaction between NOS subunits shows that iNOS subunit association involves only oxygenase domain interaction, as previously shown by Ghosh and Stuehr (1995). Contrary to this, subunit association of eNOS and nNOS involves head to head interaction of heme domains, tail to tail interaction of reductase domains, and/or head to tail interactions of heme and reductase domains (Venema *et al.*, 1997a).

It has been postulated that in nNOS, BH4 is essential for maintaining the enzyme in its dimeric form (Klatt *et al.*, 1995, 1996). However, in *E. coli*-expressed nNOS, which contains no BH4, the protein eluted from a gel filtration column behaves as a dimeric species. This is also the case with eNOS, which does not require BH4 for dimerization (Rodríguez-Crespo *et al.*, 1996). The possibility that BH4 allosterically stabilizes eNOS and/or is involved in preventing superoxide generation from eNOS (Wever *et al.*, 1997) remains to be elucidated.

24 *Bettie Sue Masters* et al.

Regions involved in dimerization were studied using progressive deletional analysis of eNOS. While deletion of the first 52 amino acids did not affect the activity, deletion of 91 and 105 residues altered dimerization equilibria and retained 20 and 12% of catalytic activity, respectively. All three mutants bind heme and exhibit the ferrous-CO difference spectral maxima at 444 nm characteristic of thiolate heme ligation (Rodríguez-Crespo *et al.*, 1997).

In search of a BH4 binding site predicated on homology-based modeling of nNOS with dihydrofolate reductase, we reported that an expressed construct of nNOS (558–721) is capable of binding L-nitroarginine, which was not enhanced by BH4 (Nishimura *et al.*, 1995). Confirming that negatively charged residues are involved in L-arginine binding, the mutation of a glutamic acid residue in corresponding regions of eNOS and iNOS (E361 and E371, respectively) abolished L-arginine binding (Chen *et al.*, 1997; Gachhui *et al.*, 1997). Thus, the glutamic acid carboxylate participates in L-arginine binding by interacting with its guanidino moiety.

Based on analysis of the resonance Raman spectra of CO-derivatives of the NOSs, Fan *et al.* (1997) proposed that eNOS and iNOS differ substantially from nNOS. They suggested that the substrate-binding pocket of nNOS is much more open compared to eNOS and iNOS and that L-arginine may be bound 1 Å further from the heme in nNOS. An elegant study by Berka *et al.* (1996) used three groups of rigid planar ligands (imidazole, pyridine, and pyrimidine) to estimate the space for the distal heme ligand ($\sim$6.3 Å $\times$ 6.7 Å). Based on EPR spectroscopy of eNOS, Tsai *et al.* (1996) proposed that the distal heme environment in eNOS resembles chloroperoxidase more than $P450_{cam}$. Our EPR spectroscopic study of eNOS shows that the binding of L-arginine analogs perturbs the environment of the high-spin ferriheme in a highly-specific manner and that four categories of high-spin complexes could be distinguished, prototypes of which are L-arginine, N-hydroxy-L-arginine, N-methyl-L-arginine, and N-nitro-L-arginine (Salerno *et al.*, 1997a).

Binding and Reduction of Artificial Electron Acceptors/Superoxide Production

The ability of different isoforms to reduce artificial electron acceptors, prototypically cytochrome *c*, differs greatly. In the presence of calmodulin, eNOS has the potential to reduce cytochrome *c* as do nNOS and iNOS, but its reducing ability reaches only 5–10% of those of nNOS and iNOS (Figure 2-3). The reduction of small electron acceptors like DCIP and FeCN is also used to address the functionality of the eNOS reductase domain. The differential ability to reduce cytochrome *c* and small electron acceptors is probably due to the differences in primary structure and overall charge of the reductase domain of eNOS, compared to nNOS or iNOS. Lucigenin, frequently used as a superoxide probe (after reduction to the lucigenin cation radical), is reduced by eNOS (Vásquez-Vivar *et al.*, 1997a). Recently, we published that binding of the notoriously cardiotoxic anticancer drug adriamycin to eNOS can divert electrons from its normal pathway, NADPH$\rightarrow$FAD$\rightarrow$FMN$\rightarrow$heme, and reduce adriamycin to the semiquinone radical. As a consequence, superoxide formation is enhanced and NO production is decreased. A consequence of eNOS-mediated reductive activation of adriamycin is the disruption of the balance between NO and superoxide production. This may lead eNOS to generate peroxynitrite and hydrogen peroxide, reactive oxygen species, in cardiomyocytes and endothelial cells (Vásquez-Vivar *et al.*, 1997b).

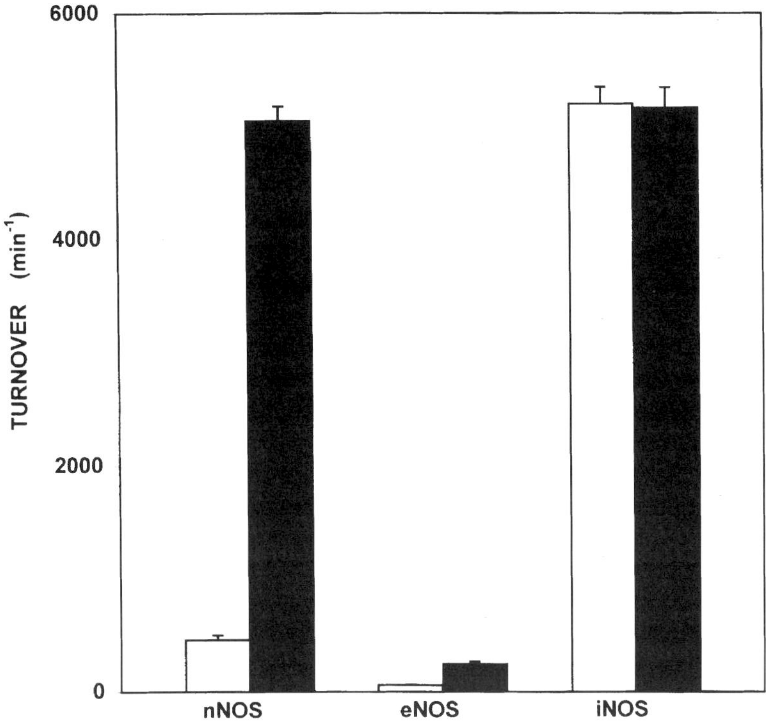

Figure 2-3. Cytochrome c reductase activity of the various NOS isoforms. Open bars = basal activity, closed bars = calmodulin-stimulated activity. Assay performed in 1 ml 50 µM Hepes, 250 µM NaCl, pH 7.6, and started with addition of NADPH. Reduction of cytochrome c is measured as the absorbance increase at 550 nm. Abbreviations: nNOS = holo-nNOS, eNOS = holo-eNOS, iNOS = holo-iNOS.

Protein–Protein Interactions

Calmodulin

Calmodulin triggers the intra-reductase domain and inter-heme-reductase domain transfer of electrons (Figure 2-2), its binding being essential for NO production (for review, see Stuehr, 1997; Michel and Feron, 1997). Endothelial NOS, as a constitutive isoform, binds calcium/calmodulin, reflecting transient elevations in intracellular calcium and providing moment-to-moment modulation of vascular tone. Constitutive isoforms contain a unique polypeptide insert in their FMN-binding domains which is not shared with iNOS. This insert probably represents an autoinhibitory control element by docking with a site on constitutive NOSs that impedes calmodulin binding and enzymatic activation (Salerno *et al.*, 1997b).

Caveolin

The interaction of eNOS with caveolins-1 and -3 has drawn much attention recently. eNOS is targeted to the caveolae in the microdomain of the plasma membrane. The major structural protein of caveolae is caveolin. Couet *et al.* (1997) have shown that the caveolin scaffolding domain (N-terminal part of caveolin protein) binds to a FxxxxFxxW motif in eNOS. This motif lies between the heme-binding and the calmodulin-binding sites and is absolutely conserved in all NOSs. The association of eNOS with the types of caveolins is tissue-specific. Endothelial NOS is present in endothelial cells where it is associated with caveolin-1, while in cardiomyocytes it is associated with caveolin-3. Caveolin and calmodulin are described as counterbalacing allosteric modulators of eNOS. Caveolin, via its scaffolding domain, directly forms an inhibitory complex with eNOS and abrogates the activation of eNOS by calmodulin (Michel *et al.*, 1997a,b). Caveolin binding to the eNOS protein prevents NO formation *in vitro* and *in vivo*, and mutation of the caveolin-binding site prevents the effect of caveolin on NO formation (García-Cardeña *et al.*, 1997; Michel *et al.*, 1997a,b; Venema *et al.*, 1997b).

Cationic Amino Acid Transporter 1 (CAT1)

Recently, it was shown that not only caveolin and eNOS but also the CAT1 protein co-localize in plasma membrane caveolae. CAT1 may be directly involved in the delivery of the substrate, L-arginine, and it has been shown that caveolar localization may be required for optimal NO production from eNOS (McDonald *et al.*, 1997). No binding site for CAT1 binding to eNOS has yet been established.

eNOS Activating Protein (ENAP)

Another protein, called ENAP-1, is also specifically bound to eNOS, but it has not yet been fully characterized. Venema *et al.* (1996a) suggested that it is involved in the phosphorylation of eNOS as a result of bradykinin stimulation.

NOSII. INDUCIBLE NITRIC OXIDE SYNTHASE (INOS)

Protein Structure

As in the other isoforms, mouse iNOS is comprised of a bidomain structure (Ghosh and Stuehr, 1995) (Figure 2-4), the major ones being the N-terminal heme or oxygenase domain (residues 1–498), which contains a cysteine thiolate-liganded heme moiety, and the C-terminal reductase domain (residues 531–1144), which contains both FAD and FMN. These two domains are bisected by a calmodulin-binding sequence (residues 499–530). Also located in the heme domain are binding sites for L-arginine and BH4, and various regulatory regions. In mouse iNOS, C194 is the proximal thiolate ligand (Sari *et al.*, 1996); in human iNOS, it is C200 (Cubberley *et al.*, 1997).

Recently, the crystal structure of a portion of the iNOS heme domain (residues 115–498) gave the world its first look at the monomeric iNOS structure (Crane *et al.*, 1997). The iNOS heme domain is an elongated curved molecule ($80 \times 57 \times 30$ Å) that the authors liken to a left-handed catcher's mitt, with the heme being held in the palm of the glove with its distal face exposed to solvent. The enzyme was co-crystallized with imidazole, which revealed that two imidazole molecules could bind, one to the heme as the sixth axial

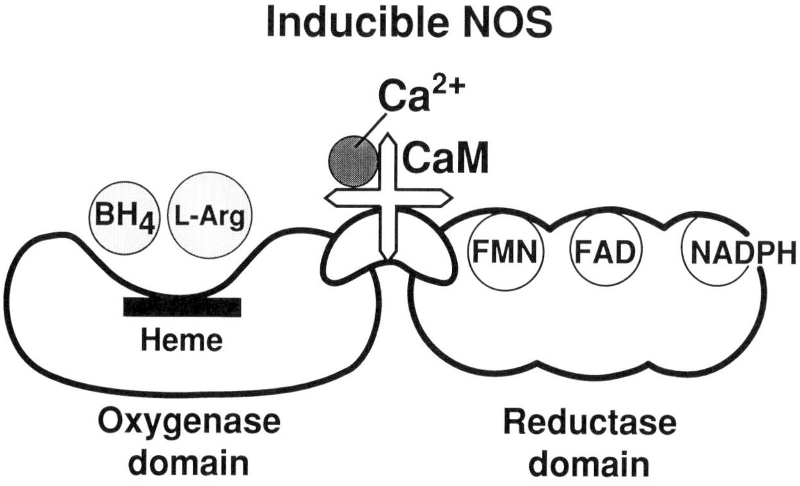

Inducible NOS

Ca^{2+} CaM

BH$_4$ L-Arg

FMN FAD NADPH

Heme

Oxygenase domain   Reductase domain

Figure 2-4. Structural representation of the inducible nitric oxide synthase. BH4, tetrahydrobiopterin; L-Arg, L-arginine; FMN, flavin mononucleotide; FAD, flavin adenine dinucleotide; CaM, calmodulin.

ligand and one hydrogen bonded with E371. Binding of aminoguanidine, an arginine analog, is also to E371, displacing the imidazole, and localizing the substrate-binding site over heme pyrrole rings A and B.

It is interesting that mouse iNOS co-crystallized with both imidazole and the arginine analog bound. Mouse iNOS reacts differently with imidazole than human iNOS and the other NOS isoforms. For mouse iNOS, imidazole is non-competitive with arginine, as shown by citrulline formation assays (Wolff and Gribin, 1994), hemoglobin capture assays, and spectral binding studies (Roman, Martásek, Miller, Salerno, and Masters, in preparation). For human iNOS (Chabin *et al.*, 1996) and all other isoforms examined (Mayer *et al.*, 1994), imidazole is competitive with arginine; *i.e.*, their binding is mutually exclusive. The catalytic sites of these enzymes must therefore be at least somewhat different than that of mouse iNOS. Future studies will illuminate the structural differences between these isoforms that give each their unique properties.

Calmodulin

The role of calmodulin in all NOS isoforms is to enable the transfer of electrons from the flavin domain to the heme moiety (Abu-Soud and Stuehr, 1993). iNOS is unique among the isoforms in that calmodulin is very tightly bound even in the absence of added calcium (Stevens-Truss and Marletta, 1995). Calmodulin copurifies with iNOS from murine macrophages and from *E. coli*, where iNOS and calmodulin have been heterologously co-expressed (Fossetta *et al.*, 1996; Wu *et al.*, 1996; Roman, Martásek, Miller, Salerno, and Masters, in preparation). Although a calmodulin-binding sequence was identified as residues 499–530 in the murine macrophage enzyme, there remains a question as to whether this sequence contains all the determinants necessary for this unusually tight interaction.

Using surface plasmon resonance studies, Zoche *et al.* (1996) concluded that this extremely tight binding is controlled by the CaM-binding segment and not by other portions of the protein. Once bound to an immobilized peptide (residues 503–528) derived from its binding site on iNOS, calmodulin does NOT dissociate. The affinity of the peptide for calmodulin was subnanomolar ($K_D < 0.1$ nM), as also seen by Anagli *et al.* (1995). In contrast, nNOS exhibits a $K_D = 5$ nM.

Other experiments, however, do not fully support that the binding sequence alone determines the calcium independence of the iNOS/calmodulin interaction. Chimeric NOS enzymes in which the mouse iNOS calmodulin-binding sequence is reciprocally exchanged with that of nNOS (Ruan *et al.*, 1996) or eNOS (Venema *et al.*, 1996b) gave intermediate results. All the chimeras produced retained some dependence on added calcium. The concentration required for half maximal activity, however, was 10- to 20-fold less than that for nNOS or eNOS, suggesting that calcium-independent calmodulin binding requires more than just the canonical binding sequence, perhaps even two adjacent or separate sites on iNOS.

Recent work by Salerno *et al.* (1997b) has identified a putative autoinhibitory control peptide present in eNOS and nNOS, but not iNOS, in the middle of the FMN binding site, which presumably functions allosterically or through binding domain overlap with calmodulin. This model explains that calmodulin binding by nNOS and eNOS versus iNOS not only is influenced by the canonical calmodulin binding sequence in each isoform, but also is modulated by the presence or absence of an autoinhibitory domain.

Given that calmodulin plays an important role in controlling the ability of the NOSs to transfer electrons from the reductase to the heme domain, can electron transfer occur in the absence of the calmodulin influence, *i.e.*, can the isolated domains, unconnected by the calmodulin binding sequence, be reconstituted to form NO? Ghosh *et al.* (1995) reported the reconstitution of purified iNOS heme and reductase domains to form NO and citrulline from N-hydroxy-L-arginine, but not arginine. The reaction was very inefficient, in that three nmoles of NADPH were required per nmole NO formed from NHA, as compared to 0.5 nmoles required by the holo-enzyme. Both the iNOS holo-enzyme and the heme domain dimer also convert NHA to NO in the presence of H_2O_2 (Pufahl *et al.*, 1995; Ghosh *et al.*, 1997). Given that the NO production by the reconstituted heme and reductase domains was uncoupled to electron transfer between them, it is feasible that little or no direct electron transfer is taking place between the separate reductase and heme domains. NO production could be entirely mediated by the superoxide/hydrogen peroxide being produced by the reductase domain (*vide infra*).

Oligomeric Status of iNOS/BH4 and L-Arginine Binding

Dimerization of iNOS is necessary for catalytic activity both *in vitro* and *in vivo* (Abu-Soud *et al.*, 1995; Tzeng *et al.*, 1995). The heme domain of iNOS contains all of the determinants necessary for dimerization. When subjected to limited proteolysis, holo-iNOS was cleaved into a heme domain, which was dimeric and contained heme and BH4, and a reductase domain, which was monomeric and contained FMN, FAD, calmodulin, and the NADPH binding site (Ghosh and Stuehr, 1995). Similarly, when residues 1–498 were expressed in *E. coli*, the resultant protein was 80% dimeric. This is consistent with a model in which the molecules of iNOS assemble into a head to head complex with the reductase domains extending independently (Ghosh and Stuehr, 1995; Venema *et al.*, 1997a). This is in contrast to nNOS and eNOS, where the dimeric complex could consist of heme/heme, heme/reductase, or reductase/reductase domain interactions (Venema *et al.*, 1997a).

Although some component(s) of the heme domain is(are) necessary for dimerization, the exact nature of the requirements is rather confusing. The involvement of BH4 binding, in particular, is controversial. In terms of actual protein sequence requirements, the deletion of amino acids 1–117 or 1–114 from the heme domain construct resulted in a protein that was monomeric or only about 25% dimeric, respectively, and deficient in BH4 and arginine binding. Inclusion of BH4 protected against cleavage at residue 117 (Ghosh *et al.*, 1997). Deletion of residues 1–65, however, had no effect on any of these properties. Thus, residues between 65 and 114 appear to be involved in dimerization and/or BH4 binding.

Several other lines of evidence also point to the involvement of BH4 in dimer formation. Human iNOS, expressed in BH4-deplete NIH3T3 cells, existed as about 20% dimeric in cytosol but was 66% dimeric when supplemented with BH4 (Tzeng *et al.*, 1995). If heme domain or holo-NOS dimers of mouse iNOS are dissociated in urea, reassociation requires BH4 and arginine (Ghosh *et al.*, 1996; Abu-Soud *et al.*, 1995). Finally, a region on the C-terminal side of the heme (residues 448–480) bears striking homology to the BH4-binding sites of amino acid hydroxylases. Mutational analysis demonstrates that residues G450 and A453 in this region are critical for BH4 binding, dimer formation, and NO production (Cho *et al.*, 1995).

Other evidence indicates that BH4 is not a requirement for dimerization. Both the heme domain and holoenzyme have been expressed in *E. coli*, which does not biosynthesize BH4, yet dimers are produced (Ghosh *et al.*, 1997; Wu *et al.*, 1996; Roman, Martásek, Miller, Salerno, and Masters, in preparation). One explanation for the apparent involvement of BH4 in dimerization is the cooperative effect seen between BH4 binding and arginine binding. Most of the dimerization/BH4-binding mutants described are also defective in L-arginine binding, so if L-arginine is required for dimerization, it might also appear that BH4 is required. A mutation of the iNOS heme domain, E371A, the residue to which aminoguanidine is bound in the heme domain crystal structure (Crane *et al.*, 1997), discounts this idea. This mutant does not bind L-arginine at all, yet is a dimer that binds BH4 (Gachhui *et al.*, 1997). Thus, there is no consistent picture from which conclusions regarding cofactor requirements for dimerization can be drawn. The cofactor requirement seen by some investigators could be a stabilization effect, *i.e.*, binding of L-arginine or BH4, by producing conformational changes reflected in the heme pocket, results in a more stable conformer which, in turn, produces a more stable dimer.

BH4

BH4 is a required cofactor for production of nitric oxide (Baek *et al.*, 1993; Cho *et al.*, 1995; Tzeng *et al.*, 1995). The exact role it plays in iNOS catalysis is unknown but it appears to have at least an allosteric function. In the absence of L-arginine and BH4, the ferric heme iron of iNOS is in the six-coordinate, low-spin state. Binding of BH4 is accompanied by a partial conversion of the heme spin state equilibrium from the low- to high-spin form. The addition of L-arginine completes this conversion (Wang *et al.*, 1995; Mayer *et al.*, 1997). The BH4 analog, 4-amino BH4, also causes a partial shift in heme spin state equilibrium and allows for dimerization of iNOS, but only BH4 permits nitric oxide production (Mayer *et al.*, 1997). Thus, allosteric effects of BH4 alone probably do not fully explain the role of BH4 in nitric oxide synthesis.

In addition to residues G450 and A453, as described above (Cho *et al.*, 1995), another residue, N-terminal to the heme-binding site, has also been implicated by mutagenesis as involved in BH4 binding. iNOS C109A has a binding constant for BH4 of 820 nM in the absence of L-arginine, as compared to 250 nM for wild type enzyme (Ghosh *et al.*, 1997). These experiments were performed with the heme domain portion, however, which did not bind BH4 in a stoichiometric manner and the mutation could still result in altered binding of L-arginine.

Gerber *et al.* (1997) proposed that BH4 binds directly above pyrrole ring D of the heme moiety or causes a constriction in the protein in this region, because BH4 markedly decreases phenyl migration to this ring. Based on the crystal structure of the iNOS heme domain solved by Crane *et al.* (1997), none of the above mentioned residues are in this area, supporting the occurence of a substantial conformational change, impinging upon this region, when BH4 binds.

L-Arginine

iNOS uses L-arginine as the primary substrate for the formation of nitric oxide, with a Km of 2.8 μM (Stuehr *et al.*, 1991). The actual binding site for L-arginine has not yet been identified, but EPR binding studies using L-arginine and various arginine analogs have demonstrated a shift in the iron spin state equilibrium to the high-spin form. Each ligand substrate perturbs the high spin ferri heme in a highly specific manner (Salerno *et al.*, 1996; Roman, Martásek, Salerno, Miller, and Masters, in preparation). Although several L-arginine and guanidine analogs bind to iNOS and shift the heme spin state equilibrium, only a few support nitric oxide synthesis (NHA, NMA, and homoarginine) (Pufahl *et al.*, 1992, 1993; Olken and Marletta, 1993; Sennequier and Stuehr, 1996). Recent experiments (discussed in the nNOS section) employing ENDOR have placed the guanidino N of L-arginine within 4 Å of the heme iron (Tierney *et al.*, 1998).

L-arginine also promotes the assembly of dimers (Baek *et al.*, 1993) and alters NADPH oxidation (Abu-Soud *et al.*, 1994). Guanidine and L-arginine analogs cause the aforementioned high-spin shift, alter NADPH oxidation, and promote dimerization but do not support nitric oxide synthesis. Since even the guanidine analogs are able to effect these results, it is reasonable to assume that these perturbations are mediated through binding of the guanidinium portion of L-arginine and not the amino acid portion.

Catalytic Activities

Superoxide Production

In arginine-depleted macrophages, superoxide appears to be generated by iNOS, along with nitric oxide (Xia and Zweier, 1997). Superoxide and nitric oxide can combine to form the potent oxidant peroxynitrite, which is also detected in these cells. Further studies with purified iNOS confirm that it is able to produce superoxide, as measured using an EPR spin-trapping technique (Xia *et al.*, submitted) or the adrenochrome assay (Roman, Martásek, Miller, Salerno, and Masters, in preparation). Superoxide synthesis by iNOS, as detected by EPR, occurs primarily through the flavin domain of the enzyme because it is sensitive to the flavoprotein inhibitor diphenyleneiodonium, but not to the heme blocker cyanide

(Xia *et al.*, 1998). Superoxide production is also greatly attenuated in the presence of 5 μM BH4, under conditions that support nitric oxide production, indicating that BH4 increases the coupling of the heme and flavin domains of iNOS (Roman, Martásek, Miller, and Masters, in preparation).

NO Production

A common strategy for NO production is used by all of the various NOS isoforms, so mechanistic details will not be discussed in this chapter. All of the enzymes use L-arginine to form NO and citrulline. One of the reaction products, NO, has a high affinity for binding to heme iron prosthetic groups in various enzymes. It seems this would be a problem for NOS, itself a heme-containing enzyme. Indeed, NO does appear to inhibit NOS activity *in vivo* under certain conditions, but nNOS and eNOS appear to be much more sensitive to this inhibition than iNOS (Griscavage *et al.*, 1995). *In vitro*, NO forms both ferrous and ferric nitrosyl complexes, which are spectrally detectable, with the heme of iNOS.

The oxidation state of the heme also plays a role in NO sensitivity; inhibition by NO is greatly decreased when the heme is in the ferrous oxidation state (Griscavage *et al.*, 1995). Both L-arginine and BH4 shift the heme to a high-spin state, which then accepts transfer of the first electron, reducing the heme from the ferric to the less NO-sensitive ferrous form. Consistent with this scheme, arginine further decreases the affinity of NO for the ferrous heme of iNOS (Hurshman and Marletta, 1995). Perhaps one role for BH4 in these enzymes is to favor formation of this state to protect against feedback inhibition. This is consistent with the previous observation that BH4 causes a shift in the heme spin state equilibrium.

REFERENCES

Abu-Soud, H.M., Feldman, P.L., Clark, P. and Stuehr, D.J. (1994) Electron transfer in the nitric-oxide synthases. Characterization of L-arginine analogs that block heme iron reduction. *J. Biol. Chem.*, **269**, 32318–32326.

Abu-Soud, H.M., Loftus, M. and Stuehr, D.J. (1995) Subunit dissociation and unfolding of macrophage NO synthase: Relationship between enzyme structure, prosthetic group binding, and catalytic function. *Biochemistry*, **34**, 11167–11175.

Anagli, J., Hofmann, F., Quadroni, M., Vorherr, T. and Carafoli, E. (1995) The calmodulin-binding domain of the inducible (macrophage) nitric oxide synthase. *European Journal of Biochemistry*, **233**, 701–708.

Anfinsen, C.B. (1973) Principles that govern the folding of protein chains. *Science*, **181**, 223–230.

Baek, K.J., Thiel, B.A., Lucas, S. and Stuehr, D.J. (1993) Macrophage nitric oxide synthase subunits: Purification, characterization, and role of prosthetic groups and substrate in regulating their association into a dimeric enzyme. *J. Biol. Chem.*, **268**, 21120–21129.

Balligand, J-L., Kelly, R.A., Marsden, P.A., Smith, T.W. and Michel, T. (1993) Control of cardiac muscle cell function by an endogenous nitric oxide signaling system. *Proc. Natl. Acad. Sci. USA*, **90**, 347–351.

Berka, V., Chen, P-F. and Tsai A-L. (1996) Spatial relationship between L-arginine and heme binding sites of endothelial nitric-oxide synthase. *J. Biol. Chem.*, **271**, 33293–33300.

Bredt, D.S. and Snyder, S.H. (1990) Isolation of nitric oxide synthetase, a calmodulin-requiring enzyme. *Proc. Natl. Acad. Sci. USA*, **87**, 682–685.

Bredt, D.S., Hwang, P.M., Glatt, C.E., Lowenstein, C., Reed, R.R. and Snyder, S.H. (1991) Cloned and expressed nitric oxide synthase structurally resembles cytochrome P-450 reductase. *Nature*, **351**, 714–718.

Brenman, J.E., Chao, D.S., Xia, H., Aldape, K. and Bredt, D.S. (1995) Nitric oxide synthase complexed with dystrophin and absent from skeletal muscle sarcolemma in Duchenne muscular dystrophy. *Cell*, **82**, 743–752.

Brenman, J.E., Chao, D.S., Gee, S.H., McGee, A.W., Craven, S.E., Santillano, D.R., Wu Z., Huang, F., Xia, H., Peters, M.F., Froehner, S.C. and Bredt, D.S. (1996) Interaction of nitric oxide synthase with the postsynaptic density protein PSD-95 and α1-syntrophin mediated by PDZ domains. *Cell*, **84**, 757–767.

Busconi, L. and Michel, T. (1993) Endothelial nitric oxide synthase: N-terminal myristoylation determines subcellular localization. *J. Biol. Chem.*, **268**, 8410–8413.

Chabin, R.M., McCauley, E., Calaycay, J.R., Kelly, T.M., MacNaul, K.L., Wolfe, G.C., Hutchinson, N.I., Madhusudanaraju, S., Schmidt, J.A., Kozarich, J.W. and Wong, K.K. (1996) Active-site structure analysis of recombinant human inducible nitric oxide synthase using imidazole. *Biochemistry*, **35**, 9567–9575.

Chang, W-J., Iannaccone, S.T., Lau, K.S., Masters, B.S.S., McCabe, T.J., McMillan, K., Padre, R.C., Spencer, M.J., Tidball, J.G. and Stull, J.T. (1996) Neuronal nitric oxide synthase and dystrophin-deficient muscular dystrophy. *Proc. Natl. Acad. Sci. USA*, **93**, 9142–9147.

Chen, P-F., Tsai, A-L., Berka, V. and Wu, K.K. (1996) Endothelial nitric oxide synthase. Evidence for bidomain structure and successful reconstitution of catalytic activity from two separate domains generated by a baculovirus expression system. *J. Biol. Chem.*, **271**, 14631–14635.

Chen, P-F., Tsai, A-L., Berka, V. and Wu, K.K. (1997) Mutation of Glu-361 in human endothelial nitric-oxide synthase selectively abolishes L-arginine binding without perturbing the behaviour of heme and other redox centers. *J. Biol. Chem.*, **272**, 6114–6118.

Cho, H.J., Martin, E., Xie, Q., Sassa, S. and Nathan, C. (1995) Inducible nitric oxide synthase: Identification of amino acid residues essential for dimerization and binding of tetrahydrobiopterin. *Proc. Natl. Acad. Sci. USA*, **92**, 11514–11518.

Couet J., Li, S., Okamoto, T., Ikezu, T. and Lisanti, M.P. (1997) Identification of peptide and protein ligands for the caveolin-scaffolding domain. *J. Biol. Chem.*, **272**, 6525–6533.

Crane, B.R., Arvai, A.S., Gacchui, R., Wu, C., Ghosh, D.K., Getzoff, E.D., Stuehr, D.J. and Tainer, J.A. (1997) The structure of nitric oxide synthase oxygenase domain and inhibitor complexes. *Science*, **278**, 425–431.

Cubberley, R.R., Alderton, W.K., Boyhan, A., Charles, I.G., Lowe, P.N. and Old, R.W. (1997) Cysteine-200 of human inducible nitric oxide synthase is essential for dimerization of haem domains and for binding of haem, nitroarginine, and tetrahydrobiopterin. *Biochem. J.*, **323**, 141–146.

Dinerman, J.L., Dawson, T.M., Schell, M.J., Snowman, A. and Snyder S.H. (1994) Endothelial nitric oxide synthase localized to hippocampal pyramidal cells: implications for synaptic plasticity. *Proc. Natl. Acad. Sci. USA*, **91**, 4214–4218.

Fan, B., Wang, J., Stuehr, D.J. and Rousseau, D.L. (1997) NO synthase isozymes have distinct substrate binding sites. *Biochemistry*, **36**, 1260–1265.

Fossetta, J.D., Niu, X.D., Lunn, C.A., Zavodny, P.J., Narula, S.K. and Lundell, D. (1996) Expression of human inducible nitric oxide synthase in *Escherichia coli*. *FEBS Lett.*, **379**, 135–138.

Frey, C., Narayanan, K., McMillan, K., Spack, L., Gross, S.S., Masters, B.S.S. and Griffith, O.W. (1994) L-Thiocitrulline. A stereospecific, heme-binding inhibitor of nitric oxide synthases. *J. Biol. Chem.*, **269**, 26083–26091.

Furchgott, R.F. (1988) Studies on relaxation of rabbit aorta by sodium nitrite: the basis for the proposal that the acid-activatable inhibitory factor from bovine retractor penis is inorganic nitrite and the endothelium-derived relaxing factor is nitric oxide. In *Vasodilation: Vascular Smooth Muscle, Peptides, Autonomic Nerves, and Endothelium*, edited by P.M. Vanhoutte, pp. 401–414. New York: Raven.

Furchgott, R.F. and Zawadski, J.V. (1980) The obligatory role of endothelial cells in the relaxation of arterial smooth muscle by acetylcholine. *Nature*, **288**, 373–376.

Gacchui, R., Presta, A., Bentley, D.F., Abu-Soud, H.M., McArthur, R., Brudvig, G., Ghosh, D.K. and Stuehr, D.J. (1996) Characterization of the reductase domain of rat neuronal nitric oxide synthase generated in the methyltrophic yeast *Pichia pastoris*. Calmodulin response is complete within the reductase domain itself. *J. Biol. Chem.*, **271**, 20594–20602.

Gachhui, R., Ghosh, D.K., Wu, C., Parkinson, J., Crane, B.R. and Stuehr, D.J. (1997) Mutagenesis of acidic residues in the oxygenase domain of inducible nitric-oxide synthase identifies a glutamate involved in arginine binding. *Biochemistry*, **36**, 5097–5103.

Galli, C., MacArthur, R., Abu-Soud, H.M., Clark, P., Stuehr, D.J. and Brudvig, G.W. (1996) EPR spectroscopic characterization of neuronal NO synthase. *Biochemistry*, **35**, 2804–2810.

García-Cardeña, G., Martásek, P., Masters, B.S.S., Skidd, P.M., Couet, L., Li, S., Lisanti, M.P. and Sessa, W.C. (1997) Dissecting the interaction between nitric oxide synthase (NOS) and caveolin. Functional significance of the NOS caveolin binding domain *in vivo*. *J. Biol. Chem.*, **272**, 25437–25440.

Gerber, N.C., Rodríguez-Crespo, I., Nishida, C.R. and Ortiz de Montellano, P.R. (1997) Active site topologies and cofactor-mediated conformational changes of nitric-oxide synthases. *J. Biol. Chem.*, **272**, 6285–90.

Ghosh, D.K., Wu, C., Pitters, E., Moloney, M., Werner, E.R., Mayer, B. and Stuehr, D.J. (1997) Characterization of the inducible nitric oxide synthase oxygenase domain identifies a 49 amino acid segment required for subunit dimerization and tetrahydrobiopterin interaction. *Biochemistry*, **36**, 10609–10619.

Ghosh, D.K., Abu-Soud, H.M. and Stuehr, D.J. (1996) Domains of macrophage NO synthase have divergent roles in forming and stabilizing the active dimeric enzyme. *Biochemistry*, **35**, 1444–1449.

Ghosh, D.K., Abu-Soud, H.M. and Stuehr, D.J. (1995) Reconstitution of the second step in NO synthesis using the isolated oxygenase and reductase domains of macrophage NO synthase. *Biochemistry*, **34**, 11316–11320.

Ghosh, D.K. and Stuehr, D.J. (1995) Macrophage NO synthase: characterization of isolated oxygenase and reductase domains reveals a head-to-head subunit interaction. *Biochemistry*, **34**, 801–807.

Greaves, M.D. and Rotello, V.M. (1997) Model systems for flavoenzyme activity. Specific recognition of flavin in a silicate sol-gel. *J. Am. Chem. Soc.*, **119**, 10569–10572.

Green, L.C., De Luzuriaga, K.R., Wagner, D.A., Rand, W., Istfan, N., Young, V.R. and Tannenbaum, S.R. (1981a) Nitrate biosynthesis in man. *Proc. Natl. Acad. Sci. USA*, **78**, 7764–7768.

Green, L.C., Tannenbaum, S.R and Goldman, P. (1981b) Nitrate synthesis and reduction in the germ-free and conventional rat. *Science*, **212**, 56–58.

Griscavage, J.M., Hobbs, A.J. and Ignarro, L.J. (1995) Negative modulation of nitric oxide synthase by nitric oxide and nitroso compounds. *Advances in Pharmacology*, **34**, 215–234.

Heinzel, B., John, M., Klatt, P., Böhme, E. and Mayer, B. (1992) Ca^{2+}/calmodulin-dependent formation of hydrogen peroxide by brain nitric oxide synthase. *Biochem. J.*, **281**, 627–630.

Hevel, J.M., White, K.A. and Marletta, M.A. (1991) Purification of the inducible murine macrophage nitric oxide synthase. Identification as a flavoprotein. *J. Biol. Chem.*, **266**, 22789–22791.

Hubbard, M.J. and Klee, C.B. (1989) Functional domain structure of calcineurin A: mapping by limited proteolysis. *Biochemistry*, **28**, 1868–1874.

Hurshman, A.R. and Marletta, M.A. (1995) Nitric oxide complexes of inducible nitric oxide synthase: Spectral characterization and effect on catalytic activity. *Biochemistry*, **34**, 5627–5634.

Ignarro, L.J., Buga, F.M., Wood, K.S., Byrns, R.E. and Chaudhuri, G. (1987) Endothelium-derived relaxing factor produced and released from artery and vein is nitric oxide. *Proc. Natl. Acad. Sci. USA*, **84**, 9265–9269.

Jaffrey, S.R. and Snyder, S.H. (1996) PIN: an associated protein inhibitor of neuronal nitric oxide synthase. *Science*, **274**, 774–777.

Janssens S.P., Shimouchi A., Quertemous T., Bloch D.B. and Bloch K.D. (1992) Cloning and expression of a cDNA encoding human endothelium-derived relaxing factor/nitric oxide synthase. *J. Biol. Chem.*, **267**, 14519–14522.

Ju, H., Zou, R., Venema, V.J. and Venema, R.C. (1997) Direct interaction of endothelial nitric-oxide synthase and caveolin-1 inhibits synthase activity. *J. Biol. Chem.*, **272**, 18522–18525.

Klatt, P., Schmidt, K. and Mayer, B. (1992) Brain nitric oxide synthase is a heme protein. *Biochem. J.*, **288**, 15-17.

Klatt, P., Schmidt, K., Lehner, D., Glatter, O., Bächinger, H.P. and Mayer, B. (1995) Structural analysis of porcine brain nitric oxide synthase reveals a role for tetrahydrobiopterin and L-arginine in the formation of an SDS-resistant dimer. *EMBO J.*, **14**, 3687–3695.

Klatt, P., Pfeiffer, S., List, B.M., Lehner, D., Glatter, O., Bächinger, H.P., Werner, E.R., Schmidt, K. and Mayer, B. (1996) Characterization of heme-deficient neuronal nitric oxide synthase reveals a role for heme in subunit dimerization and binding of the amino acid substrate and tetrahydrobiopterin. *J. Biol. Chem.*, **271**, 7336–7342.

Kwon, N.S., Nathan, C.F. and Stuehr, D.J. (1989) Reduced biopterin as a cofactor in the generation of nitrogen oxides by murine macrophages. *J. Biol. Chem.*, **264**, 20496–20501.

Liu, J. and Sessa, W.C. (1994) Identification of covalently bound amino-terminal myristic acid in endothelial nitric oxide synthase. *J. Biol. Chem.*, **269**, 11691–11694.

Marletta , M.A. (1993) Nitric oxide synthase structure and mechanism. *J. Biol. Chem.*, **268**, 12231–12234.

Marsden P.A., Schappert K.T., Chen H.S., Flowers M., Sundell C.L., Wilcox J.N., Lamas S. and Michel T. (1992) Molecular cloning and characterization of human endothelial nitric oxide synthase. *FEBS Lett.*, **307**, 287–293.

Martásek, P., Roman, L.J., Raman, C.S., Liu, Q., Salerno, J.C., Ikeda-Saito, M., Gross, S.S. and Masters, B.S.S. (1998) The Cys331Ala mutant of nNOS is defective in arginine binding. In: *The biology of nitric oxide – Part 6*, edited by S. Moncada, N. Toda, H. Maeda, E.A. Higgs. London: Portland Press, *in press*.

Masters, B.S.S. (1994) Nitric oxide synthases: Why so complex? *Annu. Rev. Nutr.*, **14**, 131–145.

Mayer, B., John, M. and Böhme, E. (1990) Purification of a Ca^{++}/calmodulin-dependent nitric oxide synthase from porcine cerebellum. Cofactor-role of tetrahydrobiopterin. *FEBS Lett.*, **277**, 215–219.

Mayer, B., John, M., Heinzel, B., Werner, E.R., Wachter, H., Schultz, G. and Böhme, E. (1991) Brain nitric oxide synthase is a biopterin- and flavin-containing multi-functional oxido-reductase. *FEBS Lett.*, **288**, 187–191.

Mayer, B., Klatt, P., Werner, E. and Schmidt, K. (1994) Identification of imidazole as L-arginine-competitive inhibitor of porcine brain nitric oxide synthase. *FEBS Lett.*, **350** 199–202.

Mayer, B., Wu, C., Gorren, A.C.F., Pfeiffer, S., Schmidt, K., Clark, P., Stuehr, D.J. and Werner, E.R. (1997) Tetrahydrobiopterin binding to macrophage inducible nitric oxide synthase: Heme spin shift and dimer stabilization by the potent pterin antagonist 4-amino-tetrahydrobiopterin. *Biochemistry*, **36**, 8422–8427.

McDonald, K.K., Zharikov, S., Block, E.R. and Kilberg, M.S. (1997) A caveolar complex between the cationic amino acid transporter 1 and endothelial nitric-oxide synthase may explain the "arginine paradox". *J. Biol. Chem.*, **272**, 31213–31216.

McMillan, K. and Masters, B.S.S. (1993) Optical difference spectrophotometry as a probe of rat brain nitric oxide synthase heme-substrate interaction. *Biochemistry*, **32**, 9875–9880.

McMillan, K. and Masters, B.S.S. (1995) Prokaryotic expression of the heme- and flavin-binding domains of rat neuronal nitric oxide synthase as distinct polypeptides: Identification of the heme-binding proximal thiolate ligand as cysteine-415. *Biochemistry*, **34**, 3686–3693.

McMillan, K., Bredt, D.S. Hirsch, D.J., Snyder, S.H., Clark, J.E. and Masters, B.S.S. (1992) Cloned, expressed rat cerebellar nitric oxide synthase contains stoichiometric amounts of heme, which binds carbon monoxide. *Proc. Natl. Acad. Sci. USA*, **89**, 11141–11145.

Michel, T. and Feron, O. (1997) Nitric oxide synthases: Which, where, how, and why? *J. Clin. Invest.*, **100**, 2146–2152.

Michel, J.B., Feron, O., Sase, K., Prabhakar, P. and Michel T. (1997a) Caveolin versus calmodulin: counter-balancing allosteric modulators of nitric oxide synthase. *J. Biol. Chem.*, **272**, 25907–25912.

Michel, J.B., Feron, O., Saks, D. and Michel, T. (1997b) Reciprocal regulation of endothelial nitric-oxide synthase by Ca^{2+}-calmodulin and caveolin. *J. Biol. Chem.*, **272**, 15583–15586.

Miller, R.T., Martásek, P., Roman, L.J., Nishimura, J.S. and Masters, B.S.S. (1997) Involvement of the reductase domain of neuronal nitric oxide synthase in superoxide anion production. *Biochemistry*, **36**, 15277–15284.

Nishida K., Harrison D.G., Navas J.P., Fisher A.A., Dockery S.P., Uematsu M., Nerem R.M., Alexander R.W. and Murphy T.J. (1992) Molecular cloning and characterization of the constitutive bovine aortic endothelial cell nitric oxide synthase. *J. Clin. Invest.*, **90**, 2092–2096.

Nishimura, J.S., Martásek, P., McMillan, K., Salerno, J.C., Liu, Q., Gross, S.S. and Masters, B.S.S. (1995) Modular structure of neuronal nitric oxide synthase: Localization of the arginine binding site and modulation by pterin. *Biochem. Biophys. Res. Commun.*, **210**, 288–294.

Olken, N.M. and Marletta, M.A. (1993) N^{G}-methyl-L-arginine functions as an alternate substrate and mechanism-based inhibitor of nitric oxide synthase. *Biochemistry*, **32**, 9677–9685.

Palmer, R.M.F., Ferrige, A.G. and Moncada, S. (1987) Nitric oxide release accounts for the biological activity of endothelium-derived relaxing factor. *Nature*, **327**, 524–526.

Palmer, R.M.J., Ashton, D.S. and Moncada, S. (1988) Vascular endothelial cells synthesize nitric oxide from L-arginine. *Nature*, **333**, 664–666.

Pollock, J.S. Förstermann, U., Mitchell, J.A., Warner, T.D., Schmidt, H.H.H.W., Nakane, M. and Murad, F. (1991) Purification and characterization of particulate endothelium-derived relaxing factor synthase from cultured and native bovine aortic endothelial cells. *Proc. Natl. Acad. Sci. USA*, **88**, 10480–10484.

Pollock, J.S., Klinghofer V., Förstermann U. and Murad, F. (1992) Endothelial nitric oxide synthase is myristylated. *FEBS Lett.*, **309**, 402–404.

Pufahl, R.A., Nanjappan, P.G., Woodard, R.W. and Marletta, M.A. (1992) Mechanistic probes of N-hydroxylation of L-arginine by the inducible nitric oxide synthase from murine macrophages. *Biochemistry*, **31**, 6822–6828.

Pufahl, R.A. and Marletta, M.A. (1993) Oxidation of N^{G}-hydroxy-L-arginine by the inducible nitric oxide synthase from murine macrophages. *Biochemistry*, **31**, 6822–6828.

Pufahl, R.A., Wishnok, J.S. and Marletta, M.A. (1995) Hydrogen peroxide-supported oxidation of N^{G}-hydroxy-L-arginine by nitric oxide synthase. *Biochemistry*, **34**, 1930–1941.

Richards, M.K., Clague, M.J. and Marletta, M.A. (1996) Characterization of C415 mutants of neuronal nitric oxide synthase. *Biochemistry*, **35**, 7772–7780.

Robinson, L.J. and Michel, T. (1995) Mutagenesis of palmitoylation sites in endothelial nitric oxide synthase identifies a novel motif for dual acylation and subcellular targeting. *Proc. Natl. Acad. Sci. USA*, **92**, 11776–11780.

Rodríguez-Crespo, I., Gerber, N.C. and Ortiz de Montellano, P.R. (1996) Endothelial nitric oxide synthase. Expression in *Escherichia coli*, spectroscopic characterization, and role of tetrahydrobiopterin in dimer formation. *J. Biol. Chem.*, **271**, 11462–11467.

Rodríguez-Crespo, I., Moenne-Loccoz, P., Loehr, T.M. and Ortiz de Montellano P.R. (1997) Endothelial nitric oxide synthase: Modulations of the distal heme site produced by progressive N-terminal deletions. *Biochemistry*, **36**, 8530–8538.

Roman, L.J., Sheta, E.A., Martásek, P., Gross, S.S., Liu, Q. and Masters, B.S.S. (1995) High-level expression of functional rat neuronal nitric oxide synthase in *Escherichia coli. Proc. Natl. Acad. Sci. USA*, **92**, 8428–8432.

Ruan, J., Xie, Q., Hutchinson, N., Cho, H., Wolfe, G.C. and Nathan, C. (1996) Inducible nitric oxide synthase requires both the canonical calmodulin-binding domain and additional sequences in order to bind calmodulin and produce nitric oxide in the absence of free Ca2+. *J. Biol. Chem.*, **271**, 22679–22686.

Sakuma, I., Stuehr, D.J., Gross, S.S., Nathan, C.F. and Levi, R. (1988) Identification of arginine as a precursor of endothelium-derived relaxing factor. *Proc. Natl. Acad. Sci. USA*, **85**, 8664–8667.

Salerno, J.C., Frey, C., McMillan, K., Williams, R.F., Masters, B.S.S. and Griffith, O.W. (1995) Characterization by electron paramagnetic resonance of the interactions of L-arginine and L-thiocitrulline with the heme cofactor region of nitric oxide synthase. *J. Biol. Chem.*, **270**, 27423–27428.

Salerno, J.C., Martásek, P., Roman, L.J. and Masters, B.S.S. (1996a) Electron paramagnetic resonance spectroscopy of the heme domain of inducible nitric oxide synthase: Binding of ligands at the arginine site induces changes in the heme ligation geometry. *Biochemistry*, **35**, 7626–7630.

Salerno, J.C., McMillan, K. and Masters, B.S.S. (1996b) Binding of intermediate, product, and substrate analogs to neuronal nitric oxide synthase: Ferriheme is sensitive to ligand-specific effects in the L-arginine binding site. *Biochemistry*, **35**, 11839–11845.

Salerno, J.C., Martásek, P., Williams, R.F. and Masters, B.S.S. (1997a) Substrate and substrate analog binding to endothelial nitric oxide synthase: Electron paramagnetic resonance as an isoform-specific probe of the binding mode of substrate analogs. *Biochemistry*, **36**, 11821–11827.

Salerno, J.C., Harris, D.E., Irizarry, K., Patel, B., Morales, A.J., Smith, S.M.E., Martásek, P., Roman, L.J., Masters, B.S.S., Jones, C.L., Weissman, B.A., Liu, Q. and Gross, S.S. (1997b) A unique autoinhibitory control element functionally defines nitric oxide synthase isoforms. *J. Biol. Chem.*, **272**, 29769–29777.

Sari, M.-A., Booker, S., Jaouen, M., Vadon, S., Boucher, J., Pompon, D. and Mansuy, D. (1996) Expression in yeast and purification of functional macrophage nitric oxide synthase. Evidence for cysteine-194 as iron proximal ligand. *Biochemistry*, **35**, 7204–7213.

Sase, K. and Michel, T. (1995) Expression of constitutive endothelial nitric oxide synthase in human blood platelets. *Life Sci.*, **57**, 2049–2055.

Sase, K. and Michel, T. (1997) Expression and regulation of endothelial nitric oxide synthase. *Trends Cardiovasc. Med.*, **7**, 28–37.

Schmidt, H.H.H.W., Nau, H., Wittfoht, W., Gerlach, J., Prescher, K.E., Klein, M-M., Niroomand, F. and Böhme, E. (1988) Arginine is a physiological precursor of endothelium-derived nitric oxide. *Eur. J. Pharmacol.*, **154**, 213–216.

Sennequier, N. and Stuehr, D.J. (1996) Analysis of substrate-induced electronic, catalytic, and structural changes in inducible NO synthase. *Biochemistry*, **35**, 5883–5892.

Sessa, W.C., Harrison, J.K., Barber, C.M., Zeng, D., Durieux M.E., D'Angelo, D.D., Lynch, K.R. and Peach, M.J. (1992) Molecular cloning and expression of a cDNA encoding endothelial cell nitric oxide synthase. *J. Biol. Chem.*, **267**, 15274–15276.

Shaul, P.W., North, A.J., Wu, L.C., Wells, L.B., Brannon, T.S., Lau, K.S., Michel, T., Margraf, L.R. and Star, R.A. (1994) Endothelial nitric oxide synthase is expressed in cultured human bronchiolar epithelium. *J. Clin. Invest.*, **94**, 2231–2236.

Sheta, E.A., McMillan, K. and Masters, B.S.S. (1994) Evidence for a bidomain structure of constitutive cerebellar nitric oxide synthase. *J. Biol. Chem.*, **269**, 15147–15153.

Silvagno, F., Xia, H. and Bredt, D.S. (1996) Neuronal nitric-oxide synthase - an alternatively spliced isoform expressed in differentiated skeletal muscle. *J. Biol. Chem.*, **271**, 11204–11208.

Song, K.S., Scherer, P.E., Tang, Z., Okamoto, T., Li, S., Chafel, M., Chu, C., Kohtz, D.S. and Lisanti, M.P. (1996) Expression of caveolin-3 in skeletal, cardiac, and smooth muscle cells. *J. Biol. Chem.*, **271**, 15160–15165.

Stevens-Truss, R. and Marletta, M.A. (1995) Interaction of calmodulin with the inducible murine macrophage nitric oxide synthase. *Biochemistry*, **34**, 15638–15645.

Stuehr, D.J. (1997) Structure-function aspects in the nitric oxide synthases. *Annu. Rev. Pharmacol. Toxicol.*, **37**, 339–359.

Stuehr, D.J. and Ikeda-Saito, M. (1992) Spectral characterization of brain and macrophage nitric oxide synthases. *J. Biol. Chem.*, **267**, 20547–20550.

Stuehr, D.J. and Marletta, M.A. (1985) Mammalian nitrate biosynthesis: Mouse macrophages produce nitrate and nitrate in response to Escherichia coli lipopolysaccharide. *Proc. Natl. Acad. Sci. USA*, **82**, 7738–7742.

Stuehr, D.J., Cho, H.J., Kwon, N.S., Weise, M.F. and Nathan, C.F. (1991) Purification and characterization of the cytokine-induced macrophage nitric oxide synthase: an FAD- and FMN-containing flavoprotein. *Proc. Natl. Acad. Sci. USA*, **88**, 7773–7777.

Tayeh, M.A. and Marletta, M.A. (1989) Macrophage oxidation of L-arginine to nitric oxide, nitrate, and nitrite. Tetrahydrobiopterin is required as a cofactor. *J. Biol. Chem.*, **264**, 19654–19658.

Tierney, D.L., Martasek, P., Doan, P.E., Masters, B.S.S. and Hoffman, B.M. (1998) The location of guanidino N of L-arginine substrate bound to neuronal nitric oxide synthase: Determination by Q-band pulsed ENDOR spectroscopy. *J. Am. Chem. Soc.*, *in press*.

Tracey, W.R., Pollock, J.S., Murad, F., Nakane, M. and Förstermann U. (1994) Identification of an endothelial-like type III NO synthase in LLC-PK1 kidney epithelial cells. *Am. J. Physiol.*, **266**, C22–C28.

Tsai, A-L., Berka, V., Chen, P-F. and Palmer, G. (1996) Characterization of endothelial nitric-oxide synthase and its reaction with ligand by electron paramagnetic resonance spectroscopy. *J. Clin. Invest.*, **271**, 32563–32571.

Tzeng, E., Billiar, T.R., Robbins, P.D., Loftus, M. and Stuehr, D.J. (1995) Expression of human inducible nitric oxide synthase in a tetrahydrobiopterin (H4B)-deficient cell line: H4B promotes assembly of enzyme subunits into an active dimer. *Proc. Natl. Acad. Sci. USA*, **92**, 11771–11775.

Vasquez-Vivar, J., Hogg, N., Pritchard, K.A., Martasek, P. and Kalyanaraman B. (1997a) Superoxide anion formation from lucigenin: an electron spin resonance spin-trapping study. *FEBS Lett.*, **403**, 127–130.

Vasquez-Vivar, J., Martasek, P., Masters, B.S.S., Pritchard, K.A. and Kalyanaraman, B. (1997b) Endothelial nitric oxide synthase-dependent superoxide generation from adriamycin. *Biochemistry*, **36**, 11293–11297.

Venema V.J., Marrero M.B., and Venema R.C. (1996a) Bradykinin-stimulated protein tyrosine phosphorylation promotes endothelial nitric oxide synthase translocation to the cytoskeleton. *Biochem. Biophys. Res. Commun.*, **226**, 703–710.

Venema, R.C., Sayegh, H.S., Kent, J.D. and Harrison, D.G. (1996b) Identification, characterization, and comparison of the calmodulin-binding domains of the endothelial and inducible nitric oxide synthases. *J. Biol. Chem.*, **271**, 6435–6440.

Venema, R.C., Ju, H., Zou, R., Ryan, J.W. and Venema, V.J. (1997a) Subunit interactions of endothelial nitric-oxide synthase. Comparisons to the neuronal and inducible nitric-oxide synthase isoforms. *J. Biol. Chem.*, **272**, 1276–1282.

Venema, V.J., Ju, H., Zou, R. and Venema, R.C. (1997b) Interaction of neuronal nitric-oxide synthase with caveolin-3 in skeletal muscle. *J. Biol. Chem.*, **272**, 28187–28190.

Wang J., Stuehr D.J. and Rousseau, D.L. (1995) Tetrahydrobiopterin-deficient nitric oxide synthase has a modified heme environment and forms a cytochrome P-420 analogue. *Biochemistry*, **34**, 7080–7087.

Wang Y., Goligorsky, M.S., Lin, M., Wilcox, J.N. and Marsden P.A. (1997) A novel, testis-specific mRNA transcript encoding an NH2-terminal truncated nitric-oxide synthase. *J. Biol. Chem.*, **272**, 11392–11401.

Wever, R.M.F., van Dam, T., van Rijn, H.J.M., de Groot, F. and Rabelink T.J. (1997) Tetrahydropterin regulates superoxide and nitric oxide generation by recombinant endothelial nitric oxide synthase. *Biochem. Biophys. Res. Commun.*, **237**, 340–344.

White, K.A. and Marletta, M.A. (1992) Nitric oxide synthase is a cytochrome P-450 type hemoprotein. *Biochemistry*, **31**, 6627–6631.

Wolff, D.J. and Gribin, B.J. (1994) Interferon-gamma-inducible murine macrophage nitric oxide synthase: Studies on the mechanism of inhibition by imidazole agents. *Arch. Biophys. Biochem.*, **311**, 293–299.

Wu, C., Zhang, J., Abu-Soud, H., Ghosh, D.K. and Stuehr, D.J. (1996) High-level expression of mouse inducible nitric oxide synthase in *Escherichia coli* requires coexpression with calmodulin. *Biochem. Biophys. Res. Commun.*, **222**, 439–444.

Xie, Q.W., Cho, H.J., Calaycay, J., Mumford, R.A., Swiderek, K.M., Lee, T.D., Ding, A., Troso, T. and Nathan, C. (1992) Cloning and characterization of inducible nitric oxide synthase from mouse macrophages. *Science*, **256**, 225–228.

Xia, Y., Roman, L.J., Masters, B.S.S. and Zweier, J.L. (1998) Inducible nitric oxide synthase generates superoxide from the reductase domain. *J. Biol. Chem.*, **273**, 22635–22639.

Xia, Y., Dawson, V.L., Dawson, T.M., Snyder, S.H. and Zweier, J.L. (1996) Nitric oxide synthase generates superoxide and nitric oxide in arginine-depleted cells leading to peroxynitrite-mediated cellular injury. *Proc. Natl. Acad. Sci. USA*, **93**, 6770–6774.

Xia, Y. and Zweier, J.L. (1997) Superoxide and peroxynitrite generation from inducible nitric oxide synthase in macrophages. *Proc. Natl. Acad. Sci. USA*, **94**, 6954–6958.

Zoche, M., Bienert, M., Beyermann, M. and Koch, K.-W. (1996) Distinct molecular recognition of calmodulin-binding sites in the neuronal and macrophage nitric oxide synthases: A surface plasmon resonance study. *Biochemistry*, **35**, 8742–8747.

3 Nitric Oxide Synthase Gene Regulation

Kenneth Kun-yu Wu

Vascular Biology Research Center and Division of Hematology, University of Texas-Houston Medical School, Houston, Texas USA, and
Institute of Biomedical Sciences, Academia Sinica, Taipei, Taiwan
E-mail: kkwu@heart.med.uth.tmc.edu; Fax: (713) 500-6812; Tel: (713) 500-6801

Three isoforms of nitric oxide synthase are encoded by discrete genes. Expressions of these genes are under tight regulation. Neuronal NOS (nNOS or NOS-1) and endothelial NOS (eNOS or NOS-3) are constitutively expressed. Their expressions are upregulated by stress. Functional studies of eNOS promoter regions reveal that binding of SP1 to sites within 100-bp of the major transcription site is critical for constitutive and lysophosphatidylcholine-induced expression. Inducible NOS (iNOS or NOS-2) is constitutively not or weakly expressed but highly induced by lipopolysaccharide (LPS) and cytokines. The promoter region of murine iNOS contains two clusters of putative binding sites and response elements. LPS induction of iNOS involves the proximal region (nucleotide −35 to −279) while interferon-γ induction involves the distal region (−900 to −1150). NF-κB sites in the proximal region and distal region are important in iNOS induction.

Key words: Nitric oxide, nitric oxide synthase, nitric oxide synthase gene regulation, neuronal nitric oxide synthase, endothelial nitric oxide synthase, inducible nitric oxide synthase.

INTRODUCTION

Nitric oxide (NO) is an important messenger molecule, mediating diverse physiological functions such as long-term potential, vascular wall integrity, vascular relaxation, platelet reactivity and defense against microorganisms and tumor proliferation (1). Excessive productions under pathological conditions lead to tissue injury and inflammatory changes (Moncada *et al.*, 1991; Marletta, 1993). NO synthesis is catalyzed by a heme-containing bifunctional enzyme, nitric oxide synthase (NOS). NOS catalyzes the conversion of L-arginine to L-citrulline and NO. The catalytic reaction requires NADPH, FAD, FMN and tetrahydrobiopterin (BH$_4$). Three isoforms of NOS have been cloned and characterized. Neuronal NOS (nNOS or NOS-1) constitutively expressed primarily in neurons and skeletal muscle cells is cytosolic and requires Ca^{+2}/calmodulin (CaM) binding for its catalytic activity. Cytokine-inducible NOS (iNOS or NOS-2), expressed in a wide variety of cells, is cytosolic. Endothelial NOS (eNOS or NOS-3), constitutively expressed primarily in vascular endothelial cells, is membrane-associated via myristoylation and palmitoylation and requires Ca^{+2}/CaM binding for its catalytic activity. Sequence comparison of the three isoforms of NOS shows a sequence identity of 60–80%.

All three isoforms of NOS have a similar bidomain structure (McMillan and Masters, 1995; Chen *et al*, 1996). The carboxyl half of the molecule contains reductase activity and the amino half, oxygenase activity. The middle of the molecule bears the CaM binding site. The reductase domain contains binding sites for NADPH, FAD and FMN. These binding sites are conserved in all three isoforms of NOS. The oxygenase domain contains heme, BH$_4$ and L-arginine binding sites. The heme ligand on the oxygenase domain has recently been identified to be Cys-184 in eNOS (Chen *et al.*, 1994) and this site is conserved in iNOS and nNOS (Richards *et al.*, 1996). We have recently identified Glu-361 of eNOS as the L-arginine binding site (Chen *et al.*, 1997), which is verified by X-ray crystallographic structure of iNOS oxygenase domain (Crane *et al.*, 1998).

All the studies have so far reported a high degree of structural and functional similarities among the three isoforms of NOS. Expression of these three isozymes is, on the other hand, controlled by distinct transcriptional and post-transcriptional regulatory mechanisms. As NOS undergoes autoinactivation during catalysis (Abu-Soud *et al.*, 1995), their distinct, regulated expressions provide an essential mechanism for maintaining a cellular level of the isozymes to carry out physiological functions. It is, hence, important to elucidate in details the mechanisms by which the expressions of these genes are regulated.

EXPRESSION OF NOS

nNOS and eNOS are constitutively expressed in neuronal and vascular endothelial cells, respectively. iNOS is not constitutively expressed but is induced by lipopolysaccharide (LPS), interferon-γ (IFN-γ), interleukin-1 (IL-1) and tumor necrosis factor α (TNF-α) in many types of cells. It has been estimated that constitutively-expressed nNOS and eNOS catalyze the synthesis of NO in nanomolar concentrations while the induced iNOS catalyzes the formation of NO in micromolar concentrations. The striking difference in the concentration of NO synthesized by the constitutive vs. inducible isoforms of NOS is probably to meet the demands for maintaining normal physiological functions by the constitutively expressed enzymes and protecting against microbial infection and tumor growth as has been suggested for iNOS.

The constitutively expressed eNOS and nNOS are inactive in quiescent cells. These constitutive enzymes become catalytically active when neuronal or endothelial cells are stimulated by physiological agonists whereby their intracellular calcium levels are elevated and binding of calcium to CaM leads to a conformational change of CaM to increase its binding affinity for nNOS or eNOS. Binding of CaM to the constitutive NOS facilitates electron transfer from NADPH via FAD and FMN to the heme reaction center with the eventual catalytic conversion of L-arginine to NO and L-citrulline. By contrast, iNOS, once induced is catalytically active without requirement for an elevation of cytosolic calcium levels, despite a similar reaction mechanism. This has attributed to a difference in the CaM binding site between iNOS and constitutive NOS in that the former has a much higher binding affinity for CaM (Venema *et al*, 1996). Hence, the catalytic activity of constitutive NOS depends on cell activation while the activity of inducible NOS is governed by the induced de novo synthesis of the enzyme in response to stimulation.

Recent studies have provided evidence to indicate that the constitutively-expressed eNOS and nNOS are also inducible. It has been shown that eNOS mRNA levels are upregulated by shear stress (Nishida *et al*, 1992), physical exercise (Sessa *et al*, 1994) and lysophosphatidylcholine (lysoPC) (Zembowicz *et al.*, 1995). The mechanisms by which chemical and physical mediators induce eNOS expression are not entirely clear. We have carried out a series of experiments to determine the transcriptional stimulation of eNOS expression. The results are discussed in the next section.

UPREGULATION OF eNOS EXPRESSION

LysoPC is a major product of LDL oxidation (Parthasarathy *et al.*, 1989). Its levels in atherosclerotic lesions are several fold higher than in normal tissue (Portman and Alexander, 1969). Several lines of evidence suggest a role of lysoPC in atherogenesis: (1) lysoPC is a chemotactant for monocytes (Nakano *et al*, 1994); (2) It induces several endothelial cell genes expressed in early atherosclerosis such as vascular cell adhesion molecule-1 (VCAM-1), intercellular adhesion molecule-1 (ICAM-1), platelet derived growth factor A and B chains and heparin-binding epidermal growth factor (Kume *et al.*, 1992; Kume and Gimbrone, 1994). To test the hypothesis that lysoPC induces endothelial cell genes that code for vasoprotective proteins, we evaluated the effect of lysoPC on eNOS expression in cultured human umbilical vein endothelial cells (HUVEC). LysoPC increased eNOS mRNA levels in a concentration and time-dependent manner (Zembowicz *et al.*, 1995). Maximal stimulation was noted when HUVECs were treated with lysoPC 100 µM in medium containing 5% fetal bovine serum for 4 h. As lysoPC binds to serum albumin in the medium, the free lysoPC concentration in the medium is much lower than 100 µM and did not have any toxic effect on HUVEC. The eNOS mRNA level estimated by densitometry of eNOS mRNA bands on Northern blots was maximally increased by lysoPC by about 5-fold over the basal level (Zembowicz *et al.*, 1995). The stimulatory effect of lysoPC lasted for up to 24 h. LysoPC-induced increase in eNOS protein levels was in accordance with eNOS mRNA increase. The eNOS protein level in HUVEC was maximally increased by lysoPC by about 5-fold over the basal level (Zembowicz *et al.*, 1995). LysoPC induced eNOS mRNA and protein increases were inhibited by actinomycin D and cycloheximide (Zembowicz *et al.*, 1995). Nuclear run-off experiments revealed a 2–3 fold increase in eNOS mRNAs in HUVEC by lysoPC treatment for 3 h. Taken together, these results indicate that lysoPC increases the transcription of eNOS.

BASAL AND LYSOPC-INDUCED eNOS PROMOTER FUNCTION

Human eNOS gene spans 21 kb in chromosome 7q 35–36 (Marsden *et al.*, 1993). It comprises 26 exons. The 5'-flanking region of eNOS gene does not contain a canonical TATA box. It is G and C rich and bears potential binding sites for Sp1, AP2, AP1, NF-1, as well as response elements for heavy metal, acute phase reactants, shear stress and corticosteroids (Marsden *et al.*, 1993). The promoter function of eNOS has recently been reported (Zhang *et al.*, 1995; Tang *et al.*, 1995). We constructed a 1.3-kb 5'-flanking fragment (nucleotides −1322 to +22) into a promoterless luciferase expression vector, pXP1 and transfected it along with pSV-LacZ into cultured HUVEC by lipofection. This 5'-flanking fragment conferred considerable promoter activity when compared to the vector control (Tang *et al.*, 1995). A series of 5'-deletion mutants were prepared and constructed into the pXP1. The promoter activity was unaltered by 5'-deletion until the deletion reached nucleotides −355 to +22 when the luciferase activity was reduced to approximately half of that of the parent fragment. The promoter activity of fragment −265 to +22 was recovered to be approximately the same as the parent fragment and this full promoter activity was conferred by fragment −198 to +22. These results suggest a suppressor element between nucleotides −355 and −265. Furthermore, the elements that are involved in basal promoter activity reside at the region between −198 and +22 (Tang *et al.*, 1995). To identify the elements in this region that are critical for basal promoter activity, DNase I footprinting was performed by incubating fragment –265 to +22 with HUVEC nuclear extracts or purified Sp1. A region –89 to –109 was clearly protected by Sp1. This region bears two overlapping binding sites (–104 GGGGCGGGG –96 and –99 GGGGCGAGG –91). Electrophoretic mobility shift assays using fragment –100 to –22 and fragment –265 to –22 revealed two shifted bands competed out by Sp1 but not by AP2 oligonucleotides (Tang *et al.*, 1995). Sp1 antibodies caused a supershift of the bands using –265/–22 as the template. By contrast, there was no band shift when fragment –94/–22 which is devoid of Sp1 binding sites was used as the template. These data indicate that the two overlapped Sp1 sites, especially the –99 to –91 site bind Sp1 or its related proteins. To establish the functional role of this Sp1 binding region, we mutated four nucleotides (–100G→T, –98G→A, –96G→T and –95G→A) in fragment –521/+22. Transient transduction experiments in HUVEC showed that the promoter activity conferred by the mutant was reduced to less than 10% of that of the wild fragment. Similar results were obtained by Zhang *et al.* (1995) who carried out the transfection experiments in bovine aortic endothelial cells (BAEC). Taken together, these findings unequivocally establish binding of Sp1 to –104/–95 to be essential for basal transcription of eNOS gene.

We then determined the site on the 5'-promoter region that is involved in lysoPC-induced transcriptional activation. The minimal promoter that is upregulated by lysoPC resides between –198 and +22. In these experiments, we identified, in addition to Sp1 binding to the –104/–95 site, binding of PEA3-related proteins to a site at –32 AAGGAA –27. LysoPC treatment of HUVEC increased Sp1 binding to the –104/–95 site. Mutation of 4Gs (–100, –98, –96 and –95) resulted in an >90% reduction in the promoter activity in response to lysoPC stimulation accompanied by diminished binding of Sp1 to the transfected 5'-flanking fragment (Cieslik *et al.*, 1998). LysoPC increased Spl binding activity by about 5-fold over the based binding activity (Cieslik *et al.*, 1998). Okadaic acid exerted a concentration-dependent abrogation of Spl binding and promoter augmentation by lysoPC while sodium vanadate had no effect (Cieslik *et al.*, 1998). These results are consistent with the involvement of protein serine/threonine phosphatases in regulating Spl binding activity and its transactivation of eNOS gene. LysoPC increased nuclear PP2A activity but had no effect on PP2B or PP2C. Furthermore, PP1 specific inhibitor did not alter Sp1 binding activity induced by lysoPC. These results are consistent with involvement of PP2A in catalyzing dephosphorylation of Sp1. In this context, it may be postulated that there must be a nuclear kinase that controls the basal Sp1 binding activity. The identity of this kinase is being investigated. We conclude from these experiments that eNOS transcription is dynamically regulated by Sp1 phosphorylation and dephosphorylation under the action of kinase and phosphatases.

EXPRESSION AND REGULATION OF nNOS GENE

Expression of nNOS in neurons is tightly regulated. The nNOS gene is constitutively expressed and is inducible. Its expression has been shown to be induced in spinal cord neurons (Wu, 1993) and dorsal root ganglion (Verge *et al.*, 1992) by avulsion and proximal transection, respectively. The mechanisms by which these traumatic events induce nNOS expressions have not been elucidated.

Human nNOS gene is mapped to chromosome 12q24 (Hall *et al.*, 1994). The 5'-flanking region is characterized by multiple transcription start sites and absence of a canonical TATA box. It has recently been proposed that basal transcription of nNOS is driven by two distinct promoters separated by about 300-bp at the 5′-flanking region (Xie *et al.*, 1995). Two distinct transcripts which have identical coding region and the 3′-end differ at the 5′-flanking region. The distinct 5′-flanking region appeared to represent two different non-coding exons. Genomic cloning and sequence analysis reveal that the unique exons separated by less than 300-bp but separated from the common exon 2 by more than 20 kb. It is postulated that these two transcripts with distinct 5′- unique exons are the products of alternate splicing. These two transcripts were shown to be expressed in cultured cells by reverse transcription-polymerase chain reaction (RT-PCR). The physiological significance is, however, unclear. It has been suggested that this may serve as a control mechanism for nNOS expression (Xie *et al.*, 1995).

It has further been shown that Oct-2 transcription factor plays an important role in nNOS gene transcription (Deans *et al.*, 1996). Oct-2 transcription factor comprises several members in the family. Oct-2.1, expressed primarily in B lymphocytes, plays an important role in B-cell activity. Oct-2.4 and Oct-2.5, on the other hand, predominantly expressed in neuronal cells, inhibit activation of the herpes simplex virus and neuronal tyrosine hydroxylase. By using an antisense approach to reduce Oct-2, it was shown that nNOS expression was suppressed when Oct-2 levels were reduced by antisense nucleotides. It was further shown that of the two promoters that drive nNOS gene expression, the downstream promoter is activated by Oct-2. Hence, it is possible that Oct-2 plays an important role in regulating the transcriptional activation of one form of nNOS transcript. The physiological relevance of this is unclear and requires further investigation.

INDUCTION OF iNOS GENE EXPRESSION

Many reports have confirmed the original observations in mouse macrophages (Stuehr and Marletta, 1985) that iNOS message expression is inducible in diverse cell types of different animal species by LPS, IFN-γ, IL-1β and, TNF-α. In mouse macrophage cell lines (J744 and RAW264.7 cells) and peritoneal macrophages, iNOS mRNA and protein were maximally induced by LPS and IFN-γ (Ding *et al.*, 1990; Weisz *et al.*, 1994; Wang *et al.*, 1994; Lorsbach *et al.*, 1993; Adler *et al.*, 1994). iNOS in other types of cells in mice (mammary adenocarcinoma cell line, EMT-6, brain endothelial cells and VEC cells) is also maximally induced by LPS and IFN-γ following a similar time course (Kilbourn and Belloni, 1990; Amber *et al.*, 1988). A number of studies have shown that iNOS in diverse rat cells (mesangial cells, hepatocytes, adult cyocytes, astroglial cells) was induced by LPS augmented by TNF, PMA, IL-1 and IFN-γ (Stevens *et al.*, 1996; Archer *et al.*, 1995; Kunz *et al.*, 1996; Hortelano *et al.*, 1992; Geller *et al.*, 1993; Balligand *et al.*, 1994; Lysiak *et al.*, 1995). Experiments with rat aortic smooth muscle cells reported iNOS induction by combined TNFα and IL-1β (Nakayama *et al.*, 1994; Perrella *et al.*, 1994; Koide *et al.*, 1993). Maximal induction of iNOS in human cells (neutrophils, astrocytoma, colorectal carcinoma cell line DLD-1, lung cancer cell line A549, lung epithelial cells, mesangial cells) requires combination of cytokines with or without LPS (Kanno *et al.*, 1994; Evans *et al.*, 1996; Mollace *et al.*, 1993; Heck *et al.*, 1992; Ansano *et al.*, 1994; Weinberg *et al.*, 1995; Saura *et al.*, 1996; Boczkowski *et al.*, 1996).

Endotoxin or LPS administered intravenously into rats induced iNOS expressions in multiple organs, including lungs, liver, heart and spleen (Wu *et al.*, 1995; Knowles *et al.*, 1990; Nagasaki *et al.*, 1996; Oguchi *et al.*, 1992). Administration of parasites to mice is accompanied by iNOS induction which was blocked by IL-4 or anti-IFN-γ antibodies (Proudfoot *et al.*, 1996). Dexamethasone has been shown to suppress iNOS induction in many cell types of different animal species including human cells (Amber *et al.*, 1988; Knoeles *et al.*, 1990; Radomski *et al.*, 1990). Acidic and basic fibroblast growth factor (FGF), epidermal growth factor (EGF), macrophage stimulating protein (MSP), transforming growth factor β (TGFβ) and macrophage deactivating factor (MDF) have been shown to suppress iNOS induction in selected cells (Ding *et al.*, 1990; Lysiak *et al.*, 1995; Perrella *et al.*, 1994; Asano *et al.*, 1994). Collectively, these findings indicate that iNOS expression is dynamically regulated by inducers notably, IFN-γ, LPS, TNF-α, IL-1β and microorganisms and suppressors such as dexamethasone and growth factors. Induction of iNOS by microorganisms is an important host defense mechanism against the invading microorganism, as evidenced by a study in iNOS gene knockout mice in which mycobacterium tuberculesis replicate faster in lungs of these animals (MacMicking *et al.*, 1997).

Several intracellular signal molecules are shown to be involved in iNOS induction. It has been demonstrated that an elevation of cellular cyclic AMP levels induces iNOS expression in rat mesangial cells and the effect of cAMP is synergistic with IL-1β (Kunz *et al.*, 1994). In addition to increasing the rate of iNOS transcription, cAMP prolonged iNOS mRNA half-life suggesting an effect on iNOS mRNA stability. A heterodimeric G protein, $G\alpha_{13}$ overexpression in mouse epithelial cells is accompanied by iNOS induction whereas α_s, α_{i1} and α_q overexpressions have no effects on iNOS expression (Kenichiro *et al.*, 1996). $G\alpha_{13}$ is coupled to thromboxane and thrombin receptors. These results indicate multiple signal pathways for iNOS induction.

The signal transduction pathway via which cytokines and LPS induce iNOS expression is not completely clear. A recent study reveals that MAP kinase, ERK1/ERK2 activation is necessary for iNOS induction by IL-1β and IFN-γ in rat cardiac myocytes and cardiac microvascular endothelial cells (Singh *et al.*, 1996). It further indicates that protein kinase C-linked activation of Ras or Raf may be involved in MAP kinase activation and iNOS induction. The exact signal transduction pathway and the mechanism by which the signal(s) induce iNOS gene expression require further investigation.

TRANSCRIPTIONAL ACTIVATION OF iNOS GENE

The human iNOS gene is mapped to chromosome 17 between p 13.1 and q 25 (Xu *et al.*, 1994). Chartrain *et al.* (1994) reported the human iNOS gene to span about 37-kb and contain 26 exons and 25 introns. A more recent report by Xu *et al.* (1996) indicates that human iNOS gene is about 40 kb in length and consists of 27 exons and 26 exons. The reason for this discrepancy is unclear at the present time. In either case, the translation start codon resides in exon 2.

There is a consensus that iNOS is encoded by a single gene. However, Block *et al.* (1995) reported cloning of three related human iNOS genes from human genomic libraries. A similar suggestion has been made regarding rat iNOS genes. As the structure of these related genes have not been elucidated, it remains uncertain whether these observations are physiologically relevant.

Multiple alternatively spliced iNOS mRNA species at the 5'-untranslated region have been reported (Chu *et al.*, 1995). These results are reminiscent of human nNOS mRNAs (Xie *et al.*, 1995). Significance of alternate splicing of iNOS or nNOS mRNAs at exon 1 in regulation of the translation of these two isozymes remains to be investigated.

The promoter structure and function of mouse iNOS gene have been extensively investigated (Xie *et al.*, 1993; Lowenstein *et al.*, 1993). A canonical TATA box is located about 30-bp upstream from the major transcription start site. Multiple putative binding sites for transcription activators or response elements for signal molecules are located within a 1500-bp fragment of the 5'-flanking region. These sites are clustered in two regions: the proximal region from nucleotide –35 to –279 bears putative binding sites or response elements for IFN-γ (γ-IRE), IFN-γ activation (γ-activated sites or GAS), nuclear factor κB (NF-κB), IFN-α-stimulation (ISRE), activated protein 1 (AP-1), tumor necrosis factor (TNF-RE) and NF-IL6 and the distal region from –900 to –1150 bears sites for ISRE, GAS, TNF-RE, NF-κB, γ-IRE, AP1 and IFN regulatory factor-1 (IRF-1). Transient transfection of progressive 5'-deletion mutants of iNOS promoter into mouse macrophages has revealed that LPS induction of iNOS involves the proximal region while IFN-γ induction requires the distal region of the iNOS gene (Xie *et al.*, 1993; Lowenstein *et al.*, 1993). Xie *et al.* (1994) have further demonstrated that an NF-κB binding site in the proximal region (–85 to –76) is essential for LPS-induced iNOS expression. Electrophoretic mobility assays indicate binding of NF-κB P50 homodimers and P50/RelA(P65) heterodimers to this NF-κB binding site. By contrast, Spink *et al.* (1995) reported an essential role of the upstream NK-κB site (nucleotides –957 to –979) for cytokine-induced iNOS transcription in rat vascular smooth muscle cells. It has further been shown *in vivo* in IRF-1 deleted mice the complete abolition of NO synthesis and iNOS induction by IFN-γ and LPS (Kamijo *et al.*, 1994). There are two putative IRF-1 binding sites in the proximal enhancer region. IRF-1 binds to these sites. This *in vivo* study suggests a critical role of IRF-1 in IFN-γ and LPS-induced iNOS expression (Kamijo *et al.*, 1994).

Human iNOS promoter differs considerably from mouse or rat iNOS promoter. Unlike murine promoter region which is located at the proximal 1 kb from the transcription start

Table 3-1. Comparison of expressions of three NOS genes.

	nNOS (NOS-1)	*eNOS (NOS-3)*	*iNOS (NOS-2)*
Expression	constitutive, regulated	constitutive, regulated	induced
Inducing Agents	neuronal trauma	lysoPC shear & mechanical stress	murine M: LPS & IFNγ human cells: TNFα IL-1β IFN-γ LPS
5'-flanking region	TATA-less, two distinct promoters	TATA-less	canonical TATA
Transcriptional Activation	Oct-2	Sp1	NF-κB IRF-1
mRNA Transcripts	multiple forms	single	multiple forms

site, the human promoter activity requires a region longer than 7.2 kb (de Vera *et al.*, 1996; Taylor *et al.*, 1998). Analysis of the promoter activity reveals that multiple NF-κB sites located between –3.8 and –7.2 kb are involved in human iNOS induction by cytokines (Taylor *et al.*, 1998).

SUMMARY AND CONCLUSION

All three isoforms of NOS are inducible. The level of induction is different: iNOS levels are higher than the two constitutive NOS. Mechanisms of transcriptional activation of three genes are also different. NF-κB is a key transcription factor in iNOS expression while Sp1 is critical for eNOS transcription. Table 3-1 summarizes the inducing agents and the transcription regulators of three isoforms of NOS.

Basal expressions of eNOS and nNOS are of fundamental importance in maintaining blood pressure and CNS long-term potentials. Induced eNOS and nNOS expressions by stress signals are considered to play an important role in augmenting defense mechanisms. Induction of iNOS is thought to be important in controlling microbial infections and tumor proliferation. However, inappropriate iNOS induction probably contributes to tissue injury, inflammation, etc., via the formation of NO-related molecules, such as peroxynitrite. Further *in vivo* studies are required to determine the pathophysiological roles in iNOS inductions in various disease states.

ACKNOWLEDGMENTS

The author wishes to thank Artur Zembowicz, Jih-Lu Tang and Pei-Feng Chen for collaboration; Susan Mitterling and Angela Wang for assistance in preparing the manuscript. The work is supported by grants from National Institutes of Health, U.S.A. (NS-23327 and HL-50675).

REFERENCES

Abu-Soud, H.M., Wang, J., Rousseau, D.L., Fukuto, J.M., Ignarro, L.J. and Stuehr, D.J. (1995) Neuronal nitric oxide synthase self-inactivates by forming a ferrous-nitrosyl complex during aerobic catalysis. *J. Biol. Chem.*, **270**, 22997–23006.

Adler, H., Peterhans, E., Nicolet, J. and Jungi, T.W. (1994) Inducible L-arginine-dependent nitric oxide synthase activity in bovine bone marrow-derived macrophages. *Biochem. Biophys. Res. Commun.*, **198**, 510–515.

Amber, I.J., Hibbs Jr., J.B., Taintor, R.R. and Vavrin, Z. (1988) Cytokines induce an L-arginine-dependent effector system in nonmacrophage cells. *J. Leukoc. Biol.*, **44**, 58–65.

Archer, S.L., Freude, K.A. and Shultz, P.J. (1995) Effect of graded hypoxia on the induction and function of inducible nitric oxide synthase in rat mesangial cells. *Circ. Res.*, **77**, 21–8.

Asano, K., Chee, C.B., Gaston, B., Lilly, C.M., Gerard, C., Drazen, J.M. *et al.* (1994) Constitutive and inducible nitric oxide synthase gene expression, regulation, and activity in human lung epithelial cells. *Proc. Natl. Acad. Sci.*, **91**, 10089–10093.

Balligand, J.L., Ungureanu-Longrois, D., Simmons, W.W., Pimental, D., Malinski-T.A., Kapturczak, M. *et al.* (1994) Cytokine-inducible nitric oxide synthase (iNOS) expression in cardiac myocytes. Characterization and regulation of iNOS expression and detection of iNOS activity in single cardiac myocytes in vitro. *J. Biol. Chem.*, **269**, 27580–27588.

Bloch, K.D., Wolfram, J.R., Brown, D.M., Roberts Jr., J.D., Zapol, D.G., Lepore, J.J. *et al.*, (1995) Three members of the nitric oxide synthase II gene family (NOS2A, NOS2B, and NOS2C) colocalize to human chromosome 17. *Genomics*, **27**, 526–530.

Boczkowski, J., Lanone, S., Ungureanu-Longrois, D., Danialou, G., Fournier, T. and Aubier, M. (1996) Induction of diaphragmatic nitric oxide synthase after endotoxin administration in rats: role on diaphragmatic contractile dysfunction. *J. Clin. Invest.*, **98**, 1550–1559.

Chartrain, N.A., Geller, D.A., Koty, P.P., Sitrin, N.F., Nussler, A.K., Hoffman, E.P. *et al.* (1994) Molecular cloning, structure, and chromosomal localization of the human inducible nitric oxide synthase gene. *J. Biol. Chem.*, **269**, 6765–6772.

Chen, P.F., Tsai, A.L., Berka, V. and Wu, K.K. (1996) Endothelial nitric-oxide synthase. Evidence for bidomain structure and successful reconstitution of catalytic activity from two separate domains generated by a baculovirus expression system. *J. Biol. Chem.*, **271**, 14631–14635.

Chen, P.F., Tsai, A.L. and Wu, K.K. (1994) Cysteine 184 of endothelial nitric oxide synthase is involved in heme coordination and catalytic activity. *J. Biol. Chem.*, **269**, 25062–25066.

Chen, P.F., Tsai, A-L., Berka, V. and Wu, K.K. (1997) Mutation of GLU-361 in human endothelial nitric oxide synthase selectively abolished L-arginine binding without perturbing the behavior of heme and other redox centers. *J. Biol. Chem.*, **242**, 6114–6118.

Chu, S.C., Wu, H.P., Banks, T.C., Eissa, N.T. and Moss, J. (1995) Structural diversity in the 5'-untranslated region of cytokine-stimulated human inducible nitric oxide synthase mRNA. *J. Biol. Chem.*, **270**, 10625–10630.

Cieslik, K., Zembowicz, A., Tang, J-L. and Wu, K.K. (1998) Transcriptional regulation of endothelial nitric-oxide synthase by lysophosphatidylcholine. *J. Biol. Chem.*, **273**, 14885–14890.

Crane, B.R., Arvai, A.S., Ghosh, D.K., Wu, C., Getzoff, E.D., Stuehr, D.J. and Tainer, J.A. (1998) Structure of nitric oxide synthase oxygenase dimer with pterin and subtrate. *Science*, **278**, 2121–2126.

Deans, Z., Dawson, S.J., Xie, J., Young, A.P., Wallace, D. and Latchman, D.S. (1996) Differential regulation of the two neuronal nitric-oxide synthase gene promoters by the oct-2 transcription factor. *J. Biol. Chem.*, **271**, 32153–32158.

Ding, A., Nathan, C.F., Graycar, J., Derynck, R., Stuehr, D.J. and Srimal, S. (1990) Macrophage deactivating factor and transforming growth factors-beta 1 -beta 2 and -beta 3 inhibit induction of macrophage nitrogen oxide synthesis by IFN-gamma. *J. Immunol.*, **145**, 940–944.

Evans, T.J., Buttery, L.D., Carpenter, A., Springall, D.R., Polak, J.M. and Cohen, J. (1996) Cytokine-treated human neutrophils contain inducible nitric oxide synthase that produces nitration of ingested bacteria. *Proc. Natl. Acad. Sci.*, **93**, 9553–9558.

Geller, D.A., Nussler, A.K., Silvio, M.D., Lowenstein, C.J., Shapiro, R.A., Wang, S.T. *et al.* (1993) Cytokines, endotoxin, and glucocorticoids regulate the expression of inducible nitric oxide synthase in hepatocytes. *Proc. Natl. Acad. Sci.*, **90**, 522–526.

Hall, A.V., Antoniou, H., Wang, Y., Cheung, A.H., Arbus, A.M., Olson, S.L. *et al.* (1994) Structural organization of the human neuronal nitric oxide synthase gene (NOS1). *J. Biol. Chem.*, **269**, 33082–33090.

Heck, D.E., Laskin, D.L., Gardner, C.R. and Laskin, J.D. (1992) Epidermal growth factor suppresses nitric oxide and hydrogen peroxide production by keratinocytes. Potential role for nitric oxide in the regulation of wound healing. *J. Biol.Chem.*, **267**, 21277–21280.

Hortelano, S., Genaro, A.M. and Bosca, L. (1992) Phorbol ester induce nitric oxide synthase activity in rat hepatocytes. antagonism with the induction elicited by lipopolysaccharide. *J. Biol. Chem.*, **267**, 24937–24940.

Kamijo, R., Harada, H., Matsuyama, T., Bosland, M. Gerecitano, J. Shapiro, D. *et al.* (1994) Requirement for transcription factor IRF-1 in NO synthase induction in macrophages. *Science*, **263**, 1612–1615.

Kanno, K., Hirata, Y., Imai, T., Iwashina, M. and Marumo, F. (1994) Regulation of inducible nitric oxide synthase gene by interleukin-1 beta in rat vascular endothelial cells. *Am. J. Physiol.*, **267** (6 Pt 2), H2318–2324.

Kenichiro, K., Singer, W.D., Star, R.A., Muallem, S. and Miller, R.T. (1996) Induction of inducible nitric-oxide synthase by the heterotrimeric G protein G_{13}. *J. Biol. Chem.*, **271**, 7412–7415.

Kilbourn, R.G. and Belloni, P. (1990) Endothelial cell production of nitrogen oxides in response to interferon gamma in combination with tumor necrosis factor, interleukin-1, or endotoxin. *J. Natl. Cancer. Inst.*, **82**, 772–776.

Knowles, R.G., Salter, M., Brooks, S.L. and Moncada, S. (1990) Anti-inflammatory glucocorticoids inhibit the induction by endotoxin of nitric oxide synthase in the lung, liver and aorta of the rat. *Biochem. Biophys. Res. Commun.*, **172**, 1042–1048.

Koide, M., Kawahara, Y., Tsuda, T. and Yokoyama, M. (1993) Cytokin-induced expression of an inducible type of nitric oxide synthase gene in cultured vascular smooth muscle cells. Cytokine-induced expression of an inducible type of nitric oxide synthase gene in cultured vascular smooth muscle cells. *FEBS.*, **318**, 213–217.

Kume, N., Cybulsky, M.I. and Gimbrone Jr., M.A. (1992) Lysophosphatidylcholine, a component of atherogenic lipoproteins, induces mononuclear leukocyte adhesion molecules in cultured human and rabbit arterial endothelial cells. *J. Clin. Invest.*, **90**, 1138–1144.

Kume, N. and Gimbrone Jr., M.A. (1994) Lysophosphatidylcholine transcriptionally induces growth factor gene expression in cultured human endothelial cells. *J. Clin. Invest.*, **93**, 907–911.

Kunz, D., Walker, G., Eberhardt, W. and Pfeilschifter, J. (1996) Molecular mechanisms of dexamethasone inhibition of nitric oxide synthase expression in interleukin 1-stimulates mesangial cells: evidence for the involvement of transcriptional and posttranscriptional regulation. *Proc. Natl. Acad. Sci.*, **93**, 255–259.

Kunz, D., Muhl, H., Walker, G. and Pfeilschifter, J. (1994) Two distinct signaling pathways trigger the expression of inducible nitric oxide synthase in rat renal mesangial cells. *Proc. Natl. Acad. Sci.*, **91**, 5387–5391.

Lorsbach, R.B., Murphy, W.J., Lowenstein, C.J., Snyder, S.H. and Russell, S.W. (1993) Expression of nitric oxide synthase gene in mouse macrophages activated for tumor cell killing. *J. Biol. Chem.*, **268**, 1908–1913.

Lowenstein, C.J., Alley, E.W., Raval, P., Snowman, A.M., Snyder, S.H., Russell, S.W. *et al.* (1993) Macrophage nitric oxide synthase gene: two upstream regions mediate induction by interferon gamma and lipopolysaccharide. *Proc. Natl. Acad. Sci.*, **90**, 9730–9734.

Lysiak, J.J., Hussaini, I.M., Webb, D.J., Glass 2nd, W.F., Allietta, M. and Gonias, S.L. (1995) Alpha 2-macroglobulin functions as a cytokine carrier to induce nitric oxide synthesis and cause nitric oxide-dependent cytotoxicity in the RAW 264.7 macrophage cell line. *J. Biol. Chem.*, **270**, 21919–21927.

MacMicking, J.D., North, R.J., LaCourse, R., Mudgett, J.S., Shah, S.K. and Nathan, C.F. (1997) Identification of nitric oxide synthase as a protective focus against tuberculosis. *Proc. Natl. Acad. Sci. USA*, **94**, 5243–5248.

Marletta, M.A. (1993) Nitric oxide synthase structure and mechanism. *J. Biol. Chem.*, **268**, 12231–12234.

Marsden, P.A., Heng, H.H., Scherer, S.W., Stewart, R.J., Hall, A.V., Shi, X.M. *et al.* (1993) Structure and chromosomal localization of the human constitutive endothelial nitric oxide synthase gene. *J. Biol. Chem.*, **268**, 17478–17488.

McMillan, K. and Masters, B.S. (1995) Prokaryotic expression of the heme- and flavin-binding domains of rat neuronal nitric oxide synthase as distinct polypeptides: identification of the heme-binding proximal thiolate ligand as cysteine-415. *Biochemistry*, **34**, 3686–3693.

Mollace, V., Colasanti, M., Rodino, P., Massoud, R., Lauro, G.M. and Nistico, G. (1993) Cytokine-induced nitric oxide generation by cultured astrocytoma cells involves Ca^{+2}-calmodulin-independent NO-synthase. *Biochem. Biophys. Res. Commun.*, **191**, 327–334.

Moncada, S., Palmer, R.M. and Higgs, E.A. (1991) Nitric oxide: physiology, pathophysiology, and pharmacology. *Pharmacol. Rev.*, **43**, 109–142.

Nagasaki, A., Gotoh, T., Takeya, M., Yu, Y., Takiguchi, M., Matsuzaki, H. *et al.* (1996) Coinduction of nitric oxide synthase, argininosuccinate synthetase, and argininosuccinate lyase in lipopolysaccharide-treated rats. RNA blot, immunoblot, and immunohistochemical analyses. *J. Biol. Chem.*, **271**, 2658–2662.

Nakano, T., Raines, E.W., Abraham, J.A., Klagsbrun, M. and Ross, R. (1994) Lysophosphatidylcholine upregulates the level of heparin-binding epidermal growth factor-like growth factor mRNA in human monocytes. *Proc. Natl. Acad. Sci.*, **91**, 1069–1073.

Nakayama, I., Kawahara, Y., Tsuda, T., Okuda, M. and Yokoyama, M. (1994) Angiotension II inhibits cytokine-stimulated inducible nitric oxide synthase expression in vascular smooth muscle cells. *J. Biol. Chem.*, **269**, 11628–11633.

Nishida, K., Harrison, D.G., Navas, J.P., Fisher, A.A, Dockery, S.P., Uematsu, *et al.* (1992) Molecular cloning and characterization of the constitutive bovine aortic endothelial cell nitric oxide synthase. *J. Clin. Invest.*, **90**, 2092–2096.

Oguchi, S., Iida, S., Adachi, H., Oshima, H. and Esumi, H. (1992) Induction of Ca^{+2}/calmodulin-dependent NO synthase in various organs of rats by Propionibacterium acnes and lipopolysaccharide treatment. *FEBS*, **308**, 22–25.

Parthasarathy, S., Quinn, M.T., Schwenke, D.C., Carew, T.E. and Steinberg, D. (1989) Oxidative modification of beta-very low density lipoprotein. Potential role in monocyte recruitment and foam cell formation. *Arteriosclerosis*, **9**, 398–404.

Perrella, M.A., Yoshizumi, M., Fen, Z., Tsai, J.C., Hsieh, C.M., Kourembanas, S. *et al.* (1994) Transforming growth factor-1, but not dexamethasone, down-regulates nitric-oxide synthase mRNA after its induction by interleukin-1 in rat smooth muscle cells. *J. Biol. Chem.*, **269**, 14595–14600.

Portman, O.W. and Alexander, M. (1969) Lysophosphatidylcholine concentrations and metabolism in aortic intima plus inner media: effect of nutritionally induced atherosclerosis. *J. Lipid Res.*, **10**, 158–165.

Proudfoot, L., Nikolaev, A.V., Feng, G.J., Wei, W.Q., Ferguson, M.A., Brimacombe, J.S. *et al.* (1996) Regulation of the expression of nitric oxide synthase and leishmanicidal activity by glycoconjugates of Leishmania lipophosphoglycan in murine macrophages. *Proc. Natl. Acad. Sci.*, **93**, 10984–10989.

Radomski, M.W., Palmer, R.M. and Moncada, S. (1990) Glucocorticoids inhibit the expression of an inducible, but not the constitutive, nitric oxide synthase in vascular endothelial cells. *Proc. Natl. Acad. Sci.*, **87**, 10043–10047.

Richards, M.K., Clague, M.J. and Marletta, M.A. (1996) Characterization of C415 mutants of neuronal nitric oxide synthase. *Biochemistry*, **35**, 7772–7780.

Saura, M., Perez-Sala, D., Canada, F.J. and Lamas, S. (1996) Role of tetrahydrobiopterin availability in the regulation of nitric-oxide synthase expression in human mesangial cells. *J. Biol. Chem.*, **271**, 14290–14295.

Sessa, W.C., Pritchard, K., Seyedi, N., Wang, J. and Hintze, T.H. (1994) Chronic exercise in dogs increases coronary vascular nitric oxide production and endothelial cell nitric oxide synthase gene expression. *Circ. Res.*, **74**, 349–353.

Singh, K., Balligand, J.L., Fischer, T.A., Smith, T.W. and Kelly, R.A. (1996) Regulation of cytokine-inducible Nitric oxide synthase in cardiac myocytes and microvascular endothelial cells. *J. Biol. Chem.*, **271**, 1111–1117.

Spink, J., Cohen, J. and Evans, T.J. (1995) The cytokine responsive vascular smooth muscle cell enhancer of inducible nitric oxide synthase. Activation by nuclear factor-kappa B. *J. Biol. Chem.*, **270**, 29541–29547.

Stevens, B.R., Kakuda, D.K., Yu, K., Waters, M., Vo, C.B. and Raizada, M.K. (1996) Induced nitric oxide synthesis in dependent on induced alternatively spliced *CAT-2* encoding l-arginine transport in brain astrocytes. *J. Biol. Chem.*, **271**, 24017–24022.

Stuehr, D.J. and Marletta, M.A. (1985) Mammalian nitrate biosynthesis: mouse macrophages produce nitric and nitrate in response to *escherichia coli* lipopolysaccharide. *Proc. Natl. Acad. Sci.*, **82**, 7738–7742.

Tang, J.L., Zembowicz, A., Xu, X.M. and Wu, K.K. (1995) Role of Sp1 in transcriptional activation of human nitric oxide synthase type III gene. *Biochem. Biophys. Res. Commun.*, **213**, 673–680.

Taylor, B.S., de Vera, M.E., Ganster, R.W., Wang, Q., Shapiro, R.A., Morris, S.M., Billiar, T.R. and Geller, D.A. (1998) Multiple NF-κB enhancer elements regulate cytokine induction of the human inducible nitric oxide synthase gene. *J. Biol. Chem.*, **273**, 15148–15156.

Venema, R.C., Sayegh, H.S., Kent, J.D. and Harrison, D.G. (1996) Identification, characterization, and comparison of the calmodulin-binding domains of the endothelial and inducible nitric oxide synthases. *J. Biol. Chem.*, **271**, 6435–6440.

De Vera, M.E., Shapiro, R.A., Nussler, A.K., Mudget, J.S., Simmons, R.L., Morris, S.M., Billiar, T.R. and Geller, D.A. (1996) Transcriptional regulation of human inducible nitric oxide synthase gene by cytokines: initial analysis of the human NOS2 promoter. *Proc. Natl. Acad. Sci. USA*, **93**, 1054–1059.

Verge, V.M., Xu, Z., Xu, X.J., Wiesenfeld, Hallin Z. and Hokfelt, T. (1992) Marked increase in nitric oxide synthase mRNA in rat dorsal root ganglia after peripheral axotomy: *in situ* hybridization and functional studies. *Proc. Natl. Acad. Sci.*, **89**, 11617–11621.

Wang, M.H., Cox, G.W., Yoshimura, T., Sheffler, L.A., Skeel, A. and Leonard, E. (1994) Macrophage-stimulating protein inhibits induction of nitric oxide production by endotoxin-or cytokine-stimulated mouse macrophages. *J. Biol. Chem.*, **269**, 14027–14031.

Weinberg, J.B., Misukonis, M.A., Shami, P.J., Mason, S.N., Sauls, D.L. and Dittman, W.A. (1995) Human mononuclear phagocyte inducible nitric oxide synthase (iNOS): analysis of iNOS mRNA, iNOS protein, biopterin, and nitric oxide production by blood monocytes and peritoneal macrophages. *Blood*, **86**, 1184–1195.

Weisz, A., Oguchi, S., Cicatiello, L. and Esumi, H. (1994) Dual Mechanism for the control of inducible-type NO synthase gene expression in macrophages during activation by interferon-and bacterial lipopolysaccharide. transcriptional and post-transcriptional regulation. *J. Biol. Chem.*, **269**, 8324–8333.

Wu, C.C., Croxtall, J.D., Perretti, M., Bryant, C.E., Thiemermann, C., Flower, R.J. *et al.* (1995) Lipocortin 1 mediates the inhibition by dexamethasone of the induction by endotoxin of nitric oxide synthase in the rat. *Proc. Natl. Acad. Sci.*, **92**, 3473–3477.

Wu, W. (1993) Expression of nitric-oxide synthase (NOS) in injured CNS neurons as shown by NADPH diaphorase histochemistry. *Exp. Neurol.*, **120**, 153–159.

Xie, J., Roddy, P., Rife, T.K., Murad, F. and Young, A.P. (1995) Two closely linked but separable promoters for human neuronal nitric oxide synthase gene transcription. *Proc. Natl. Acad. Sci.*, **92**, 1242–1246.

Xie, Q.W., Whisnant, R. and Nathan, C. (1993) Promoter of the mouse gene encoding calcium-independent nitric oxide synthase confers inducibility by interferon gamma and bacterial lipopolysaccharide. *J. Exp. Med.*, **177**, 1779–1784.

Xie, Q.W., Kashiwabara, Y. and Nathan, C. (1994) Role of transcription factor NF-kappa B/Rel in induction of nitric oxide synthase. *J. Biol. Chem.*, **269**, 4705–4708.

Xu, W., Charles, I.G., Moncada, S., Gorman, P., Sheer, D., Liu, L. *et al.* (1994) Mapping of the genes encoding human inducible and endothelial nitric oxide synthase (NOS2 and NOS3) to the pericentric region of chromosome 17 and to chromosome 7, respectively. *Genomic*, **21**, 419–422.

Xu, W., Charles, I.G., Liu, L., Moncada, S., Emson, P. (1996) Molecular cloning and structural organization of the human inducible nitric oxide synthase gene (NOS2). *Biochem. Biophys. Res. Commun.*, **219**, 784–788.

Zembowicz, A., Tang, J.L. and Wu, K.K. (1995) Transcriptional induction of endothelial nitric oxide synthase type III by lysophosphatidylcholine. *J. Biol. Chem.*, **270**, 17006–17010.

Zhang, R., Min, W. and Sessa, W.C. (1995) Functional analysis of the human endothelial nitric oxide synthase promoter. Sp1 and GATA factors are necessary for basal transcription in endothelial cells. *J. Biol. Chem.*, **270**, 15320–15326.

4 Localization and Subcellular Targeting of Nitric Oxide Synthases

Guillermo García-Cardeña and William C. Sessa

*Department of Pharmacology, Yale University School of Medicine,
Boyer Center for Molecular Medicine, 295 Congress Avenue,
New Haven CT 06536-0812, USA*

Key words: nitric oxide synthase, cellular trafficking, immunocytochemistry, cell biology, fatty acylation

INTRODUCTION

Nitric oxide (NO) is produced by the nitric oxide synthase family of proteins: neuronal (nNOS or NOS1), inducible (iNOS or NOS 2) and endothelial (eNOS or NOS 3) (Sessa, 1994). NOSs are cytochrome P-450 hemoproteins that utilize L-arginine as the substrate, NADPH and molecular oxygen as co-substrates and calmodulin as a critical regulator. Embedded as prosthetic groups in the proteins are heme iron, flavin adenine dinucleotide and flavin mononucleotide (Griffith and Stuehr, 1995). NOSs exhibit approximate 40–50% sequence identity across the family and each isoform is highly homologous across various species (80–94% identity) (Knowles and Moncada, 1994). Based on findings accumulating over the past several years, all NOSs are catalytically competent to produce NO only as homodimers and for iNOS and to lesser extent nNOS and eNOS, tetrahydrobiopterin (BH_4) is an essential co-factor for dimerization *in vitro* and *in vivo* (Klatt *et al.*, 1996; Lee *et al.*, 1995; Nathan and Xie, 1994; Tzeng *et al.*, 1995).

As alluded to above, the production of NO can be regulated by a variety of factors (oxygen, L-arginine, calcium, calmodulin and BH_4) that directly impinge on the catalytic heart of NOS. However, in the past several years a novel concept has emerged suggesting that cell type specific subcellular targeting of the enzyme provides a dynamic mechanism to control NO synthesis. Therefore, in this review we will highlight the recent advances in tissue specific NOS expression and subcellular targeting.

NEURONAL NITRIC OXIDE SYNTHASE

Tissue Distribution

Initial mapping of nNOS in brain revealed the highest amounts of immunoreactive protein in cerebellar granule cells and olfactory bulb. Subsequent immunohistochemical studies demonstrated that nNOS is expressed in certain areas of the spinal cord (Dun *et al.*, 1992), sympathetic ganglia, adrenal glands (Dun *et al.*, 1993), and peripheral nitrergic nerves (Sheng *et al.*, 1992), in epithelial cells of lung, uterus and stomach (Schmidt *et al.*, 1992), in pancreatic islet cells (Schmidt *et al.*, 1992), juxtaglomerular cells of the kidney and in skeletal muscle (Kobzik *et al.*, 1994).

Subcellular Targeting

nNOS was purified as 150 kDa protein from cytosolic fractions of rat and porcine cerebellum (Bredt and Snyder, 1990; Schmidt *et al.*, 1991). In contrast to nNOS purified from brain, human and rat nNOS isolated from skeletal muscle is mostly membrane associated (>80%). After the cloning of the three prototypical NOS isoforms, it was clear that the amino terminus of nNOS was markedly different from that of iNOS and eNOS and may contribute to the unusual differences in nNOS fractionation between cytosol and membranes. nNOS has a unique, 210 amino acid N-terminal extension that contains a PDZ motif, named after the three proteins that contain repeats of this domain, post-synaptic density protein-95 (**PSD**-95), *Drosophila* disc large protein (**D**lg) and zona occludens protein-1 (**Z**O-1) (Hendriks, 1995). In the core of the nNOS PDZ motif are the amino acids GLGF embedded in a sequence of 100 flanking amino acids. The PDZ motif permits nNOS to interact biochemically with other PDZ motifs found in PSD-95 and a related protein, PSD-93, in brain extracts (Brenman *et al.*, 1996). PSD-95 also interacts with other neural signaling proteins such as the NMDA receptor and the Shaker family of K^+ channels via the carboxy terminal amino acid motifs ESDV or ETDV (Gomperts, 1996). Interestingly, peptides derived from these regions of the NMDA receptor and Shaker block the interaction of nNOS with PSD-95 suggesting that they compete for the similar binding sites (Brenman *et al.*, 1996). The precise functional reasons for these molecular interactions of nNOS with PDZ motif containing proteins are not known but likely represents another example of a growing theme in biology; i.e. compartmentalization of signal transduction to defined regions of the cell. In nerves, it is possible that nNOS interacts with other PSD-95 associated proteins such as the NMDA receptor and a voltage dependent K^+ channel, and that locally produced nitric oxide can influence NMDA receptor, K-channel gating or the assembly of proteins into the post-synaptic density of neurons. It is not known if nNOS bound to PSD-95 is more or less active than the substantial amount of active enzyme in the cytosol of nerves, however the specific docking of the enzyme highly suggests that regulation of both pools will be different and perhaps fine tune post-synaptic signaling and neurotransmission.

In the periphery, PDZ domain interactions also mediate binding of nNOS to the skeletal muscle protein α1-syntrophin, a dystrophin associated protein. This interaction is responsible for the targeting of nNOS to the sarcolemma of fast-twitch fibers of skeletal muscle. Interestingly, *mdx* mice and humans with Duchenne muscular dystrophy (DMD) demonstrate a selective loss of nNOS protein and catalytic activity from muscle membranes, suggesting that aberrant regulation of nNOS may contribute to preferential degeneration of fast-twitch muscle fibers in DMD (Brenman *et al.*, 1995; Chang *et al.*, 1996). In addition, nNOS in sarcolemmal membranes may influence excitation-contraction coupling as evidenced by the ability of NOS inhibitors to increase the amplitude of muscle contraction (Kobzik *et al.*, 1994). Further elucidation of specific amino acids required for nNOS interaction with PSD proteins and α1-syntrophin will permit detailed analysis of the functional importance of these interactions.

INDUCIBLE NITRIC OXIDE SYNTHASE

Tissue Distribution

iNOS can be induced in many cells types in response to cytokines, bacterial products such as lipopolysacharride and several other immunostimulants (Moncada *et al.*, 1991). In general, the expression of iNOS is pleiotropic, occurring in mostly all nucleated cells that are cytokine responsive. Immunohistochemical studies of iNOS in rats treated with *Propionibacterium acnes* and LPS demonstrated the enzyme in macrophages, neutrophils and eosinophils in the spleen, Kupffer cells, endothelial cells and hepatocytes in liver, alveolar macrophages in lung, macrophages and endothelial cells in adrenal glands, and eosinophils, mast cells and endothelial cells in colon (Bandaletova *et al.*, 1993). iNOS immuoreactivity has been reported in pancreatic islets of diabetic BB rats, where the immunoreactivity was restricted to areas of islet infiltration by macrophages (Kleeman *et al.*, 1993). Kobzik *et al.*, 1993) found strong iNOS labeling of macrophages from LPS-treated, but not untreated rats. Human alveolar macrophages were occasionally iNOS immunreactive, especially in areas of inflammation. iNOS immunoreactivity has been shown in alveolar macrophages from patients with acute bronchopneumonia, whereas no immunoreactivity was detected in grossly normal human lung tissue (Tracey *et al.*, 1994). Recently, iNOS has been shown in foam cells of atherosclerotic lesions (Buttery *et al.*, 1996) and in human neutrophils, either activated *in vitro* (Evans *et al.*, 1996) or isolated from patients with urinary tract infections (Wheeler *et al.*, 1997).

Subcellular Targeting

iNOS was purified as a 133 kDa protein from the cytosol of activated murine macrophages. In serially passaged murine macrophages, the majority of iNOS is clearly cytosolic. Recently, both in primary cultures of murine macrophages and in activated human neutrophils isolated from patients with ongoing urinary tract infections, a large percentage of the total iNOS activity resides within the particulate fraction (Evans *et al.*, 1996; Wheeler *et al.*, 1997). The reasons why iNOS localization varies in different cell lines are not obvious. Based on the cDNA cloning, iNOS does not contain a PDZ motif or is it fatty acylated suggesting a novel mechanism for subcellular targeting into membrane. More importantly, the significance of two pools of iNOS are not clear. In cytokine stimulated human neutrophils, iNOS colocalizes with myeloperoxidase within neutrophil primary granules (Evans *et al.*, 1996). In primary cultures of murine macrophages and in neutrophils isolated from patients with urinary tract infections (Wheeler *et al.*, 1997) iNOS localizes in the perinuclear region of the cells. The precise nature of that perinuclear staining remains to be defined, but appears to be regions of the Golgi-complex and in an unidentified vesicle population (Vodovotz *et al.*, 1995). As suggested by these authors, perhaps the localization of iNOS in vesicles will facilitate a higher local concentration of NO in the vicinity of a phagocytized pathogen.

ENDOTHELIAL NITRIC OXIDE SYNTHASE

Tissue Distribution

eNOS was first identified as the isoform responsible for the production of endothelium-derived relaxing factor in endothelial cells (Moncada *et al.*, 1991). Immunohistochemical studies demonstrated the presence of eNOS in several types of arterial and venous endothelial cells in many tissues, including human tissues (Pollock *et al.*, 1993). eNOS immunoreactivity has also been detected in syncytiotrophoblasts of human placenta (Myatt *et al.*, 1993), LLC-PK1 kidney tubular epithelial cells (Tracey *et al.*, 1994), platelets (Sase and Michel, 1995) and cardiac myocytes (Balligand *et al.*, 1995). Interestingly eNOS immunoreactivity has also been described in neurons of the CA1 region of the hippocampus and in other brain regions (Dinerman *et al.*, 1994).

Subcellular Targeting

eNOS was originally purified as a 135 kDa membrane protein from both freshly isolated and cultured bovine aortic endothelial cells (Pollock *et al.*, 1991). eNOS is membrane associated by virtue of co-translational N-myristoylation at glycine 2 and post-translational cysteine palmitoylation at positions 15 and 26 (Liu *et al.*, 1995; Liu and Sessa, 1994). In cultured endothelial cells and intact blood vessels, eNOS is localized on membranes of the Golgi complex and on plamalemma caveolae of endothelial cells (García-Cardeña *et al.*, 1996; Sessa *et al.*, 1995; Shaul *et al.*, 1996). Mutation of the N-myristoylation site inhibits the subsequent palmitoylation and prevents Golgi and caveolae targeting (García-Cardeña *et al.*, 1996; Liu *et al.*, 1995) (Liu *et al.*, 1997). Interestingly, mutation of the palmitoylation sites blocks caveolae targeting suggesting that palmitoylation is a "molecular zip code" for eNOS (García-Cardeña *et al.*, 1996). In both circumstances, inhibition of proper targeting results in a 50–70% reduction in stimulated NO production demonstrating that cellular compartmentalization is critical for optimal NOS activation (Liu *et al.*, 1996; Sessa *et al.*, 1995). What is the functional consequence of impaired localization, *in vivo*? A recent demonstration using a paradigm of memory (long-term potentiation, LTP) in CA1 region neurons of the hippocampus (rich in eNOS) shows that inhibition of eNOS N-myristoylation blocks membrane targeting and attenuates LTP providing support for the concept that compartmentalization is critical for NO production (Kantor *et al.*, 1996). Thus, it is likely that targeting onto the cytoplasmic face of the Golgi via N-myristoylation and then into caveolae via cysteine palmitoylation are important events necessary for placing eNOS into a milieu rich in NOS substrates (L-arginine, NADPH, oxygen) cofactors (Ca^{2+}, tetrahydrobiopterin and flavins) and NOS regulatory proteins (calmodulin and caveolin).

By analogy to nNOS, palmitoylation of eNOS plays a role in its targeting near to other signaling proteins to form a highly efficient signal transduction cascade in discrete regions of endothelial cells (García-Cardeña *et al.*, 1996). Interestingly, we and others have showed that eNOS interacts in endothelial cells with caveolin-1 (Feron *et al.*, 1996; García-Cardeña *et al.*, 1996), and in cardic myocytes with caveolin-3 (Feron *et al.*, 1996). This interaction is through two cytoplasmic domains of caveolin-1, the scaffolding domain (amino acids 61–101) and the carboxy terminal tail (amino acids 153–178). The scaffolding domain of caveolin is also the docking site for Ha-Ras, c-Src and G-protein α subunits, suggesting that in addition to its role as a structural protein for caveolae, specific protein-protein interactions between caveolin and other resident proteins could regulate signal transduction (Li *et al.*, 1996; Li *et al.*, 1995; Song *et al.*, 1996). Incubation of pure eNOS with peptides derived from the scaffolding domains of caveolin-1 and -3 result in inhibition of eNOS, iNOS and nNOS activities (García-Cardeña *et al.*, 1997). These results suggest a common mechanism and site of inhibition. The region of eNOS that interacts with caveolin was mapped using a yeast two-hybrid system and GST-fusion proteins to the oxygenase domain (Ju *et al.*, 1997), between amino acids 310 and 570 (García-Cardeña *et al.*, 1997). Site directed mutagenesis of the predicted caveolin binding motif within eNOS blocked the ability of caveolin-1 to supress NO release in co-transfection experiments (García-Cardeña *et al.*, 1997). These results suggests that caveolin negatively regulates NO production and that additional mechanisms of regulation must prevail during stimulation of NO release by shear stress, growth factors and calcium-mobilizing agonists.

Conclusion and Perspectives

The availability of high quality antibodies has allowed investigators to anatomically map the NOS isoforms in normal and disease tissues. However the appearance or disappearance of a given NOS isoform still does answer the question of whether NO is beneficial or detrimental to the disease process. Better understanding of the physiological and patho-physiological roles of NO may be facilitated by the generation of isoform selective knockout mice (Huang and Fishman, 1996). With respect to subcellular targeting, we are viewing the tip of the iceberg. Clearly, several aspects of complex cell signaling require compartmentalization of proteins for local activation and inactivation of biochemical pathways. All the NOS isoforms display specific subcellular targeting suggesting that mutiple protein-protein or protein-lipid interactions are yet to be discovered. Elucidation of such interactions will increase our understanding of the molecular mechanisms of NO generation. Finally, it is important for us as a investigators in this exciting field to appreciate the subtle differences in NOS targeting and to unravel the significance of crude fractionation of NOS into cytosol and membranes.

ACKNOWLEDGMENTS

This work was supported by NIH grants R29-HL51948 and RO-1HL57665 and a AHA Grants (National and Connecticut Affiliate). WCS is an Established Investigator of the American Heart Association.

REFERENCES

Balligand, J.L., Kobzik, L., Han, X., Kaye, D.M., Belhassen, L., O'Hara, D.S., Kelly, R.A., Smith, T.W. and Michel, T. (1995) Nitric oxide-dependent parasympathetic signaling is due to activation of constitutive endothelial (type III) nitric oxide synthase in cardiac myocytes. *J. Biol. Chem.*, **270**, 14582–6.

Bandaletova, T., Brouet, I., Bartsch, H., Sugimura, T., Esumi, H. and Ohshima, H. (1993) Immunohistochemical localization of an inducible form of nitric oxide synthase in various organs of rats treated with propioni-bacterium-acnes and lipopolysaccharide. *APMIS*, **101**, 330–336.

Bredt, D.S. and Snyder, S.H. (1990) Isolation of nitric oxide synthetase, a calmodulin-requiring enzyme. *Proc. Natl. Acad. Sci. USA*, **87**, 682–685.

Brenman, J.E., Chao, D.S., Gee, S.H., McGee, A.W., Craven, S.E., Santillano, D.R., Wu, Z., Huang, F., Xia, H., Peters, M.F., Froehner, S.C. and Bredt, D.S. (1996) Interaction of nitric oxide synthase with the postsynaptic density protein PSD-95 and a1-syntrophin mediated by PDZ domains. *Cell*, **84**, 757–67.

Brenman, J.E., Chao, D.S., Xia, H., Aldape, K. and Bredt, D.S. (1995) Nitric oxide synthase complexed with dystrophin and absent from skeletal muscle sarcolemma in Duchenne muscular dystrophy. *Cell*, **82**, 743–52.

Buttery, L.D., D.R., S., Chester, A.H., Evans, T.J., Standfield, E.N., Parums, D.V., Yacoub, M.H. and Polak, J.M. (1996) Inducible nitric oxide synthase is present within human atherosclerotic lesions and promotes the formation and activity of peroxynitrite. *Lab. Invest.*, **75**, 77–85.

Chang, W.J., Iannaccone, S.T., Lau, K.S., Masters, B.S., McCabe, T.J., McMillan, K., Padre, R.C., Spencer, M.J., Tidball, J.G. and Stull, J.T. (1996) Neuronal nitric oxide synthase and dystrophin-deficient muscular dystrophy. *Proc. Natl. Acad. Sci. USA*, **93**, 9142–9147.

Dinerman, J.L., Dawson, T.M., Schell, M.J., Snowman, A. and Snyder, S.H. (1994) Endothelial nitric oxide synthase localized to hippocampal pyramidal cells: Implications for synaptic plasticity. *Proc. Natl. Acad. Sci. USA*, **91**, 4214–4218.

Dun, N.J., Dun, S.L., Wu, S.Y. and Forstermann, U. (1993) Nitric oxide synthase immunoreactivity in rat superior cervical ganglia and adrenal glands. *Neurosci. Lett.*, **158**, 51–54.

Dun, N.L., Dun, S.L., Fostermann, U. and Tseng, L.F. (1992) Nitric oxide synthase immunoreactivity in rat spinal cord. *Neurosci. Lett.*, **147**, 217–220.

Evans, T.J., Buttery, L.D.K., Carpenter, A., Springall, D.R., Polak, J.M. and Cohen, J. (1996) Cytokine-treated human neutrophils contain inducible nitric oxide synthase that produces nitration of ingested bacteria. *Proc. Natl. Acad. Sci. USA*, **93**, 9553–9558.

Feron, O., Belhassen, L., Kobzik, L., Smith, T.W., Kelly, R.A. and Michel, T. (1996) Endothelial nitric oxide synthase targeting to caveolae. Specific interactions with caveolin isoforms in cardiac myocytes and endothelial cells. *J. Biol. Chem.*, **271**, 22810–22814.

García-Cardeña, G., Fan, R., Stern, D.F., Liu, J. and Sessa, W.C. (1996) Endothelial nitric oxide synthase is regulated by tyrosine phosphorylation and interacts with caveolin-1. *J. Biol. Chem.*, **271**, 27237–27240.

García-Cardeña, G., Martasek, P., Masters, B.S.S., Skidd, P.M., Couet, J., Li, S., Lisanti, M.P. and Sessa, W.C. (1997) Dissecting the interaction between nitric oxide synthase (NOS) and caveolin. Vol. 272, 25437–25440.

García-Cardeña, G., Oh, P., Liu, J., Schnitzer, J.E. and Sessa, W.C. (1996) Targeting of nitric oxide synthase to endothelial cell caveolae via palmitoylation; implications for nitric oxide signaling. *Proc. Natl. Acad. Sci. USA*, **93**, 6448–6453.

Gomperts, S.N. (1996) Clustering membrane proteins: it's all coming together with the PSD-95/SAP90 protein family. *Cell*, **84**, 659–662.

Griffith, O.W. and Stuehr, D.J. (1995) Nitric oxide synthases: properties and catalytic mechanism. [Review]. *Annu. Rev. Physiol.*, **57**, 707–36.

Hendriks, W. (1995) Neuronal nitric oxide synthase contains a discs-large homologous region (DHR) sequence motif. *Biochem. J.*, **305**, 687–688.

Huang, P.L. and Fishman, M.C. (1996) Genetic analysis of nitric oxide synthase isoforms: targeted mutation in mice. *J. Mol. Med.*, **74**, 415–421.

Ju, H., Zou, R., Venema, V. and Venema, R. C. (1997) Direct interaction of endothelial nitric oxide synthase and caveolin-1 inhibits synthase activity. *J. Biol. Chem.*, **272**, 18522–18525.

Kantor, D.B., Lanzrein, M., Stary, S.J., Sandoval, G.M., Smith, W.B., Sullivan, B.M., Davidson, N. and Schuman, E.M. (1996) A role for endothelial NO synthase in LTP revealed by adenovirus-mediated inhibition and rescue. *Science*, **274**, 1744–1748.

Klatt, P., Pfeiffer, S., List, B.M., Lehner, D., Glatter, O., Bachinger, H.P., Werner, E.R., Schmidt, K. and Mayer, B. (1996) Characterization of heme-deficient neuronal nitric oxide synthase reveals a role for heme in subunit dimerization and binding of the amino acid substrate and tetrahydrobiopterin. *J. Biol. Chem.*, **271**, 7336–7342.

Kleeman, R., Rothe, H., Kolb-Bachofen, V., Xie, Q. W., Nathan, C., Martin, S. and Kolb, H. (1993) Transcription and translation of inducible nitric oxide synthase in the pancreas of prediabetic BB rats. *FEBS Lett.*, **328**, 9–12.

Knowles, R.G. and Moncada, S. (1994) Nitric oxide synthases in mammals. *Biochem. J.*, **298**, 249–258.

Kobzik, L., Bredt, D.S., Lowenstein, C.J., Drazen, J., Gaston, B., Sugarbaker, D. and Stamler, J.S. (1993) Nitric oxide synthase in human and rat lung: Immunocytochemical and histochemical localization. *Am. J. Respir. Cell. Mol. Biol.*, **9**, 371–377.

Kobzik, L., Reid, M.B., Bredt, D.S. and Stamler, J.S. (1994) Nitric oxide in skeletal muscle. *Nature*, **372**, 546–588.

Lee, C.M., Robinson, L.J. and Michel, T. (1995) Oligomerization of endothelial nitric oxide synthase. Evidence for a dominant negative effect of truncation mutants. *J. Biol. Chem.*, **270**, 27403–6.

Li, S., Couet, J. and Lisanti, M.P. (1996) Src tyrosine kinases, Ga sununits and H-Ras share a common membrane-anchored scaffolding protein, caveolin. *J. Biol. Chem.*, **271**, 29182–29190.

Li, S., Okamoto, T., Chun, M., Sargiacomo, M., Casanova, J.E., Hansen, S.H., Nishimoto, I. and Lisanti, M.P. (1995) Evidence for a regulated interaction between heterotrimeric G proteins and caveolin. *J. Biol. Chem.*, **270**, 15693–701.

Liu, J., García-Cardeña, G. and Sessa, W.C. (1995) Biosynthesis and palmitoylation of endothelial nitric oxide synthase: mutagenesis of palmitoylation sites, cysteines-15 and/or -26, argues against depalmitoylation-induced translocation of the enzyme. *Biochemistry*, **34**, 12333–40.

Liu, J., García-Cardeña, G. and Sessa, W.C. (1996) Palmitoylation of endothelial nitric oxide synthase is necessary for optimal stimulated release of nitric oxide: implications for caveolae localization. *Biochemistry*, **35**, 13277–13281.

Liu, J., Hughes, T.E. and Sessa, W.C. (1997) The first 35 amino acids and fatty acylation sites determine the molecular targeting of endothelial nitric oxide synthase into the golgi region of cells: A green fluorescent protein study. *J. Cell Biol.*, **137**, 1525–1535.

Liu, J. and Sessa, W.C. (1994) Identification of covalently bound amino-terminal myristic acid in endothelial nitric oxide synthase. *J. Biol. Chem.*, **269**, 11691–4.

Moncada, S., Palmer, R.M. and Higgs, E.A. (1991) Nitric oxide: physiology, pathophysiology and pharmacology. [Review]. *Pharmacol. Rev.*, **43**, 109–42.

Myatt, L., Brockman, D.E., Eis, A. and Pollock, J.S. (1993) Immunohistochemical localization of nitric oxide synthase in the human placenta. *Placenta*, **14**, 487–495.

Nathan, C. and Xie, Q.W. (1994) Nitric oxide synthases: roles, tolls and controls. [Review]. *Cell*, **78**, 915–8.

Pollock, J.S., Forstermann, U., Mitchell, J.A., Warner, T.D., Schmidt, H.H., Nakane, M. and Murad, F. (1991) Purification and characterization of particulate endothelium-derived relaxing factor synthase from cultured and native bovine aortic endothelial cells. *Proc. Natl. Acad. Sci. USA*, **88**, 10480–4.

Pollock, J.S., Nakane, M., Buttery, L.D., Martinez, A., Springall, D., Polak, J.M., Forstermann, U. and Murad, F. (1993) Characterization and localization of endothelial nitric oxide synthase using specific monoclonal antibodies. *Am. J. Physiol.*, **265**, C1379–C1387.

Sase, K. and Michel, T. (1995) Expression of constitutive endothelial nitric oxide synthase in human blood platelets. *Life Sciences*, **57**, 2049–2055.

Schmidt, H., Gagne, G.D., Nakane, M., Pollock, J.S., Miller, M.F. and Murad, F. (1992) Mapping of neuronal nitric oxide synthase in the rat suggests frequent co-localization with NADPH diaphorase but not with soluble guanylyl cyclase and novel paraneural functions for nitrinergic signal transduction. *J. Histochem. Cytochem.*, **40**, 1439–1456.

Schmidt, H., Pollock, J.S., Nakane, M., Gorsky, L.D., Fostermann, U. and Murad, F. (1991) Purification of a soluble isoform of guanylyl cyclase-activating-factor synthase. *Proc. Natl. Acad. Sci. USA*, **88**, 365–369.

Schmidt, H., Warner, T.D., Ishii, K., Sheng, H. and Murad, F. (1992) Insulin secretion from pancreatic B cells caused by L-arginine-derived nitrogen oxides. *Science*, **255**, 721–723.

Sessa, W.C. (1994) The nitric oxide synthase family of proteins. *J. Vasc. Res.*, **31**, 131–43.

Sessa, W.C., García-Cardeña, G., Liu, J., Keh, A., Pollock, J.S., Bradley, J., Thiru, S., Braverman, I.M. and Desai, K.M. (1995) The Golgi association of endothelial nitric oxide synthase is necessary for the efficient synthesis of nitric oxide. *J. Biol. Chem.*, **270**, 17641–17644.

Shaul, P.W., Smart, E.J., Robinson, L.J., German, Z., Yuhanna, I.S., Ying, Y., Anderson, R.G.W. and Michel, T. (1996) Acylation targets endothelial nitric-oxide synthase to plasmalemmal caveolae. *J. Biol. Chem.*, **271**, 6518–22.

Sheng, H., Schmidt, H., Nakane, M., Mitchell, J.A., Pollock, J.S., Fostermann, U. and Murad, F. (1992) Characterization and localization of nitric oxide synthase in non-adrenergic non-cholinergic nerves from bovine retractor penis muscles. *Br. J. Pharmacol.*, **106**, 768–773.

Song, K.S., Li, S., Okamoto, T., Quilliam, L.A., Sargiacomo, M. and Lisanti, M.P. (1996) Co-purification and direct interaction of Ras with caveolin, an integral membrane protein of caveolae microdomains. *J. Biol. Chem.*, **271**, 9690–9697.

Tracey, W.R., Pollock, J.S., Murad, F., Nakane, M. and Forstermann, U. (1994) Identification of a type III (endothelial-like) particulate nitric oxide synthase in LLC-PK1 kidney tubular epithelial cells. *Am. J. Physiol.*, **266**, C22–C26.

Tracey, W.R., Xue, C., Klinghofer, V., Barlow, J., Pollock, J.S., Forstermann, U. and Johns, R.A. (1994) Immunochemical detection of inducible NO synthase in human lung. *Am. J. Physiol.*, **266**, L722–L727.

Tzeng, E., Billiar, T.R., Robbins, P.D., Loftus, M. and Stuehr, D.J. (1995) Expression of human inducible nitric oxide synthase in a tetrahydrobiopterin (H4B)-deficient cell line: H4B promotes assembly of enzyme subunits into an active dimer. *Proc. Natl. Acad. Sci. USA*, **92**, 11771–5.

Vodovotz, Y., Russell, D., Xie, Q., Bogdan, C. and Nathan, C.F. (1995) Vesicle membrane association of nitric oxide synthase in primary mouse macrophages. *J. Immunol.*, **154**, 2914–2925.

Wheeler, M.A., Smith, S.D., García-Cardeña, G., Nathan, C.F., Weiss, R.M. and Sessa, W.C. (1997) Bacterial infection induces nitric oxide synthase in human neutrophils. *J. Clin. Invest.*, **99**, 110–116.

5 Molecular Mechanisms of Peroxynitrite Reactivity

A.J. Gow and H. Ischiropoulos

IFEM, University of Pennsylvania, 1 John Morgan Building, 3620 Hamilton Walk, Philadelphia, PA 19104–6068, USA

Nitric oxide, like molecular oxygen, is capable of forming reactive species and although nitric oxide may not be essential for life, it is as important as oxygen in sustaining normal physiological function. Like molecular oxygen, nitric oxide poses a paradox as it can contribute to cellular oxidative stress. Partial reduction of oxygen by one or two electrons results in the formation of superoxide and hydrogen peroxide respectively. Again, similar to oxygen, nitric oxide is capable of forming reactive species, such as peroxynitrite and nitrosonium ion. It is probable that these oxidized forms of nitric oxide, and in particular peroxynitrite, form the basis of nitric oxide-mediated toxicity. Both nitric oxide and superoxide are free radicals and therefore they react rapidly, forming peroxynitrite:

$$\cdot NO + O_2^- \rightarrow ONOO^-$$

The second order rate constant of this reaction is 6.7×10^9 M^{-1} sec^{-1} (Huie and Padjama, 1993). This is an extremely fast rate that is approximately thirty times faster than the reaction of nitric oxide with oxyhemoglobin or guanylate cyclase and three times faster than the reaction of superoxide with superoxide dismutase (2.9×10^9 M^{-1} sec^{-1}). The goal of this chapter is to examine whether the formation of peroxynitrite by the reaction between nitric oxide and superoxide provides a reasonable biochemical explanation for the cyto-toxicity of nitric oxide.

NITRIC OXIDE AND TISSUE INJURY

Recent publications have provided evidence that nitric oxide, and the production of nitric oxide-derived oxidants, is part of the pathogenic mechanism associated with tissue injury in major organs (Table 5-1). The development of pulmonary injury after lungs were challenged with immune-complexes, or smoke, conditions that activate inflammatory cells (macrophages, neutrophils), was inhibited by a competitive inhibitor of nitric oxide synthase (NOS) (Mulligan *et al.*, 1991; Ischiropoulos *et al.*, 1994). Nitric oxide-derived oxidants are implicated in the pathogenic mechanisms of lung injury derived from sepsis, respiratory distress syndrome and ischemia-reperfusion (Haddad *et al.*, 1994; Kooy *et al.*, 1995; Wizemann *et al.*, 1994; Ischiropoulos *et al.*, 1995a). Inhibition of nitric oxide synthesis was shown to be protective in a global myocardial ischemia-reperfusion piglet model (Matheis *et al.*, 1992) and to reduce infarct size in the *in situ* rabbit heart (Patel *et al.*, 1993). Moreover, nitric oxide-derived oxidants were detected in rat heart after

Correspondence: H. Ischiropoulos, IFEM, University of Pennsylvania, 1 John Morgan Building, 3620 Hamilton Walk, Philadelphia, PA 19104–6068.

ischemia reperfusion injury (Wang and Zweir, 1996). A clear indication for the involvement of nitric oxide in ischemic injury was provided by Huang *et al.*, who showed that mice lacking the neuronal NOS were protected from ischemic injury (Huang *et al.*, 1994) and by data showing that antisense oligoncleotides targeting the inducible form of NOS protected rat kidney against ischemia (Noiri *et al.*, 1996). Neuronal injury from hypoxia-reoxygenation and NMDA-receptor activation was blocked by NOS inhibition (Dawson *et al.*, 1991; Cazevielle *et al.*, 1993; Beckman, 1991; Lipton *et al.*, 1993). Neurotoxicity derived from injections of 1-methyl-4-phenyl-1,2,3,6-tetrahydropyridine (MPTP), a model of Parkinson's disease, 3-nitropropionic acid and malonate are associated with nitric oxide-derived oxidants (Schulz *et al.*, 1995a, b; 1996). Recently, it was shown that apoptotic death of PC12 cells following downregulation of Cu,Zn superoxide dismutase proceeds via the formation of nitric oxide-derived oxidants (Troy *et al.*, 1996) and that nitric oxide-derived oxidants inhibit DOPA synthesis in the same cell system (Ischiropoulos, 1995b). In addition, upregulation of nitric oxide synthesis by endotoxin has been associated with tissue injury in liver ischemia reperfusion (Ma *et al.*, 1995).

Although nitric oxide is a free radical it is a weak one electron oxidant and, therefore, does not react with as many biological targets as one might have thought. Nitric oxide toxicity has been proposed to occur via its autooxidation. However, this reaction is dependent upon the square of the nitric oxide concentration and is thus slow at physiologically relevant concentrations of nitric oxide (approximately 5×10^{-9} Ms^{-1} at micromolar concentrations of nitric oxide) (Pryor *et al.*, 1982; Wink *et al.*, 1994; Kharitonov *et al.*, 1995). Biological molecules that can outcompete oxygen as a reactant for nitric oxide include, iron (both heme and non-heme bound), oxygen bound to heme, thiols, and other free radicals (Stadlar *et al.*, 1993; Eich *et al.*, 1996; Gow *et al.*, 1996a; Darley-Usman *et al.*, 1995). As stated previously, the most rapid of all these reactions is the combination with radicals such as superoxide to produce peroxynitrite. This reaction converts the relatively weak oxidants nitric oxide and superoxide into a powerful oxidizing and nitrating agent. Under physiological conditions the relative concentrations of the two reactants are low, however, since this is a second order reaction a small increase in the reactant concentrations produce large increases in the rate of product formation. Therefore, it can be envisaged that increases in the reactant concentrations, as would occur in many pathological conditions such as ischemia-reperfusion and sepsis, will result in peroxynitrite formation. Activated rat alveolar macrophages, human neutrophils, endothelial and smooth muscle cells have been shown to generate peroxynitrite (Ischiropoulos *et al.*, 1992; Carreras *et al.*, 1994; Kooy and Royall, 1994; Phelps *et al.*, 1995; Boota *et al.*, 1996) thus providing evidence for the formation of ONOO$^-$ *in vivo*. Furthermore, the above published studies (Table 5-1) have provided evidence that peroxynitrite may be the critical mediator of nitric oxide-derived toxicity. In order to understand the molecular mechanisms of peroxynitrite-mediated toxicity we need to examine the biochemistry of peroxynitrite. These studies have revealed unique reactivities of peroxynitrite and have lead to the development of methodologies for its detection in biological systems. The major findings regarding the biochemistry of peroxynitrite are outlined below.

OXIDATIVE BIOCHEMISTRY OF PEROXYNITRITE

Peroxynitrite exists in both a protonated, peroxynitrous acid, and an unprotonated form, peroxynitrite anion. The reactivities of these two forms differ greatly and, therefore, the

Table 5-1. Published Data Implicating ·NO-derived Oxidants with Tissue Injury.

Physiological *function*	DOUBLE EDGED ROLE OF NITRIC	*Pathological* *mechanism*
	ORGAN	
Vasodilation	*Heart*	Ischemia-reperfusion injury
Prevention of platelet and neutrophil adherence to endothelium	*Vasculature*	Sepsis, Hemorrhagic shock
Vasodilation Bronchodilation	*Lung*	Injury mediated by activated inflammatory cells, Sepsis and ARDS Ischemia-reperfusion injury
Vasodilation Long term potentiation	*Brain*	Glutamate, MPTP, 3-NP Ischemic injury
Vasodilation	*Kidney*	Ischemia-reperfusion injury Septic chronic failure Rejection of allografts

degree of protonation is critical. The pKa of peroxynitrtie protonation is 6.8 and thus at the physiological pH of 7.4, approximately 25% will be protonated. In the anionic form peroxynitrite is very stable and can be stored for days under alkaline conditions. However, peroxynitrous acid rapidly isomerizes to nitrate at a rate of 0.6 sec^{-1} in phosphate buffer and 37°C (Beckman *et al.*, 1990). Peroxynitrous acid is a strong one and two electron oxidant and is capable of interacting with a large variety of biological molecules forming products similar to those generated by either the hydroxyl radical (Beckman *et al.*, 1990) or nitrogen dioxide, although it does not appear to physically separate to form hydroxyl radical and nitrogen dioxide (·OH ···NO_2). Protonation, and hence hydroxyl radical-like reactivity, of peroxynitrite may not be important *in vivo* because the rate of peroxynitrite dissociation to hydroxyl radical like reactivity is relatively slow (k $\approx$ 1–2 s^{-1}). The second order rate constants for the other interactions of peroxynitrite anion with biological molecules vary from 10^5 $M^{-1}s^{-1}$ for the oxidation of Zn/S and Fe/S centers to 10^3 $M^{-1}s^{-1}$ for thiol oxidation. The actual targets of peroxynitrite oxidation within a biological system will depend upon the relative concentrations of the potential targets and the rate constants of their reaction with peroxynitrite. As a result where the peroxynitrite is generated within a cell may also be critical. For instance within the area of the mitochondria the potential for interaction with metal containing proteins is much higher, whilst in the cytosol protein thiols may form one of the main targets. One of the major biological reactants for peroxynitrite may be CO_2 as it has a relatively high concentration, approximately 1–5 mM in biological systems, and a moderate reaction rate, 10^4 $M^{-1}s^{-1}$ (Denicola *et al.*, 1996). However, reaction with CO_2 does not result in a terminal product but in an intermediate that is capable of nitrating tyrosine residues (Gow *et al.*, 1996c). Therefore, 3-nitrotyrosine may represent a major product of peroxynitrite formation in biological systems.

Nucleic acids also form an oxidative target for peroxynitrite at the level of both the sugar and the base. Oxidation of the deoxyribose moiety can result in strand breaks within DNA strands, whilst the guanine nucleotide is susceptible to both hydroxylation and nitration. 8-nitroguanine has been shown to be formed in a dose-dependent manner upon exposure of calf thymus DNA to peroxynitrite (Yermilov *et al.*, 1995). Carbon dioxide has been shown to be critical in the effects of peroxynitrite upon DNA. Incubation of DNA with peroxyntirite in bicarbonate buffer increased the yield of 8-nitroguanine six-fold and inhibited the formation of strand breaks (Yermilov *et al.*, 1996). Therefore, it would appear that the product of the CO_2/peroxynitrite reaction is capable of nitrating DNA but is inhibited from performing a hydroxyl radical like reaction with deoxyribose. Similarly, Denicola *et al.*, described the inhibition of one and two electron oxidations by the reaction of CO_2 and peroxynitrite (Denicola *et al.*, 1996).

The rapid and direct reactions of the peroxynitrite anion with zinc- and iron-thiolate centers, along with nitration of tyrosine residues, can result in the inactivation of key enzymes *in vivo* (Crow *et al.*, 1995; Ischiropoulos *et al.*, 1992; Bechman *et al.*, 1992). In addition, peroxynitrite oxidation of cellular thiols, which occurs 1,000 times faster than H_2O_2 mediated oxidation, and lipids (Radi *et al.*, 1991a,b) will deplete cells of antioxidant defenses (Radi *et al.*, 1991a). Recent data is starting to elucidate other potential intracellular targets of peroxynitrite including proteins, antioxidant defense pathways (such as ascorbate, a-tocopherol, and glutathione), transcription factors (such as AP-1 and NFκB and DNA (Yemilov *et al.*, 1995; Ischiropoulos, 1995; Bonfoco *et al.*, 1995; Estévez *et al.*, 1995; Lin *et al.*, 1995; Szabo *et al.*, 1996; Salgo *et al.*, 1995; Roussyn *et al.*, 1996). Overall, the net result of peroxynitrite reactivity is the induction of oxidative stress and ultimately cell death. The nature of peroxynitrite-induced cell death is dependent upon the concentration of peroxynitrite and the cell type; high levels of peroxynitrite result in necrotic death whereas relatively low levels result in the induction of apoptosis (Ischiropoulos, 1995; Bonfoco *et al.*, 1995; Estévez *et al.*, 1995; Lin *et al.*, 1995). Peroxynitrite-mediated delayed cell death has a different time profile in endothelial, epithelial, and neuronal cells. Exposure of type II pneumocytes to peroxynitrite significantly decreased cyanide-sensitive O_2 consumption (Hu *et al.*, 1994). Peroxynitrite exposure resulted in the inactivation of succinate-cytochrome c reductase and cytochrome c oxidase of the mitochondrial electron transport chain in neuronal cells but not in astrocytes (Bolãnos *et al.*, 1995). These data indicate that peroxynitrite can establish a diffusion gradient that will allow its diffusion inside cells where it can react with cellular targets such as mitochondria.

MITOCHONDRIA AS A POTENTIAL CELLULAR TARGET OF PEROXYNITRITE

In addition to studies utilizing intact cells (Hu *et al.*, 1994; Bolaños *et al.*, 1995), published data using isolated mitochondria preparations have revealed important mecahnistic information regarding the interaction of peroxynitrite with mitochondria. Cassina and Radi reported recently that while NO induces reversible injury to mitochondrial electron transport components, peroxynitrite induces irreversible injury that is consistent with the changes observed in intact cells (Cassina and Radi, 1996). These observations have lead to the speculation that peroxynitrite is the ultimate species responsible for mitochondrial dysfunction under nitric oxide mediated pathological conditions. Another event that is associated with the reaction of peroxynitrite with isolated mitochondria is the uncoupling of the electron transport chain. Uncoupling of the mitochondrial electron transport chain and an increase in hydrogen peroxide production has been shown to occur after exposure of isolated rat heart mitochondria to peroxynitrite (Radi *et al.*, 1994). Peroxynitrite mediated damage to mitochondria will result in disruption of intracellular redox balance and release of calcium stored within the mitochondria. This has been demonstrated in isolated liver mitochondria, which are depolarized and release calcium upon peroxynitrite exposure (Packer and Murphy, 1995; Schweizer and Richter, 1996). It may be that mitochondrial disruption represents an early event in peroxynitrite mediated delayed cell death, as loss of mitochondrial reductive capacity occurs within six hours of exposure of endothelial cells whilst cell death does not occur until sixteen hours post exposure. The increase in mito-chondria-derived reactive species may be further amplified as peroxynitrite can inactivate Mn superoxide dismutase that is strategically located inside the mitochondria to account for the production of superoxide (Ischiropoulos *et al.*, 1992). Recently Mn superoxide dismutase was found to be nitrated and inactivated in rejected human transplanted kidney tissues (MucMillan-Crow *et al.*, 1996).

PEROXYNITRITE-MEDIATED NITRATION OF TYROSINE

The major product of the spontaneous reaction of peroxynitrite with tyrosine is 3-nitrotyrosine. Low molecular weight metal catalysts such as Fe^{+3}-EDTA or Cu^{+2}, metal-containing enzymes like Cu, Zn superoxide dismutase and CO_2 can catalyze the nitration of phenolic compounds and protein tyrosine residues to give nitrophenols and nitrotyrosine (Beckman *et al.*, 1990; Denicola *et al.*, 1996; Ischiropoulos *et al.*, 1992; Beckman *et al.*, 1992). The reaction of peroxynitrite with CO_2 may provide the necessary nitrating agent that can explain the formation of nitrotyrosine *in vivo*. Recent data from our laboratory as well as from others independently provided experimental evidence that CO_2 reacts in a catalytic manner with peroxynitrite to form a potent nitrating agent. Not only does the yield of protein nitration increase over two fold by the reaction of peroxynitrite with CO_2 but all other reactivities of peroxynitrite such as oxidation of cysteine and tryptophan are partially inhibited (Beckman *et al.*, 1990; Denicola *et al.*, 1996). In addition, CO_2 reverses the inhibitory effects of small molecular weight antioxidants, such as ascorbate, upon peroxynitrite mediated oxidation. Endogenous tyrosine nitration is almost certainly derived via enzymatically produced nitric oxide although nitric oxide itself is not a nitrating agent. Potential *in vivo* nitrating agents include; peroxynitrous acid, metal nitrates, alkyl and acyl nitrates, nitryl helides, nitrogen oxides, and acid catalyzed reactions of nitrite. Using human plasma as a model, nitrotyrosine could not be detected even at concentrations of nitric oxide which resulted in up to 1 mM nitrogen dioxide or after exposure to nitric oxide in the presence of other biological oxidants such as H_2O_2 and metal catalysts (Denicola *et al.*, 1996). Therefore, under plausible pathophysiological concentrations the intermediate formed by the reaction of peroxynitrite with CO_2, appears to be the most reasonable nitrating agent.

However, there is no single mechanism for the nitration of tyrosine but a continuum of reactions that are a function of the reactive nitrating species and the nitration conditions. Recently, Van der Vliet *et al.* (1997) demonstrated a potential nitration reaction mechanism involving the oxidation of nitrite by myeloperoxidase. This reaction mechanism is unlikely to occur to a greater extent than a peroxynitrite mediated pathway as the formation of nitrite is a slow process and often production of nitric oxide is accompanied by the production of superoxide. In addition, this process is both slow and relatively inefficient, approximately

10 fold less efficient than CO_2 catalyzed peroxynitrite nitration. Moreover neutropenia (>90% elimination of circulating neutrophils) did not change the levels of nitrotyrosine measured in a rat model of CO-mediated oxidative stress (Ischiropoulos *et al.*, 1996). Therefore, it seems improbable that this reaction mechanism contributes significantly to nitration observed under conditions of acute inflammation. However, it may contribute significantly in the processes of chronic inflammation. The mechanism of nitration by peroxynitrite, and other nitrating agents, is complicated by the presence of more than one oxidizing species in the reaction mixture. However, in the presence of CO_2 the reaction of peroxynitrite with proteins will preferentially yield nitrotyrosine. Consistent with these data is the detection of extensive tyrosine nitration in pathologic conditions known to increase the levels of both bicarbonate/CO_2 and peroxynitrite, such as inflammation and ischemia-reperfusion injury (Table 5-2).

The direct evidence for peroxynitrite formation in human atherosclerotic plaques is consistent with the observations that peroxynitrite, but not NO, induced lipid oxidation of β-VLDL and LDL (Bechman *et al.*, 1994b; White *et al.*, 1994) and that hypercholesterolemia induced the release of both nitrogen oxides and superoxide from rabbit aorta (Minor *et al.*, 1990; Ohara *et al.*, 1993). The evidence for peroxynitrite formation in human lungs is consistent with the detection of nitrotyrosine in animal models of these diseases and the detection of nitrotyrosine in inflammatory disease is consistent with the ability of macrophages and neutrophils to generate peroxynitrite (Ischiropoulos *et al.*, 1992a; Carreras *et al.*, 1994). Moreover, in all the animal models of injury the increase in the levels of nitrotyrosine were prevented by inhibition of ·NO synthesis indicating that nitrotyrosine is in part derived from nitric oxide. This is confirmed in brain models of injury in which the use of nNOS knockout mice and mice overexpressing Cu/Zn superoxide dismutase has indicated the necessity of nitric oxide and superoxide production to both nitration and neurotoxicity. A direct relation

Table 5-2. Detection of Nitrotyrosine in Human and Animal Diseases.

HUMAN:
  Atherosclerotic plaques of coronary vessels (Beckman 1994b)
  LDL isolated from atherosclerotic lesions (Leeuwenburgh *et al.*, 1997)
  Lungs of infants with sepsis and/or respiratory disease (Haddad *et al.*, 1994; Kooy *et al.*, 1995)
  Synovial fluid of patients with arthritis (Kaur and Halliwell, 1994)
  Multiple sclerosis plaques (Basarga *et al.*, 1995)
  Chronic renal failure in septic patients (Fukuyama *et al.*, 1997)
  Rejected renal allografts (MacMillan-Crow *et al.*, 1996)
  Amyotrophic Lateral Sclerosis (Beckman *et al.*, 1994a)
ANIMAL:
  Rabbit lungs following exposure to hyperoxia (Haddad *et al.*, 1994)
  Lungs of endotoxin treated rats (Wizemann *et al.*, 1994)
  Ischemia reperfusion injured rat lungs (Ischiropoulos *et al.*, 1995a)
  Ischemia reperfusion injured rat heart (Wang and Zweir, 1996)
  Aorta of septic rats (Szabo *et al.*, 1995)
  Inflammatory bowel disease (Miller *et al.*, 1995)
  CO poisoned rats (Ischiropoulos *et al.*, 1996)
  Rat skeletal muscle SERC2a isoform (Viner *et al.*, 1996)
  Brain lesions in MPTP, 3-nitropropionic acid and malonate neurotoxicity (Schulz *et al.*, 1995a,b 1996)
  Neurofilament L in ALS models (Beckmann *et al.*, 1994a)

between nitration of tyrosine residues and toxicity has not been established as yet. However, published work has indicated that peroxynitrite-mediated nitration of proteins leads to inactivation (MacMillan-Crow *et al.*, 1996) and may interfere with tyrosine phosphorylation and signal transduction (Gow *et al.*, 1996b).

REFERENCE

Basarga, O., Michaels, F.H., Zheng. Y.M., Borboski, L.E., Spitsin, S.V., Fu, Z.F., Tawadros, R. and Koprowski, H. (1995) Activation of the inducible form of nitric oxide synthase in the brains of patients with multiple sclerosis. *Proc. Natl. Acad. Sci. USA*, **92**, 12041–12045.

Beckman, J.S. (1991) The double-edged role of nitric oxide in brain function and superoxide-mediated injury. *J. Devol. Physiol.*, **15**, 53–59.

Beckman, J.S., Beckman, T.W., Chen, J., Marshall, P.A. and B.A. Freeman. (1990) Apparent hydroxyl radical production by peroxynitrite: Implications for endothelial injury from nitric oxide and superoxide. *Proc. Natl. Acad. Sci. USA*, **87**, 1620–1624.

Beckman, J.S., Chen, J., Crow, J.P. and Ye, Y.Z. (1994a) Reactions of nitric oxide and peroxynitrite with superoxide dismutase in neurodegeneration. *Prog. Brain Res.*, **103**, 371–380.

Beckman, J.S., Ischiropoulos, H., Zhu, L., van der Woerd, M., Smith, C.D., Harrison, J., Martin, J.C. and Tsai, J-H.M. (1992) Kinetics of superoxide dismutase and iron catalyzed nitration of phenolics by peroxynitrite. *Arch. Biochem. Biophys.*, **298**, 438–445.

Beckman, J.S., Ye, Y.-Z., Anderson, P.G., Chen, J, Accavitti, M.A., Tarpey, M.M. and White, C.R. (1994b) Extensive nitration of protein tyrosine in human atherosclerosis detected by immunohistochemistry. *Biol. Chem. Hoppe-Seyler*, **375**, 81–88.

Bolanos, J.P., Heals, S.J.R., Land, J.M. and Clark, J.B. (1995) Effect of peroxynitrite on the mitochondrial respiratory chain: differential susceptibility of neurones and astrocytes in primary culture. *J. Neurochem.* **64**, 1965–1972.

Bonfoco, E., Krainc, D., Ankarrona, M., Nicotera, P. and Lipton, S.A. (1995) Apoptosis and necrosis: two distinct events induced, respectively, by mild and intense insults with N-methyl-D-aspartate or nitric oxide/superoxide in cortical cell cultures. *Proc. Natl. Acad. Sci. USA*, **92**, 1044–1048.

Boota, A., Zar, H., Kim, Y-M., Johnson, B., Pitt, B. and Davies, P. (1996) IL-1β stimulates superoxide and delayed peroxynitrite production by pulmonary vascular smooth muscle cells. *Am. J. Physiol.*, **271**, L932–L938.

Carreras, M.C., Pargament, G.A., Catz, S.D., Poderoso, J.J. and Boveris, A. (1994) Kinetics of nitric oxide and hydrogen peroxide production and formation of peroxynitrite during the respiratory burst of human neutrophils. *FEBS Lett.*, **341**, 65–68.

Cassina, A. and Radi, R. (1996) Differential inhibitory action of nitric oxide and peroxynitrite on mitochondrial electron transport. *Arch. Biochem. Biophys.*, **328**, 309–316.

Cazevieille, C., Muller, A., Meynier, F. and Bonne, C. (1993) Superoxide and nitric oxide cooperation in hypoxia/reoxygenation-induced neuronal injury. *Free Rad. Biol. Med.*, **14**, 389–395.

Crow, J.P., Beckman, J.S. and McCord, J.M. (1995) Sensitivity of the essential zinc-thiolate moiety of yeast alcohol dehydrogenase to hypoclorite and peroxynitrite. *Biochemistry*, **34**, 3544–3522.

Darley-Usmar, V., Wiseman, H. and Halliwell, B. (1995) Nitric oxide and oxygen radicals: a question of balance. *FEBS Lett.*, **369**, 131–135.

Dawson, V., Dawson, T., London, E., Brent, D. and Snyder, S. (1991) Nitric oxide mediates glutamate neurotoxicity in primary cortical cultures. *Proc. Natl. Acad. Sci., USA*, **88**, 6368–6371.

Denicola, A., Trujillo, M., Freeman, B.A. and Radi, R. (1996) Peroxynitrite reaction with carbon dioxide/bicarbonate: Kinetics and influence on peroxynitrite-mediated oxidation reactions. *Arch. Biochem. Biophys.*, **333**, 49–58.

Eich, R.F., Li, T., Lemon, D.D., Doherty, D.H., Curry, S.R., Aitken, J.F., Mathews, A.J.X., Johnson, K.A., Smith, R.D., Phillips, G.N., Jr.and Olson, J.S. (1996) Mechanism of NO-induced oxidation of myoglobin and hemoglobin. *Biochem.*, **35**, 6976–6983.

Estévez, A.G., Radi, R., Barbeito, L., Shin, J.T., Thompson, J.A. and Beckman, J.S. (1995) Peroxynitrite-induced cytotoxicity in PC12 cells: Evidence for an apoptotic mechanism differentially modulated by neurotrophic factors. *J. Neurochem.*, **65**, 1543–1550.

Fukuyama, N., Takebayashi, Y., Hida, M., Ishida, H., Ichimori, K. and Nakazawa, H. (1997) Clinical evidence of peroxynitrite formation in chronic renal failure patients with septic shock. *Free Rad. Biol. Med.*, **22**, 771–774.

Gow A.J., Buerk D.G. and Ischiropoulos, H. (1996a) A novel reaction mechanism for the formation of S-nitrosothiol *in vivo*. *J. Biol. Chem.*, **272**, 2841–2844.

Gow, A., Duran, D., Malcom, S. and Ischiropoulos, H. (1996b) Effects of peroxynitrite induced modifications to signal transduction and protein degradation. *FEBS Lett.*, **385**, 63–66.

Gow, A., Duran, D., Thom, S.R. and Ischiropoulos, H. (1996) Carbon dioxide catalyzed protein tyrosine nitration by peroxynitrite. *Arch. Biochem. Biophys.*, **333**, 42–48.

Haddad, I.Y., Pataki, G., Hu, P., Galliani, C., Beckman, J.S. and Matalon S. (1994) Quantitation of nitrotyrosine levels in lung sections of patients and animals with acute lung injury. *J. Clin. Invest.*, **94**, 2407–2413.

Hu, P., Ischiropoulos, H., Beckman, J.S. and Matalon, S. (1994) Peroxynitrite inhibition of oxygen consumption and sodium transport in alveolar Type II cells. *Am. J. Physiol.*, **266**, L628–L634.

Huang, Z., Huang, P.L., Panahian, N., Dalkara, T., Fishman, M.C. and Moskowitz, M.A. (1994) Effects of cerebral ischemia in mice deficient in neuronal nitric oxide synthase. *Science*, **265**, 1883–1885.

Huie, RE and Padjama, S. (1993) The reaction of NO with superoxide. *Free Rad. Res. Comm.*, **18**, 195–199.

Ischiropoulos, H. (1995) Exposure of endothelial cells to peroxynitrite inhibits tyrosine phosphorylation and induces apoptosis. *Endothelium*, **3s**, 47–48.

Ischiropoulos, H., Al-Mehdi, A. B. and Fisher, A.B. (1995a) Reactive species in rat lung injury: contribution of peroxynitrite. *Am. J. Physiol.*, L185–L164.

Ischiropoulos, H., Beers, M.F., Ohnishi, S.T., Fisher, D., Garner, S.E. and Thom, S.R. (1996) Nitric oxide production and perivascular tyrosine nitration in brain following carbon monoxide poisoning in the rat. *J. Clin. Invest.*, **97**, 2260–2267.

Ischiropoulos, H., Duran, D. and Horwitz, J. (1995b) Peroxynitrite-mediated inhibition of DOPA synthesis in PC12 cells. *J. Neurochem.*, **65**, 2366–2372.

Ischiropoulos, H., Mendiguren, I., Fisher, D., Fisher, A.B. and Thom, S.R. (1994) Role of neutrophils and nitric oxide in lung alveolar injury from smoke inhalation. *Am. J. Resp. Crit. Care Med.*, **150**, 337–341.

Ischiropoulos, H., Zhu, L. and Beckman, J.S. (1992a) Peroxynitrite formation from macrophage-derived nitric oxide. *Arch. Biochem. Biophys.*, **298**, 446–45.

Ischiropoulos, H., Zhu, L., Chen, J., Tsai, J-H.M., Martin, J.C., Smith, C.D. and Beckman, J.S. (1992) Peroxynitrite-mediated tyrosine nitration catalyzed by superoxide dismutase. *Arch. Biochem. Biophys.*, **298**, 431–437.

Kaur, H. and Halliwell, B. (1994) Evidence for nitric oxide-mediated oxidative damage in chronic inflammation. Nitrotyrosine in serum and synivial fluid from rheumatoid patients. *FEBS Lett.*, **350**, 9–12.

Kharitonov, V.G., Sundquist, A.R. and Sharma, V.S. (1995) Kinetics of nitrosation of thiols by nitric oxide in the presence of oxygen. *J. Biol. Chem.*, **270**, 28158–28164.

Kooy, N.W. and Royall, J.A. (1994) Agonist-induced peroxynitrite production from endothelial cells. *Arch. Biochem. Biophys.*, **310**, 352–359.

Kooy, N.W., Royall, J.A., Ye, Y-Z., Kelly, D.R. and Beckman J.S. (1995) Evidence for *in vivo* peroxynitrite production in human acute lung injury. *Am. J. Resp. Crit. Care Med.*, **151**, 1250–1254.

Leeuwenburgh, C., Hardy, M.M., Hazen, S.L., Wagner, P., Oh-ishi, S., Steinbrecher, U.P. and Heinecke, J.W. (1997) Reactive nitrogen intermediates promote low density lipoprotein oxidation in human atherosclerotic intima. *J. Biol. Chem.* **272**, 1433–1436.

Lin, K.-T., Xue, J.-Y., Nomen, M., Spur, B. and Wong, Y.-K. P. (1995) Peroxynitrite-induced apoptosis in HL-60 cells. *J. Biol. Chem.*, **70**, 16487–16490.

Lipton, S.A., Choi, Y-B., Pan, Z-H., Lei, S.Z., Chen, H-S.V., Sucher, N.J., Loscalzo, J., Singel, D.J. and Stamler, J.S. (1993) A redox-based mechanism for the neuroprotective and neurodestructive effects of nitric oxide and related nitroso-compounds. *Nature (Lond)*, **364**, 626–632.

Ma, T.T., Ischiropoulos, H. and Brass, C.A. (1995) Endotoxin stimulated nitric oxide production increases injury and reduces rat liver chemiluminescence during reperfusion. *Gastroenterology*, **108**, 463–469.

MacMillan-Crow, L.A., Crow, J.P., Kerby, J.D., Beckman, J.S. and Thompson, J.A. (1996) Nitration and inactivation of manganese superoxide dismutase in chronic rejection of human renal allografts. *Proc. Natl. Acad. Sci. USA*, **93**, 11853–11858.

Matheis, G., Sherman, M.P., Buckberg, G.D., Haybron, D.P., Young, H.H. and Ignarro, L.J. (1992) Role of L-arginine-nitric oxide pathway in myocardial reoxygenation injury. *Am. J. Physiol.*, **262**, H616–620.

Miller, M.J.S., Thompson, J.H., Zhang, X.-J., Saodowska-Krowicka, H., Kakkis, J.L., Munshi, U.K., Sandoval, M., Rossi, J.L., Eloby-Childress, S., Beckman, J.S., Ye, Y.Z., Rodi, C.P., Manning, P.T., Currie, M.G. and Clark, D.A. (1995) Role of inducible nitric oxide synthase expression and peroxynitrite formation in the guinea pig ileitis. *Gastroenterology*, **109**, 1475–1483.

Minor, R.L., Myers, P.R., Guerra, R. Bates, J.N. and Harrison, D.G. (1990) Diet-induced atherosclerosis increases the release of nitrogen oxides from rabbit aorta. *J. Clin. Invest.*, **86**, 2109–2116.

Mulligan, M.S., Hevel, J.M., Marletta, M.A. and Ward, P.A. (1991) Tissue injury caused by deposition of immune complexes is L-arginine dependent. *Proc. Natl. Acad. Sci. USA*, **88**, 6338–6342.

Noiri, E., Peresleni, T., Miller, F. and Goligorsky, M.S. (1996) In vivo targeting of inducible NO synthase with oligodeoxynucleotides protects rat kidney against ischemia. *J. Clin. Invest.*, **97**, 2377–2383

Ohara, Y., Peterson, T.E. and Harrison, D.G. (1993) Hypercholesterolemia increases endothelial superoxide anion production. *J. Clin. Invest.*, **91**, 2546–2551.

Packer, M.A. and Murphy, M.P. (1995) Peroxynitrite formed by simultaneous nitric oxide and superoxide generation causes a cyclosporin A-sensitive mitochondrial calcium efflux and depolarisation. *Eur. J. Biochem.*, **234**, 231–239.

Patel, V.C., Yellon, D.M., Singh, K.J., Neild, G.H. and Woolfson, R.G. (1993) Inhibition of nitric oxide limits infarct size in the in situ rabbit heart. *Biochem. Biophys. Res. Comm.*, **194**, 234–238.

Phelps, D.T., Ferro, T.J., Higgins, P.J. Shankar, R., Parker, D.M. and Johnson, A. (1995) Tumor necrosis factor-induces the peroxynitrite-mediated depletion of lung endothelial glutathione via protein kinase C activation. *Am. J. Physiol.*, **269**, L551–L559.

Pryor, W.A., Church, D.F., Govindian, C.K. and Crank, G. (1982). Oxidation of thiols by nitric oxide and nitrogen dioxide: Synthetic utility and toxicological implications. *J. Org. Chem.*, **47**, 157–161.

Radi, R., Beckman, J.S., Bush, K.M. and Freeman, B.A. (1991b) Peroxynitrite-induced membrane lipid peroxidation: The cytotoxic potential of superoxide and nitric oxide. *Arch. Biochem. Biophys.*, **288**, 481–487.

Radi, R., Beckman, J.S., Bush, K.M. and Freeman, B.A. (1991a) Sulfhydryl oxidation by peroxynitrite: the cytotoxic potential of superoxide and nitric oxide. *J. Biol. Chem.*, **266**, 4244–4250.

Radi, R., Rodriguez, M., Castro, L. and Telleri, R. (1994) Inhibition of mitochondrial transport chain by peroxynitrite. *Arch. Biochem. Biophys.*, **308**, 89–95.

Roussyn, I., Briviba, K., Masumoto, and Sies, H. (1996) Selenium-containing compounds protect DNA from single-strand breaks caused by peroxynitrite. *Arch Biochem. Biophys.*, **330**, 216–218.

Salgo, M.G., Stone, K., Squadrito, G.L., Battista, J.R. and Pryor, W.A. (1995) Peroxynitrite causes DNA nicks in plasmid pBR322. *Biochem. Biophys. Res. Comm.*, **210**, 1025–1030.

Schulz, J.B., Huang, P.L., Matthews R.T., Pasov, D., Fishman, M. C. and Beal M.F. (1996) Striatal malonate lesions are attenuated in nitric oxide oxide synthase knockout mice. *J. Neurochem.*, **67**, 430–433.

Schulz, J.B., Matthews R.T., Muqit M.K., Browne S.E. and Beal M.F. (1995a) Inhibition of neuronal nitric oxide synthase by 7-nitroindazole protects against MPTP-induced neurotoxicity in mice. *J. Neurochem.*, **64**, 936–939.

Schulz, J.B., Matthews, Jenkins, B.G., Ferrante, R.J., Siwek, D., Henshaw, D.R., Cipolloni, P.B., Mecocci, P., Kowall, N.W. Rosen, B.R. and Beal, M.F. (1995b) Blockade of neuronal nitric oxide synthase protects against excitotoxicity *in vivo. J. Neuroscience*, **15**, 8419–8429.

Schweizer, M. and Richter, C. (1996) Peroxynitrite stimulates the pyridine nucleotide-linked calcium release from intact rat liver mitochondria. *Biochemistry*, **35**, 4524–4528.

Stadler, J., Bergonia, H.A., Di Silvio, M., Sweetland, M.A., Billiar, T.R., Simmons, R.L. and Lancaster, J.R., Jr. (1993) Nonheme iron-nitrosyl complex formation in rat hepatocytes: detection by electron paramagnetic resonance spectroscopy. *Arch. Biochem. Biophys.*, **302**, 4–11.

Szabo, C. Salzman, A.L. and Ischiropoulos, H. (1995) Endotoxin triggers the expression of an inducible isoform of nitric oxide synthase and the formation of peroxynitrite in the rat aorta *in vivo. FEBS Lett.*, **363**, 235–238.

Szabo, C., Zignarelli, B., O'Connor, M. and Salzman, A.L. (1996) DNA strand breakage, activation of poly(ADP-ribose) synthetase, and cellular energy depletion are involved in the cytotoxicity in macrophages and smooth muscles cells exposed to peroxynitrite. *.Proc. Natl. Acad. Sci. USA*, **93**, 1753–1758.

Troy, C.M., Derossi, D., Prochiantz, A., Greene, L.A. and Shelanski, M. (1996) Down-regulation of copper/zinc superoxide dismutase leads to cell death via the nitric oxide-peroxynitrite pathway. *J. Neuroscience*, **16**, 253–261.

van der Vliet, A., Eiserich, J.P., Halliwell, B. and Cross, C.E. (1997) Formation of reactive nitrogen species during peroxidase-catalyzed oxidation of nitrite. *J. Biol. Chem.*, **272**, 7617–7625.

Viner, R.I., Ferrington, D.A., Huhmer, A.F.R., Bigelow, D.J. and Schoneich, C. (1996) Accumulation of nitrotyrosine on the SERCA2a isoform of SR Ca-ATPase of rat skeletal muscle during aging: a peroxynitrite-mediated process? *FEBS Lett.*, **379**, 286–290.

Wang, P. and Zweir, J.L. (1996) Measurement of nitric oxide and peroxynitrite generation in the postischemic heart. *J. Biol. Chem.*, **271**, 29223–29230.

White, C.R., Brock, T.A., Chang, L.-Y. Crapo, J., Brisco, P., Ku, D., Bradley, A.W., Gianturco, S.H., Gore, J., Freeman, B.A. and Tarpey, M.M. (1994) Superoxide and peroxynitrite in atherosclerosis. *Proc. Natl. Acad. Sci. USA*, **91**, 1044–1048.

Wink, D.A., Nims, R.W., Darbyshire, J.F., Christodoulou, D., Hanbauer, I., Cox, G.W., Laval, F., Laval, J., Cook, J.A., Krishna, M.C., DeGraff, W.G. and Mitchell, J.B. (1994) Reaction kinetics for nitrosation of cysteine and glutathione in aerobic nitric oxide solutions at neutral pH. Insights into the fate and physiological effects of intermediates generated in the NO/O_2 reaction. *Chem. Res. Toxicol.*, **7**, 519–525.

Wizemann, T.M., Gardner, C.R., Laskin, J.D., Quinones, S., Durham, K.D., Golle, N.L., Ohnishi, S.T. and Laskin, D.L. (1994) Production of nitric oxide and peroxynitrite in the lung during acute endotoxemia. *J. Leuk. Biol.*, **56**, 759–768.

Yermilov, V., Rubio, J. and Ohshima, H. (1995) Formation of 8-nitroguanine in DNA treated with peroxynitrite *in vitro* and its rapid removal from DNA by depurination. *FEBS Lett.*, **376**, 207–210.

Yermilov, V., Yoshie, Y., Rubio, J. and Ohshima, H. (1996) Effects of carbon dioxide/bicerbonate on induction of DNA single-strand breaks and formation of 8-nitroguanine, 8-oxoguanine and base-propenal mediated by peroxynitrite. *FEBS Lett.*, **399**, 67–70.

6 Nitric Oxide, Peroxynitrite and Poly (ADP-Ribose) Synthetase: Biochemistry and Pathophysiological Implications

Csaba Szabó

Children's Hospital Medical Center, Division of Critical Care, 3333 Burnet Avenue, Cincinnati, Ohio 45229, USA

Nitric oxide (NO) and superoxide rapidly react to yield peroxynitrite. Peroxynitrite is a potent oxidant which reacts with proteins, lipids, and DNA. Peroxynitrite (but not NO) is a potent initiator of DNA single strand breakage, which is an obligatory stimulus for the activation of the nuclear enzyme poly (ADP ribose) synthetase (PARS). Rapid activation of PARS can deplete the intracellular concentration of its substrate, NAD^+, slowing the rate of glycolysis, electron transport, and, therefore, ATP formation. This process, as demonstrated in macrophages, smooth muscle cells, endothelial cells, epithelial cells, neurons and islet cells, can result in acute cell dysfunction and cell death. Accordingly, inhibitors of PARS protect against cell death under these conditions. In addition to the direct cytotoxic pathway regulated by peroxynitrite and subsequent PARS activation, PARS also appears to regulate the process of the expression of the inducible isoform of NO synthase (iNOS), so that pharmacological inhibitors of PARS suppress the process of iNOS induction. Taken together, the experimental evidence outlined in the current chapter supports the view that the peroxynitrite–PARS pathway contributes to cellular injury in a number of pathophysiological conditions including circulatory shock, stroke, myocardial and intestinal ischemia-reperfusion injury, and diabetes mellitus. Pharmacological inhibition of PARS may be a promising approach for the experimental therapy of these conditions.

Key words: free radical, peroxynitrite, nitric oxide, superoxide, septic shock, endotoxin, inflammation, stroke, diabetes, mitochondrial respiration, nicotinamide, 3-aminobenzamide, poly (ADP-ribose) synthetase.

THE PEROXYNITRITE–PARS PATHWAY

Peroxynitrite: Formation, Decomposition and Reactivity

Nitric oxide and superoxide rapidly combine to form a toxic reaction product, peroxynitrite anion ($ONOO^-$) (Beckman *et al.*, 1990; Pryor and Squadrito, 1995). The ratio of superoxide and NO is important in determining the reactivity of peroxynitrite: excess NO reduces the oxidation elicited by peroxynitrite (Rubbo *et al.*, 1994; Szabó *et al.*, 1995a). The oxidant reactivity of peroxynitrite is mediated by an intermediate with biological activity of hydroxyl radical. However, this product does not appear to be hydroxyl radical *per se*, but it is peroxynitrous acid or its activated isomer (Pryor and Squadrito, 1995).

Peroxynitrite is a highly reactive species. Its oxidant and cytotoxic activities include rapid oxidation of sulfhydryl groups and thioethers, as well as nitration and hydroxylation of aromatic compounds, such as tyrosine, tryptophan and guanine (Ischiropoulos *et al.*, 1992; Pryor and Squadrito, 1995). While the reaction with the sulfhydryl groups is likely to represent a direct reaction of peroxynitrite, the nitration reactions probably occur through a NO_2^+-like intermediate. These reactions, when occurring during the reaction of peroxynitrite with enzymes, macromolecules and lipids, markedly influence various cellular functions.

For instance, tyrosine nitration may lead to dysfunction of nitrated proteins, as has been shown or suggested in the case of superoxide dismutase (Ischiropoulos *et al.*, 1992), cytoskeletal actin (Selden *et al.*, 1995), and neuronal tyrosine hydroxylase (Ischiropoulos *et al.*, 1995a). Oxidation of critical sulfhydryl groups results in a potent inhibition of aconitase (Hausladen and Fridovich, 1994; Castro *et al.*, 1994) and other critical enzymes of cellular energy generation (Mohr *et al.*, 1994; Radi *et al.*, 1994; Rubbo *et al.*, 1994; Bolanos *et al.*, 1995) and disruption of the zinc-thiolate center at the active site of certain enzymes (Crow *et al.*, 1995). Furthermore, peroxynitrite can interfere with membrane transport systems, such as the membrane Na^+/K^+ ATP-ase (Hu *et al.*, 1994; Guzman *et al.*, 1995), membrane sodium channels (Bauer *et al.*, 1992) and sarcoplasmatic calcium pumps (Viner *et al.*, 1996; Elliott, 1996). In addition to the above effects, peroxynitrite is a potent initiator of lipid peroxidation (Radi *et al.*, 1991), and injures DNA via a number of independent mechanisms (see below).

Considering the highly cytotoxic nature of peroxynitrite, it is not surprising that, under physiological conditions, the generation of peroxynitrite is very limited. In fact, a number of endogenous factors act as scavengers/neutralizers of peroxynitrite cytotoxicity. Scavengers of peroxynitrite include uric acid, cysteine, glutathione, ascorbic acid, melatonin, desferoxamine, Vitamin E and synthetic manganese-mesoporphyrins, but not mannitol or dimethylsulfoxide (Bartlett *et al.*, 1995; Salgo *et al.*, 1995a; Van Der Vliet *et al.*, 1994; Denicola *et al.*, 1995; Shi *et al.*, 1994; Hogg *et al.*, 1994; DeGroot *et al.*, 1993; Pryor *et al.*, 1994; Kooy *et al.*, 1995; Szabó *et al.*, 1996a; Gilad *et al.*, 1997). The presence of superoxide dismutase also plays an important role in the protection against peroxynitrite-induced cytotoxicity, as demonstrated in PC12 cells (Troy *et al.*, 1996). As mentioned above, excess NO can also act as an inhibitor of peroxynitrite-triggered oxidative processes, probably by terminating peroxynitrite-triggered radical chain propagation processes, and, possibly also via a direct interaction with peroxynitrite (Rubbo *et al.*, 1994; Szabó *et al.*, 1995a). It appears that a delicate balance exists between peroxynitrite-mediated oxidant processes and endogenous antioxidant pathways, which limit the reactivity of peroxynitrite (Darley-Usmar *et al.*, 1995). Although endogenous antioxidants play an important role in limiting peroxynitrite toxicity (as evidenced, for example, by the enhancement of peroxynitrite-induced cytotoxicity in cells depleted of endogenous glutathione [Fici *et al.*, 1997; Szabó *et al.*, 1997a]), the reactions of peroxynitrite with antioxidants may also give rise to secondary, oxidant species. Such species include ascorbyl radical, albumin-thyil radical, and an uric acid-derived radical (Vasquez-Vivar *et al.*, 1996), which may initiate secondary oxidant processes.

The reactivity and decomposition pathways of peroxynitrite are strongly influenced by the chemical environment. In solutions containing carbonate, peroxynitrite forms an adduct with carbonate, which then may decompose to yield the toxic HCO_3 radical (Lymar and Hurst, 1995). It appears that the reaction of CO_2 with peroxynitrite importantly modifies the biological actions of peroxynitrite: a reaction the biological implications of which are incompletely understood. In the presence of plasma, proteins, glucose or glutathione, peroxynitrite can form intermediates which act as NO donors (Moro *et al.*, 1994; Lymar and Hurst, 1995). As a NO donor, peroxynitrite has some (relatively low) potency for activation of cytosolic guanylyl cyclase (Tarpey *et al.*, 1995; Mayer *et al.*, 1996). It appears that peroxynitrite, at extremely low concentrations, can act as a NO-donor, or a NO-like species, in that it inhibits neutrophil adhesion and inhibits the up-regulation of cell surface adhesion receptors (Lefer *et al.*, 1997). Such an effect would be consistent with the lack of cytotoxicity when peroxynitrite is being produced in low levels, in normally functioning organisms, as a by-product of the production of NO (from constitutive isoforms of NO synthases), and of superoxide (as a by-product of normal cellular processes). However, in various pathophysiological conditions, when peroxynitrite is formed at higher concentrations, peroxynitrite can lead to depletion of endogenous antioxidants (Phelps *et al.*, 1995; Szabó *et al.*, 1996a; Vatassery, 1996) and can initiate oxidant processes. It appears that under these conditions it is superoxide, rather than NO which acts as a rate limiting factor in the generation of peroxynitrite (Szabó *et al.*, 1995; Ischiropoulos *et al.*, 1996; Szabó, 1996). The peroxynitrite-triggered oxidant processes importantly contribute to the pathophysiology of circulatory shock, ischemia-reperfusion injury of the brain, heart, gut and skeletal muscle, inflammatory pancreatic islet cell destruction and possibly a variety of other pathophysiological conditions (see below).

Comparative Toxicity of NO and Peroxynitrite

In various experimental conditions, NO, when produced in excess, has been proposed to act as an important cytotoxic mediator. This conclusion is, in general, based on the following lines of evidence: (i) NO production can be demonstrated in response to stimulation of the experimental system; (ii) cytotoxic effects are observed, which parallel with the production of NO, (iii) inhibition of NO synthase (NOS) prevents cytotoxicity, and (iv) exogenous administration of NO in the form of NO donor compounds mimicks the toxic effects of endogenously produced NO. A primary example of the above mentioned line of thought is the suppression of mitochondrial respiration in response to the induction of iNOS in macrophages and various other cell types (Nathan, 1992; Szabó, 1995). However, recent data challenged the prevailing dogma that NO is independently toxic. Accumulating evidence suggests that much of the NO-related injury may be, in significant part, due to the generation of peroxynitrite. This proposition is supported by recent observations demonstrating that (i) peroxynitrite is more cytotoxic than NO or superoxide in a variety of experimental systems (DeGroot *et al.*, 1993; Denicola *et al.*, 1993; Hausladen and Fridovich, 1994; Brunelli *et al.*, 1995; Castro *et al.*, 1994; Szabó and Salzman, 1995; Bolanos *et al.*, 1995), (ii) that oxygen radical scavengers protect against NO-mediated cell injury in various cells (Dawson *et al.*, 1993; Cazevieille *et al.*, 1993; Burkart *et al.*, 1995; Zingarelli *et al.*, 1997a), and (iii) by theoretical considerations regarding the reactivity of NO, superoxide and peroxynitrite, as compared to the levels of superoxide and the rate of peroxynitrite formation in biological systems (Squadrito and Pryor, 1995; Augusto *et al.*, 1994). Careful *in vitro* studies have demonstrated that, in some systems (*in vitro* experiments using isolated mitochondrial and cytosolic aconitase, for example), peroxynitrite is a potent inhibitor of aconitase activity, whereas NO is without significant inhibitory effect (Hausladen and Fridovich, 1994; Castro *et al.*, 1994). While NO may cause reversible inactivation of mitochondrial enzymes, peroxynitrite causes a permanent suppression of mitochondrial function (Brunelli *et al.*, 1995; Castro *et al.*, 1994; Szabó and Salzman, 1995; Bolanos *et al.*, 1995). Similarly, peroxynitrite, and not NO, is a potent initiator of DNA strand breakage (see below). It appears that even when "pure" NO donors are applied to some biological systems, there may be sufficient amounts of superoxide (produced by the mitochondria, for example) present, to form peroxynitrite and lead to cytotoxic effects. The proposal that peroxynitrite is a major cytotoxic mediator would also provide a cogent explanation for previous data demonstrating the protective role of both NOS inhibitors and superoxide neutralizing strategies in the treatment of stroke, inflammation, and shock (see below).

DNA Injury Caused by Peroxynitrite

Single strand breakage

DNA single strand breakage is an obligatory trigger of activation of the nuclear enzyme poly (ADP-ribose) synthetase (PARS) (see below). The key pathophysiologically relevant triggers of DNA single strand breakage are hydroxyl radical (see below) and peroxynitrite. In 1992, King and colleagues demonstrated that potassium peroxynitrite (ONOOK) causes DNA cleavage in solutions of end-labelled DNA restriction fragments (King *et al.*, 1992). Over the last three years, several groups have independently demonstrated the occurrence of DNA single strand breakage in various types of intact cells upon exposure to peroxynitrite. For instance, in calf thymus DNA, and in the bacteriophage PM2, DNA strand breakage has been reported after exposure to authentic peroxynitrite or to SIN-1, a sydnonimine compound that simultaneously generates NO and superoxide (Salgo *et al.*, 1995a; Inonue *et al.*, 1995; Epe *et al.*, 1996). Moreover, DNA single strand breakage has been demonstrated in our laboratory in cultured macrophages and smooth muscle cells exposed to peroxynitrite (Szabó *et al.*, 1996a,b). Recently, the phenomenon of peroxynitrite-induced DNA single strand breakage has also been described in human pancreatic islet cells (Delaney *et al.*, 1996). The mechanism of the DNA strand breakage is probably related to abstraction of hydrogen atoms from the ribose of the DNA moiety, thereby opening the sugar ring (Salgo *et al.*, 1995). A similar phenomenon (abstraction of hydrogen atoms) has been observed in conjunction with peroxynitrite-induced lipid peroxidation.

The above mentioned studies all utilized exposure of living cells or DNA to authentic peroxynitrite or to agents that generate peroxynitrite. The physiological importance of such experiments, however, remained undefined. In a pathophysiologically more relevant situation, in immunostimulated macrophages and smooth muscle cells (which simultaneously produce NO and superoxide, and thus peroxynitrite from endogenous sources) (Ischiropoulos *et al.*, 1992; Zingarelli *et al.*, 1996a; Szabó *et al.*, 1996a), DNA single strand breakage has been demonstrated, and the time course of the strand breakage paralleled the time course of NO and peroxynitrite production (Zingarelli *et al*, 1996a; Szabó *et al.*, 1996a). Similarly, results by Zhang and co-workers (1994) suggested that in cerebellar slices, activation of NMDA receptors (a trigger for enhanced NO, superoxide and peroxynitrite production) may lead to peroxynitrite-mediated DNA single strand breakage (Snyder, 1996).

Taken together, the current data, in contrast to the earlier suggestions implicating a role for NO *per se* in the genesis of DNA single strand breakage (Zhang *et al.*, 1994; Radons *et al.*, 1994), point towards the central importance of peroxynitrite as the critical species responsible for the initiation of DNA single strand breakage: "pure" NO donors, even at extremely high concentrations, are extremely weak inducers of DNA single strand breakage, when compared to the effect of peroxynitrite (see: Szabó *et al.*, 1996a; Szabó, 1996b; Snyder, 1996).

It is noteworthy, nevertheless, that, although NO per se does not directly cause DNA single strand breaks, several hypotheses have been put forward for indirect mechanisms. Unrepaired abasic sites (due to NO-mediated injury) may lead to the development of DNA single strand breaks (Tamir *et al.*, 1996). The development of these DNA strand breaks may involve AP endonucleases, excision repair, topoisomerase-mediated repair, Ca2+/Mg2+ dependent endonucleases (Tamir *et al.*, 1996). Furthermore, it has been proposed that inhibition of ribonucleotide reductase by NO may reduce the supply of deoxyribonucleotides for DNA synthesis and repair, which, in turn, may cause a delay in the repair process and lead to the prevalence of DNA single strand breaks (Zhang and Steiner, 1995). Similarly, it is possible that inhibition by NO of a variety of DNA repair processes, as demonstrated, for example, for the Fpg protein and for O6-methylguanine-DNA-methyltransferase (Wink and Laval, 1994; Laval and Wink, 1994), may increase the degree of DNA single strand breakage in cells challenged with oxidants. Additional mechanisms of NO-related injury may involve the production of oxyradicals by the mitochondrial chain. In such scenario, NO may first inhibit the activity of mitochondrial enzymes, which subsequently triggers increased oxyradical generation ('leak') from the mitochondria (Ponderosa *et al.*, 1996). One can hypothesize that this process may lead to the generation of peroxynitrite, and subsequent DNA single strand breakage. An example of such mechanism has recently been described in relation to tumor necrosis factor induced inhibition of mitochondrial respiration, oxyradical production, DNA injury and PARS activation in L929 cells (Shoji *et al.*, 1995).

Oxidative DNA base modification

In addition to DNA single strand breakage, peroxynitrite can initiate a number of DNA modifications. However, these modifications do not lead to the activation of PARS in the nucleus. Among others, the production of 8-hydroxydeoxyguanosine, 8-nitroguanineoxazolone, 8-oxo-7,8-dihydro-2'deoxyguanosine and 8-oxo-7,8-dihydro-2'deoxyadenosine (Yermilow *et al.*, 1995; Inonue and Kawanishi, 1995; Epe *et al.*, 1996) have been described in various experimental systems after peroxynitrite exposure. Peroxynitrite can also cause DNA cleavage in solutions of end-labelled DNA restriction fragments (King *et al.*, 1992), and can initiate DNA nicking in the supercoiled plasmid pBR322 (Salgo *et al.*, 1995). Peroxynitrite exposure of E coli and human AD293 cells transfected with a pSP189 plasmid resulted in a significant increase in mutation frequency, with the majority occurring at G:C base pairs, predominantly involving G:C-T:A transversions (Juedes and Wogan, 1996).

A variety of DNA base modfications have also been described in immunostimulated macrophages which produce oxyradicals, NO, as well as peroxynitrite. For instance, the production of 5-hydromethyl)uracil; 2,6-diamino-4-hydroxy-5-formamidopyrimidine and 8-oxoguanine formation have been reported, indicative of both oxidative and deaminative DNA injury (Rojas-Walker *et al.*, 1995). Thus, it appears that in immunostimulated cells, DNA injury can occur via several mechanisms, only one of them being peroxynitrite formation and single strand breakage. Another mechanism of DNA injury is likely to involve the reaction of NO with molecular oxygen, yielding N_2O_3 with subsequent nitrosation of secondary amines and the formation of N-nitrosoamines or nitrosation of primary amines and nucleic acid bases (Tannenbaum *et al.*, 1994).

Cytotoxicity Due to PARS Activation

PARS: an endogenous cytotoxic effector

DNA single strand breakage is an obligatory trigger for the activation of PARS (Ueta and Hayashi, 1985; Lautier *et al.*, 1993). PARS is a protein-modifying and nucleotide-polymerizing enzyme which is abundantly present in the nucleus (Ueta and Hayashi, 1985; Lautier *et al.*, 1993). The activation of PARS results in the cleavage of NAD$^+$ into ADP-ribose and nicotinamide. In turn, PARS covalently attaches ADP-ribose to various nuclear proteins. PARS then extends the initial ADP-ribose group into a nucleic acid-like polymer, poly (ADP-ribose) (Ueta and Hayashi, 1985; Lautier *et al.*, 1993). The major acceptors of poly (ADP-ribose) are PARS itself (automofification domain), topoisomerase I and II, DNA polymerases alpha and beta, and DNA ligase 2. For most of these enzymes, ADP-ribosylation results in a decrease in their catalytic activities (Lautier *et al.*, 1993). It is also noteworthy that the presence of ADP-ribose on histones (mainly histone H1) causes chromatin relaxation (Lautier *et al.*, 1993).

The physiological function of PARS has been the subject of debate for at least two decades, the 'classical' proposition being that PARS is a DNA repair enzyme. According to the more recent evidence, PARS is not effective as a DNA repair enzyme, since cells from a PARS knock-out mice have normal DNA repair characteristics (Wang *et al.*, 1995). It has been suggested that the physiological role of PARS is to regulate gene expression and cellular differentiation, transformation and division (see below) and/or to slow cellular metabolism as an adaptive response to altered environmental conditions.

Pronounced activation of PARS can rapidly deplete the intracellular concentration of its substrate, NAD^+, slowing the rate of glycolysis, electron transport, and, therefore, ATP formation resulting in cell dysfunction and cell death. Accordingly, inhibitors of PARS protect against cell death under these conditions. This mechanism, known as the "PARS suicide hypothesis", has previously been characterized in relation to H_2O_2-induced oxidant damage and radiation injury (Berger *et al.*, 1986; Schraufstatter *et al.*, 1988; Schraufstatter *et al.*, 1986; Thies and Autor, 1991; Berger, 1991; Cochrane, 1991). The relative contribution of PARS to the cellular metabolic changes and cell injury is dependent on the cell type studied. In endothelial cells, epithelial cells, and fibroblasts PARS appears to play a greater role in the oxidant damage, whereas in hepatocytes inhibition of PARS has little influence on oxidant-induced cell damage (Junod *et al.*, 1989; Yamamoto *et al.*, 1993; Kirkland, 1991; Spragg, 1991).

Acute cellular metabolic derangements in response to PARS activation

The above described pathway of rapid energy depletion can lead to acute (within minutes to hours) cellular metabolic derangements and even death in *in vitro* systems. In 1994, two groups, working on the pathogenesis of diabetes and neuronal injury, independently proposed a role of DNA strand breakage and PARS activation in the NO-related cellular injury (Zhang *et al.*, 1994; Radons *et al.*, 1994). In these studies, when pancreatic islet cells or rat cerebellar slices were incubated with high concentrations of NO gas or high concentration of NO donors, and an increase in PARS activity, and a consequent decrease in cellular NAD^+ levels activity was observed, leading to cytotoxic effects (Zhang *et al.*, 1994; Radons *et al.*, 1994). In a pathophysiologically more relevant setting, studies in brain slices have demonstrated that NMDA-receptor activation increases NO production from endogenous sources, which, in turn, decreases cellular viability. Inhibitors of NOS, as well as inhibitors of PARS, ameliorated the NMDA-activation mediated decreases in cell viability (Zhang *et al.*, 1994).

We have subsequently recently demonstrated that in macrophages, smooth muscle cells, epithelial cells and endothelial cells exposed to authentic peroxynitrite, development of DNA strand breaks occurs, which, in turn, results in the activation of PARS with consequent reduction of intracellular NAD^+, ATP, and mitochondrial respiration (Szabó *et al*, 1996a,b,c, Szabó *et al.*, 1997b; Kennedy *et al.*, 1997). The metabolic changes, but not the development of DNA strand breaks, can be ameliorated by pharmacological inhibition of PARS (Szabó *et al.*, 1996a.b). The finding that 3-aminobenzamide did not inhibit the development of DNA strand breaks (the trigger of PARS activation) is consistent with the proposal that 3-aminobenzamide is not a direct scavenger of peroxynitrite. Similar to the results observed with authentic peroxynitrite, we found that immunostimulation of J774 cells and vascular smooth muscle cells in culture resulted in DNA strand breakage and PARS activation, and these changes paralleled the onset of NO and peroxynitrite production (Zingarelli *et al.*, 1996a; Szabó *et al.*, 1996a). Pharmacological inhibitors of PARS activity inhibited the depression of mitochondrial respiration in these cells. We obtained similar results utilizing three different, structurally-unrelated PARS inhibitors, suggesting that the mechanism by which these agents exerted protective effects is, indeed, *via* the inhibition of PARS activity (Zingarelli *et al.*, 1996a; Szabó *et al.*, 1996a). Taken together, these data suggested that activation of PARS, caused by DNA strand breakage due to endogenously produced peroxynitrite, contributes to acute cell damage (Figures 6-1 and 6-2a). Recent flow cytometry studies demonstrate that the mode of cell death induced by peroxynitrite which is beneficially affected by inhibition of PARS is necrosis, rather than apoptosis (see also below) (Virag *et al.*, 1988).

Acute functional changes in cells and tissues in response to peroxynitrite exposure and PARS activation

Peroxynitrite-induced PARS activation has important functional consequences in various *in vitro* and *in vivo* experimental systems. In *in vitro* experiments in pulmonary and intestinal epithelial monolayers, exposure of the cells to authentic peroxynitrite, or to SIN-1, a compound that generates peroxynitrite (see above), leads to an increase in epithelial permeability, and effect, which can be ameliorated by pharmacological inhibition of PARS (Szabó *et al.*, 1997c; Kennedy *et al.*, 1997). The importance of the peroxynitrite-PARS pathway in mediating intestinal hyperpermeability has also been confirmed *in vivo*, in a rodent model of splanchnic artery ischemia/reperfusion (Cuzzocrea *et al.*, 1997; see below).

In thoracic aortic rings exposed to peroxynitrite, both the development of vascular hyporeactivity (Szabó *et al.*, 1996a) and the development of endothelial dysfunction (reduced relaxant responsiveness to endothelium-dependent, but not to endothelium-independent vasorelaxants) (Szabó *et al*, 1997c) has been observed, in accordance with similar results obtained in perfused hearts after peroxynitrite exposure (Villa *et al.*, 1994). These vascular alterations, which are common features of various forms of shock and reperfusion injury (see below), were, in a significant part, ameliorated by pharmacological inhibitors of PARS (Szabó *et al.*, 1996a; Szabó *et al.*, 1997b,c). Similar results were observed in *ex vivo* experiments, with blood vessels from animals subjected to endotoxic shock (see below).

Apoptosis in response to peroxynitrite: the potential role of PARS activation

Several reports demonstrated that NO and peroxynitrite cause necrosis or apoptosis in a variety of cell types (Denicola *et al.*, 1993; Salgo *et al.*, 1995c; Bonfoco *et al.*, 1995; Lin *et al.*, 1995; Estevez *et al.*, 1995). It appears that sustained exposure or low levels of NO or peroxynitrite cause apoptosis whereas sudden exposure to high concentrations of peroxynitrite or NO results in cell necrosis. As demonstrated in CHO-Em9 cells defective in their ability to repair DNA single-strand breaks, unrepaired single strand breaks lead to the formation of double strand breaks, with eventual cell death (Tamir *et al.*, 1996).

                    Csaba Szabó

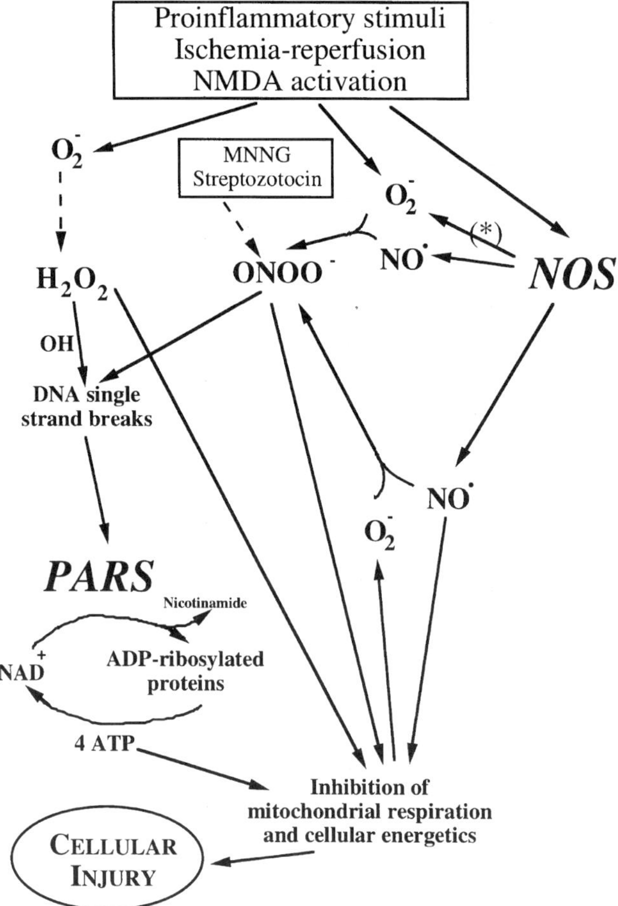

Figure 6-1. Proposed scheme of PARS-dependent and PARS-independent cytotoxic pathways involving nitric oxide (NO·), hydroxyl radical (OH·) and peroxynitrite (ONOO⁻). Proinflammatory conditions induce the expression of the inducible NO synthase (iNOS), whereas NMDA receptor ligands activate the constitutive neuronal NOS (bNOS). NO, in turn, combines with superoxide to yield peroxynitrite. During myocardial or splanchnic ischemia-reperfusion injury, NO derived from the constitutive endothelial NO synthase (ecNOS) combines with superoxide to produce peroxynitrite. Hydroxyl radical (produced from superoxide via the iron-catalyzed Haber-Weiss reaction – indicated with a dashed line in the top left part of the figure) and peroxynitrite or peroxynitrous acid induce the development of DNA single strand breakage, with consequent activation of PARS. Depletion of the cellular NAD⁺ leads to inhibition of cellular ATP-generating pathways leading to cellular dysfunction. NO alone does not induce DNA single strand breakage, but may combine with superoxide (produced from the mitochondrial chain or from other cellular sources) to yield peroxynitrite. Under conditions of low cellular L-arginine (*), NOS may produce both superoxide *and* NO, which then combine to form peroxynitrite (Xia *et al.*, 1996). There is experimental evidence suggesting that the DNA damaging agents N-methyl-N'-nitrosoguanidine (MNNG) and streptozotocin also induce DNA injury *via* the generation of peroxynitrite (indicated by a dashed line in the top middle part of the figure). There are PARS-independent, parallel pathways of cellular metabolic inhibition, and these pathways can be activated by NO, hydroxyl radical, superoxide and by peroxynitrite (alone or in combination or synergy). The relative importance of the PARS-dependent and PARS-independent component of oxidant cytotoxicity is cell-type dependent. Even in cells in which PARS activation plays an important role in the oxidant injury, activation of PARS-independent pathways of cytotoxicity may induce overwhelming cellular injury at extremely high oxidant concentrations. See the text for further explanations.

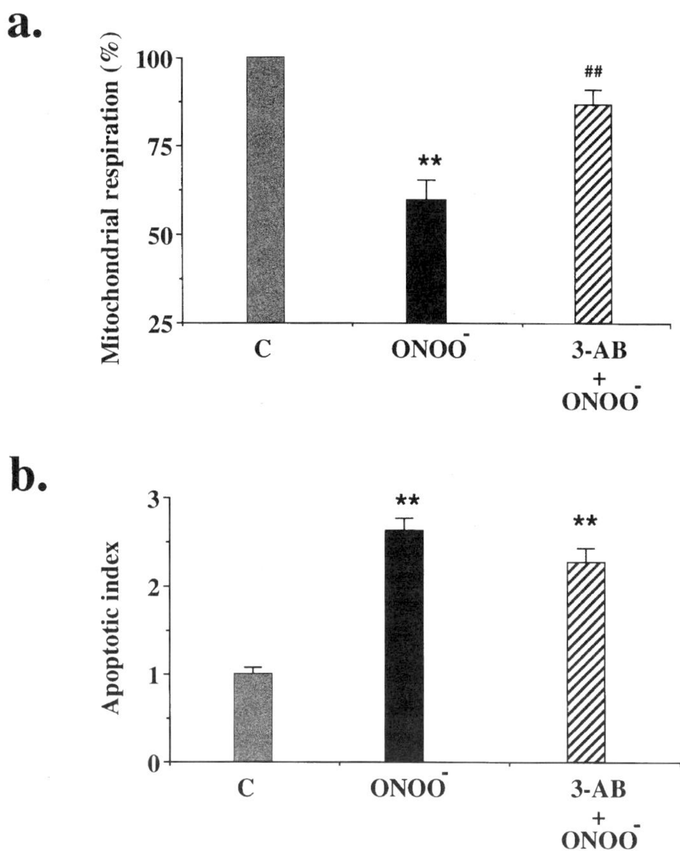

Figure 6-2. Effect of peroxynitrite (ONOO⁻, 500 μM) on (a) mitochondrial respiration, as measured at 1h after exposure and on (b) apoptotic index, as measured at 24h after exposure in cultured smooth muscle cells. (C: control value; ONOO: value after peroxynitrite exposure; ONOO+3–AB; effect of the PARS inhibitor 3-aminobenzamide, given as a 10 min pretreatment). Note that the PARS inhibitor reduced the acute suppression of the mitochondrial respiration, but did not affect the delayed apoptosis. (Methods as described in Szabó *et al.*, 1996a and Bingisser *et al.*, 1996). **P < 0.01 represents significant changes elicited by ONOO when compared to unstimulated controls; ##P < 0.01 represents significant protection by 3-AB against the peroxynitrite-induced alterations. Data are expressed as means ± standard error of the mean of n = 6–9 wells.

Although a role for PARS in the development of apoptosis has previously been demonstrated in a variety of cell types (Monti *et al.*, 1994; Monti *et al.*, 1995; Nicholson *et al.*, 1995), the role of PARS in the mediation of NO-induced apoptosis is undefined. In human leukemia cells, 3-aminobenzamide and nicotinamide reduced NO-induced apoptosis (Kuo *et al.*, 1996), whereas these inhibitors were ineffective in blocking apoptosis in RAW murine macrophages (Messmer and Brune, 1996). It is clear that the NO-induced apoptosis (which is associated with proteolytic cleavage of PARS [Messmer *et al.*, 1996]) involves mechanisms different from the apoptosis triggered by potent DNA single strand breaking agents, such as peroxynitrite and hydroxyl radical. Indeed, it appears that neither the hydrogen peroxide-induced apoptosis in intestinal epithelial cells (Watson *et al.*, 1995), nor the peroxynitrite- or immunostimulation-induced apoptosis in cultured rat aortic smooth muscle cells (Figure 6-2b) is significantly altered by pharmacological inhibition of PARS. In thymocytes exposed to peroxynitrite, both necrosis and apoptosis can be detected. Inhibition of PARS in these cells prevents necrosis, but increases the apoptotic cell population (Virag *et al.*, 1998). This latter finding is consistent with the notion that apoptosis is a process that requires cellular energy (Payne *et al.*, 1995; Richter *et al.*, 1996).

PARS cleavage during NO-induced apoptosis

In a recent line of investigations, a novel interaction between NO and PARS has been described. In this process, NO donors, or immunostimulation cause apoptosis, as demonstrated in murine macrophages (Messmer *et al.*, 1996). This process is also associated with proteolytic cleavage of PARS, a process, which, similar to the process of apoptosis, can be blocked by pharmacological inhibitors of NOS in these immunostimulated cells. Overexpression of the anti-apoptotic mammalian protein Bcl-2 in these cells has been shown to block the NO-induced apoptosis, and the NO-induced proteolytic cleavage of PARS (Messmer *et al.*, 1996). Based on these results, the hypothesis has been put forward that PARS cleavage (with a consequent inhibition of the catalytic activity of PARS) is a process involved in endonuclease activation and apoptosis in NO-treated cells (Messmer *et al.*, 1996). However, this proposal is somewhat in contrast with the finding that pharmacological inhibitors of PARS do not affect the course of immunostimulation-induced (and NO-dependent) apoptotic process in macrophages (Messmer and Brune, 1996) or in smooth muscle cells (Figure 6-2b).

ROLE OF PEROXYNITRITE–PARS PATHWAY IN VARIOUS PATHOPHYSIOLOGICAL CONDITIONS

Circulatory Shock

Various forms of circulatory shock are associated with a reduced responsiveness of arteries and veins to exogenous or endogenous vasoconstrictor agents (vascular hyporeactivity), myocardial dysfunction and disrupted intracellular energetic processes, culminating in multiple organ failure and death. Some of these alterations have previously been suggested to be related to NO overproduction, due to the activation of the endothelial isoform of NOS (ecNOS) in the early stage and expression of a distinct inducible isoform of NOS (iNOS) in the late stage (Szabó and Thiemermann, 1994; Szabó, 1995). These previous studies have reached these conclusions based, mainly, on results obtained with NOS inhibitors, and, therefore, did not or could not distinguish between the effects of NO and peroxynitrite. Recent data demonstrate, that authentic peroxynitrite is capable of mimicking many the pathophysiological alterations associated with shock (endothelial dysfunction, vascular hyporeactivity, and cellular dysfunction) (see above).

In circulatory shock, pro-inflammatory cytokines invoke the stimulation of oxygen centered free radicals production. Therefore, it is not surprising that various forms of circulatory shock are associated with the production of peroxynitrite. The production of peroxynitrite (evidenced as increased nitrotyrosine immunoreactivity or increased oxidation of the fluorescent probe dihydrorhodamine 123 to rhodamine 123) has recently been demonstrated in endotoxin shock and in hemorrhagic shock (Szabó *et al.*, 1995a,b; Wizemann *et al.*, 1994; Zingarelli *et al.*, 1997b) (Figure 6-1). Interestingly, a NOS-inhibitor inhibitable oxidation of the fluorescent probe dihydrorhodamine 123 to rhodamine 123 preceded the onset of iNOS induction (Szabó *et al.*, 1995b), and this finding has been suggested to be due to the early formation of peroxynitrite produced in response to the combination of superoxide and NO from constitutive NOS isoforms, such as the endothelial isoform, ecNOS.

Vascular contractile failure in circulatory shock

The vascular contractile failure associated with circulatory shock is closely related to overproduction of NO within the blood vessels. Expression of iNOS within the vascular smooth muscle cells has been implicated in the pathogenesis of vascular hyporeactivity during various forms of shock (See: Rees, 1995; Szabó, 1995). In these investigations, the evidence for the role of NO in the development of vascular hyporeactivity was based on experiments where pharmacological inhibitors of NOS restored the contractility of the blood vessels. However, recent studies have demonstrated that a superoxide dismutase mimetic, similar to a NOS inhibitor, offers a significant protection against the suppression of the vascular contractility of the thoracic aorta in a rat model of endotoxic shock (Zingarelli *et al.*, 1997a). This is an effect which is in marked contrast with the previously described NO-superoxide interactions in normal blood vessels, where superoxide dismutase is known to prolog the half-life of NO and thereby enhances NO-mediated relaxant responses (Gryglewski *et al.*, 1986; Rubanyi *et al*, 1986). The most obvious explanation for the protective effect of both NOS inhibitors and superoxide dismutase mimetics against the endotoxin-induced vascular hyporeactivity is related to peroxynitrite generation. Indeed, peroxynitrite exposure can cause vascular hyporeactivity in vascular rings (see above).

What, then, is the mechanism of the peroxynitrite-induced vascular hyporeactivity in endotoxic shock? In studies in anesthetized rats, inhibition of PARS with 3-aminobenzamide and nicotinamide were able to reduce the suppression of the vascular contractility of the thoracic aorta in *ex vivo* experiments (Szabó *et al.*, 1996a; Zingarelli *et al.*, 1996b). These findings are similar to the *in vitro* results with authentic peroxynitrite, which also causes a vascular hyporeactivity in thoracic aortic rings, and which is also reduced by pharmacological inhibition of PARS.

At present, it is not clear, to which extent the "pure" NO-mediated (guanylyl cyclase related) vasorelaxant mechanisms versus the peroxynitrite-mediated (and, in part, PARS-related) vasorelaxant mechanisms contribute to the vascular hyporeactivity in various forms of shock, and in various phases of shock. In also remains to be investigated whether there are interactions between the cGMP-related and the peroxynitrite-related pathways.

Endothelial dysfunction in circulatory shock

Peroxynitrite production has been suggested to contribute to endothelial injury in ischemia-reperfusion, circulatory shock and atherosclerosis (Crow and Beckman, 1995; White *et al.*, 1994; Zingarelli *et al.*, 1997a; White *et al.*, 1996). In fact, Villa and co-workers (1994) directly demonstrated in isolated perfused hearts that infusion of authentic peroxynitrite into these hearts results in an impairment of the endothelium-dependent relaxations. In light of these previous studies, the question arises as to whether the endothelial dysfunction associated with endotoxic shock is also related to peroxynitrite formation, and, if so, whether the PARS pathway plays a role in the process.

There is now good evidence for a role for peroxynitrite formation in the pathogenesis of the endothelial dysfunction in endotoxic shock. For instance, we have recently observed that a manganese-mesoporphyrin cell-permeable superoxide dismutase analog and peroxynitrite scavenger has been shown to protect against the development of endothelial dysfunction in endotoxic shock (Zingarelli *et al.*, 1997a). Similarly, selective inhibition of the inducible isoform of NO synthase by canavanine (Fatehi-Hassanabad *et al.*, 1996) or by guanidinoethyldisulfide (Szabó *et al.*, 1996d) provided protection against the endothelial dysfunction and the endothelial morphological damage in vessels obtained from rats subjected to endotoxin shock. One possible rationalization of the above mentioned results is that iNOS-derived NO combines with superoxide (the latter derived from activated neutrophils, and other sources) to form peroxynitrite, which, in turn, causes endothelial injury.

Current data, demonstrating protective effects of 3-aminobenzamide against the development of endothelial dysfunction in vascular rings obtained from rats with endotoxic shock (Szabó *et al.*, 1996b) suggest that DNA strand breakage and PARS activation occur in endothelial cells during shock and that the subsequent energetic failure reduces the ability of the cells to generate NO in response to acetylcholine-induced activation of the muscarinic receptors on the endothelial membrane. Indeed, several lines of *in vitro* data demonstrate DNA injury, PARS activation and consequent cytotoxicity in endothelial cells exposed to hydroxyl radical generators (Spragg, 1991; Thies and Autor, 1991; Junod *et al.*, 1989), or in response to peroxynitrite (Szabó *et al.*, 1997c). The relative contribution of peroxynitrite vs. hydroxyl radical in the PARS activation and endothelial injury in shock remains to be further investigated, since both species are known to be produced in shock, and the available scavengers (such as superoxide dismutase analogs) would be expected to reduce the production of both peroxynitrite and hydroxyl radical. The data with iNOS inhibitors demonstrating protection against the endothelial injury in endotoxic shock (Fatehi-Hassanabad *et al.*, 1996; Szabó *et al.*, 1996 x) would favor the contribution of an NO-derived species, such as peroxynitrite. Considering the existence of synergistic cytotoxic interactions of hydrogen peroxide and peroxynitrite (Szabó, 1997), it is also conceivable that a similar synergism may also exist *in vivo*, and both peroxynitrite and hydroxyl radical contribute to the activation of PARS and the subsequent endothelial injury *in vivo*.

Cellular energetic failure and organ injury in circulatory shock

There is now good evidence suggesting that NO (or a related species, such as peroxynitrite) plays a role in the cellular energetic changes and the related organ dysfunction associated with endotoxic shock. This conclusion is chiefly based on the results of pharmacological studies, in which inhibition of NO synthesis, especially by agents that are selective towards the inducible isoform of NOS (iNOS), reduce cellular injury and improve organ function in shock (Southan and Szabó, 1996). It is noteworthy that in the same experimental models of rodent endotoxic shock, the cell-permeable superoxide dismutase analog MnIII tetrakis (4-benzoic acid) porphyrin (Szabó *et al.*, 1996e) also reduced the endotoxin-induced depression of mitochondrial respiration in peritoneal macrophages *ex vivo* (Zingarelli *et al.*, 1997a), thereby suggesting that peroxynitrite, rather than NO per se plays a role in these alterations.

Peroxynitrite-induced activation of the PARS pathway has also been implicated in the pathophysiology of the cellular energetic failure associated with endotoxin shock by demonstration of increased DNA strand breakage, decreased intracellular NAD$^+$ and ATP levels and mitochondrial respiration in peritoneal macrophages obtained from rats subjected to endotoxin shock (Zingarelli *et al.*, 1996a,b). This cellular energetic failure was reduced by pretreatment of the animals with the PARS inhibitors 3-aminobenzamide or nicotinamide (Zingarelli *et al.*, 1996a,b).

In contrast to these encouraging results in peritoneal macrophages, it appears that the PARS pathway only plays a limited role in the liver dysfunction associated with endotoxin shock. In an endotoxic shock model in the rat, inhibition of PARS with 3-aminobenzamide and nicotinamide did not affect the alterations in most parameters of liver injury, whereas inhibition of PARS with 1,5-dihydroxyisoquinoline resulted in a marginal protective effect (Thiemermann *et al.*, unpublished data, 1997). These observations are perhaps not surprising when considering the fact that in *in vitro* studies, the oxidant-induced injury in cultured hepatocytes is not prevented by pharmacological inhibition of PARS (Yamamoto *et al.*, 1993). The exact reason why inhibition of PARS does not affect the course of the oxidant injury in hepatocytes remains to be further investigated. Although hepatic injury does not appear to be largely PARS mediated in endotoxic shock, pharmacological inhibition of PARS, either with 3-aminobenzamide (Szabó *et al.*, 1996a) or with the potent, novel potent PARS inhibitor (Bauer *et al.*, 1996) 5-iodo-6-amino-1,2,-benzopyrone (Szabó *et al.*, 1997d) improves survival rate in mice challenged with high dose endotoxin. Based on these observations, one may suggest that, in response to pharmacological inhibition of PARS, the improved hemodynamic status due to improved vascular function, and possibly, the improved cellular energetic status in some organs, does result in an overall survival benefit in this condition.

Reperfusion Injury

A growing body of evidence supports the role of the peroxynitrite in the neuronal injury associated with ischemia-reperfusion injury in a variety of organ systems. In ischemia-reperfusion injury, superoxide, produced in the reperfusion phase, rapidly reacts with NO and forms peroxynitrite. This phenomenon has been demonstrated or suggested in the heart (Matheis *et al.*, 1992; Schulz and Wambolt, 1995; Naseem *et al.*, 1995, Wang and Zweier, 1996), liver (Ma *et al.*, 1995), kidney (Yu *et al.*, 1994) intestine (Szabó *et al.* 1995b), brain (Fagni *et al.*, 1994; Gunasekar *et al.*, 1995) and lung (Ischiropoulos *et al.*, 1995b). In these conditions, prevention of peroxynitrite generation by inhibition of NO biosynthesis markedly reduces reperfusion injury, as shown by reduced pulmonary lipid peroxidation in the lung (Ischiropoulos *et al.*, 1995b) or improved mechanical performance of the heart (Matheis *et al.*, 1992; Schulz and Wambolt, 1995; Naseem *et al.*, 1995, Wang and Zweier, 1996).

Reperfusion injury in the brain

The original proposition (Beckman, 1991) that peroxynitrite (and not NO or superoxide, independently), is a major cytotoxic mediator in the neuronal injury during stroke and NMDA receptor activation, was based on theoretical considerations and previous evidence showing that reperfusion injury in the central nervous system is associated with activation of NMDA receptors which then triggers the production of superoxide and NO. There is now indirect evidence which suggests that NMDA receptor activation is associated with a marked increase in a hydroxyl-radical like reactivity in the brain (blocked by inhibition of NOS) (Hammer *et al.*, 1993), which is presumably due to peroxynitrite generation. The involvement of superoxide and the protective effect of superoxide neutralizing strategies (Fagni *et al.*, 1994; Gunasekar *et al.*, 1995; Lafon-Cazal *et al.*, 1993) as well as the involvement of NO and the protective effect of NOS inhibition (Dawson *et al.*, 1993; Schulz *et al.*, 1995a,b) have been well established in various forms of central nervous system injury, including stroke. Although it is likely that peroxynitrite, produced during the reperfusion of the brain, initiates a multitude of interrelated cytotoxic processes (see above), it appears that PARS activation is one of these cytotoxic pathways.

For instance, in cerebellar granule cells, glutamate induces a rapid increase in poly (ADP-ribose) immunoreactivity (Cosi *et al.*, 1994). Furthermore, the early phase of stroke is associated with DNA injury (MacKenzie *et al.*, 1993). *In vitro* evidence clearly suggests the involvement of DNA single strand breakage and PARS activation in the neuronal damage associated with peroxynitrite production in response to NMDA receptor activation (Zhang *et al.*, 1994; Cosi *et al.*, 1994; Wallis *et al.*, 1993: Zhang *et al.*, 1995; Zhang and Steiner, 1995; Didier *et al.*, 1996; Snyder, 1996), and in response to traumatic brain injury (Wallis *et al.*, 1996). Interestingly, NMDA-receptor activation related cytotoxicity in the periphery also appears to involve peroxynitrite formation and PARS activation, as demonstrated in the case of NMDA-induced pulmonary edema in perfused rat lungs (Said *et al.*, 1996). Recent studies provided clear *in vivo* demonstration of the neuroprotective effect of PARS inhibition in murine models of stroke (Eliasson *et al.*, 1997; Endres *et al.*, 1997).

Reperfusion injury of the heart

As described above, reperfusion of the ischemic heart results in the generation of peroxynitrite, and prevention of peroxynitrite generation by inhibition of NO biosynthesis markedly reduces reperfusion injury and improves myocardial mechanical performance (Matheis *et al.*, 1992; Schulz and Wambolt, 1995; Naseem *et al.*, 1995, Wang and Zweier, 1996; Xie and Wolin, 1996). Recent studies have indicated that activation of PARS may be responsible, at least in part, for the development of reperfusion injury in the heart. In a rabbit model of myocardial infarction, pharmacological inhibitors of PARS, such as nicotinamide and 3-aminobenzamide, given immediately prior to the reperfusion of the ischemic myocardium, caused a massive reduction of the infarct size, while the structurally related, but inactive agent 3-aminobenzoic acid was without protective effect (Thiemermann *et al.*, 1997).

In an independent line of experiments in anesthetized rats, 3-aminobenzamide also exerted cardioprotective effects in a rat model of myocardial infarction elicited by occlusion and reperfusion of the left coronary artery (Zingarelli *et al.*, 1997c). In this latter study, the following functional parameters were measured: 1) infarct size; 2) histological damage; 3) plasma creatine kinase activity; 4) cardiac myeloperoxidase; 5) cardiac ATP levels and 6) pressure-rate index. There was a reduction in the infarct size, plasma creatine kinase activity and the improved histological picture of the reperfused myocardium after pharmacological inhibition of PARS (Zingarelli *et al.*, 1997c). The reduced cardiac myeloperoxidase activity after 3-aminobenzamide treatment is consistent with reduced infiltration of neutrophil granulocytes into the reperfused myocardium. The modest preservation of the myocardial ATP levels and the lack of significant effect of 3-aminobenzamide on the pressure-rate index (Zingarelli *et al.*, 1997c), however, supports the view that the pharmacological inhibitor of PARS used in the current study only offers a limited degree of protection against the myocardial metabolic changes and against the changes in myocardial oxygen consumption during reperfusion. Taken together, 3-aminobenzamide has protective effects on many, but not all pathophysiological alterations associated with myocardial ischemia-reperfusion injury.

Although 3-aminobenzamide is not a direct scavenger of peroxynitrite and does not inhibit the synthesis or action of its precursors, the PARS inhibitor caused a marked reduction in the degree of oxidation of dihydrorhodamine 123 and in the degree of tyrosine nitration in the reperfused myocardium in the 3-aminobenzamide-treated animals (Zingarelli *et al.*, 1997c). These observations suggest that inhibition of PARS, in an indirect way, influences the amounts of reactive peroxynitrite produced. Although the mechanism of this action clearly requires further work, the following possibility, related to self-amplifying, positive feedback circles should be considered. Since 3-aminobenzamide reduced neutrophil influx into the reperfused myocardium, it is expected that a subsequent reduced production of neutrophil-derived oxidants (including superoxide and peroxynitrite) occurs. It is well established that myocardial ischemia and reperfusion is associated with neutrophil accumulation with subsequent burst of oxygen free radical production, activation of inflammation, excessive calcium entry, and ultimately, cell death (Lucchesi, 1990; Ferrari *et al.*, 1992). Activation and accumulation of polymorphonuclear cells is one of the initial events of tissue injury, which triggers the release of oxygen free radicals, arachidonic acid metabolites and lysosomal proteases, with subsequent more pronounced infiltration of neutrophils into the reperfused tissues, excessive calcium entry, and ultimately, cell death (Lucchesi, 1990; Ferrari, 1994). Peroxynitrite and hydroxyl radical are known to exert cytotoxic effects to the vascular endothelium, and the mechanism of this injury, at least in part, is mediated by PARS activation (see above). Since endothelium-derived NO is a potent inhibitor of both neutrophil aggregation and adherence, an improvement of the endothelial function by 3-aminobenzamide would reduce the infiltration of neutrophils during reperfusion, thus, resulting in reduced peroxynitrite formation and protection against the tissue injury. In other words, the following positive feedback cycle may be present in myocardial ischemia-reperfusion: early hydroxyl radical and peroxynitrite production >> PARS-related endothelial injury >> neutrophil infiltration >> more hydroxyl and peroxynitrite production. Inhibition of PARS would intercept this cycle at the level of endothelial injury. This model would explain the reduction of nitrotyrosine immunoreactivity and dihydrorhodamine oxidation during reperfusion in the 3-aminobenzamide-treated rats: reduced neutrophil infiltration leads to reduced peroxynitrite generation. It is presumable that various oxyradical scavengers, which also have been found to block the increase in myocardial myeloperoxidase activity during reperfusion (Mehta *et al.*, 1989; Campo *et al.*, 1994; Hoshida *et al.*, 1994), interrupt a similar positive feedback cycle. Recent studies, using genetically engineered mice lacking PARS have confirmed the role of PARS both in mediating myocardial necrosis and promoting neutrophil recruitment in myocardial ischemia-reperfusion (Zingarelli *et al.*, 1998).

Splanchnic reperfusion injury

A recent set of studies implicated the role of the peroxynitrite-PARS pathway in the pathophysiological changes in the reperfused gut (Cuzzocrea *et al.*, 1997). In a rat model of splanchnic occlusion shock, which was induced in rats by clamping both the superior mesenteric artery and the celiac trunk for 45 min, followed by release of the clamp (reperfusion), there was a marked increase in the oxidation of dihydrorhodamine 123 to rhodamine (a marker of peroxynitrite-induced oxidative processes) in the plasma of the SAO shocked rats after reperfusion, but not during ischemia alone. Immunohistochemical examination demonstrated a marked increase in the immunoreactivity to nitrotyrosine, a specific "footprint" of peroxynitrite, in the necrotic ileum in shocked rats. In addition, in *ex vivo* studies in aortic rings from shocked rats, there was a reduction in the contractions to noradrenaline and impaired responsiveness to a relaxant effect to acetylcholine (vascular hyporeactivity and endothelial dysfunction, respectively). Splanchnic artery ischemia and reperfusion also resulted in a marked increase in epithelial permeability. 3-aminobenzamide treatment significantly reduced ischemia/reperfusion injury in the bowel as evaluated by histological examination and significantly improved mean arterial blood pressure, improved contractile responsiveness to noradrenaline, enhanced the endothelium-dependent relaxations and reduced the reperfusion-induced increase in epithelial permeability (Cuzzocrea *et al.*, 1997). 3-aminobenzamide also prevented the infiltration of neutrophils into the reperfused intestine, as evidenced by reduced myeloperoxidase activity, improved the histological status of the reperfused tissues, reduced the production of peroxynitrite during reperfusion, and improved survival (Cuzzocrea *et al.*, 1997). These results demonstrate that the PARS inhibitor 3-aminobenzamide exerts multiple protective effects in splanchnic artery occlusion/reperfusion shock, and suggest that peroxynitrite and/or hydroxyl radical, produced during the reperfusion phase, trigger DNA strand breakage, PARS activation and subsequent cellular dysfunction. The vascular endothelium is likely to represent an important cellular site of protection by 3-aminobenzamide in SAO shock. The reduced neutrophil infiltration and the reduced nitrotyrosine staining after inhibition of PARS may be related to the interruption of positive feedback cycles similar to the ones described in relation to the reperfused heart (see above).

Pancreatic Islet Cell Damage–Diabetes

The importance of the PARS pathway in the pathogenesis of pancreatic islet cell injury has been initially put forward by Yamamoto in 1981, based on studies in streptozotocin-treated islet cells. Inhibition of PARS inhibited NAD^+ depletion and the suppression of proinsulin synthesis without modifying the extent of DNA damage in streptozotocin-treated islet cells (Yamamoto *et al.*, 1981; Uchigata *et al.*, 1982). Moreover, *in vivo* experiments demonstrated that PARS inhibition by 3-aminobenzamide or nicotinamide prevented the onset of streptozotocin- and alloxan-induced diabetes (Uchigata *et al.*, 1983; Masiello *et al.*, 1985; 1990). These above observations should be re-evaluated based on the current evidence demonstrating that (i) streptozotocin generates NO in aqueous solutions (Turk *et al.*, 1993; Kwon *et al.*, 1994) and (ii) scavenging NO protects against streptozotocin-induced DNA strand breakage (Kroncke *et al.*, 1995). Coupling these previous data with the recent evidence showing that peroxynitrite, (and not NO) is the potent trigger of DNA strand breakage (see above) it is logical to propose that PARS activation due to peroxynitrite formation underlies the pathogenesis of islet cell damage in response to streptozotocin or NO donor compounds. A recent report showing that Vitamin E inhibits DNA strand breakage, PARS activation and cytotoxicity in response to NO donor drugs (Burkart *et al.*, 1995), suggesting the involvement of oxygen-derived reactive species, most likely peroxynitrite in this toxicity. In this respect it is also noteworthy that heat shock and heat shock proteins also protects against peroxynitrite-induced cytotoxicity as well as against streptozotocin-induced islet cell injury (Bellmann *et al.*, 1995, Szabó *et al.*, 1996c; Bellmann *et al.*, 1996).

The studies into the role of PARS in diabetes gained new momentum by a recent line of investigations demonstrating showing increased poly (ADP-ribosylation) in islet cells exposed to oxidants and NO donors, and protection against the associated cellular injury by inhibition of PARS (Radons *et al.*, 1994; Inada *et al.*, 1995; Heller *et al.*, 1995; Eizirik *et al.*, 1996). Most notably, Heller and co-workers (1995) demonstrated that islet cells of the PARS$^{-/-}$ knockout mouse are resistant against injury in response to NO generator and oxyradical generator compounds. Interestingly, the protection waned when islets are challenged with extremely high concentrations of the oxidants (Heller *et al.*, 1995), demonstrating that at very high levels of oxidant stress, the PARS related pathway of cellular injury may be replaced by PARS-independent cytotoxic mechanism.

Although the evidence is solid regarding the effectiveness of PARS inhibitors against oxidant-induced islet cell injury, the relevance of these findings in relation to autoimmune diabetes is unclear. The relevance of the chemically induced diabetes models has frequently been questioned, and it appears that experimental models of *spontaneous autoimmune* diabetes (such as the NOD mice) are more relevant to the pathophysiology of the human disease. In this latter experimental model, the development of the disease is governed by the intraislet production of pro-inflammatory cytokines. These cytokines, in turn, induce the expression of the inducible isoform of NO synthase (iNOS) in the pancreatic islet cells of autoimmune diabetes-prone nonobese diabetic mice, with consequent overproduction of NO (Corbett *et al.*, 1993; Suarez-Pinzon *et al.*, 1994; Rabinovitch *et al.*, 1996). In a recent set of studies, we have recently shown the massive presence of nitrotyrosine in pancreatic islet cells of the diabetes-prone nonobese diabetic mice (Suarez-Pinzon *et al.*, 1997). The potential role of PARS activation in the autoimmune model of diabetes *in vivo* requires further investigations.

POTENTIAL REGULATION OF iNOS EXPRESSION BY PARS

It appears that PARS plays an important role in the regulation of gene expression and cell differentiation (Nagao *et al.*, 1991; Smulson *et al.*, 1995). Under basal conditions, PARS is closely associated to DNA (especially at regions of cruciform DNA, bent DNA, and in A-T rich regions) (Sastry *et al.*, 1989), and regulates histone shuttling and nucleosomal unfolding (Althaus *et al.*, 1994). PARS appears to be more frequently associated with transcriptionally active regions of chromatin (Hough *et al.*, 1994; DeMurcia *et al.*, 1988). In fact, a recent report proposes that PARS acts as a functional component of the positive cofactor 1 activity, its function being the enhancement activator-dependent transcription processes (Meisterernst *et al.*, 1997).

Using pharmacological inhibitors of PARS, it has been demonstrated that the activity of PARS is required for the expression of the major histocompatibility complex class II gene (Hiromatsu *et al.*, 1992; Taniguchi *et al.*, 1993; Qu *et al.*, 1994), ras, c-myc (Bauer *et al.*, 1996; Nagao *et al.*, 1991), DNA methyltransferase gene (Bauer *et al.*, 1996), and protein kinase C (Bauer *et al.*, 1996). Moreover, in several independent lines of investigations, it has been demonstrated that pharmacological inhibition of PARS (with nicotinamide,

3-aminobenzamide and 5-iodo-6-amino-1,2-benzopyrone [INH$_2$BP]) suppresses the expression of mRNA of iNOS (Hauschildt *et al.*, 1992; Pellat-Seceunyk *et al.*, 1994; Zingarelli *et al.*, 1996a; Szabó *et al.*, 1997d). In studies using INH$_2$BP, inhibition of iNOS expression in RAW macrophages was indicated by the inhibition of nitrite production, iNOS mRNA expression and iNOS protein expression (Szabó *et al.*, 1997d). The regulation appeared to occur in the early stage of iNOS induction, since INH$_2$BP gradually lost its effectiveness when applied at increasing times *after* the stimulus for iNOS induction. Interestingly, in trasnfection studies using INH$_2$BP in murine RAW macrophages, it was found that this PARS inhibitor suppressed the transcription of iNOS, when cells transfected with the full length (–1592 bp) promoter construct with INH$_2$BP. However, similar co-treatment of cells transfected with the –367 bp deletional construct did not significantly reduce the LPS-mediated increase in luciferase activity (Szabó *et al.*, 1997d). The regu-lation by INH$_2$BP of iNOS induction also occurred in whole animals challenged with LPS: pre-treatment, but not post-treatment of the animals with the inhibitor suppressed the LPS-induced increase in plasma nitrite/nitrate concentrations and reduced the LPS-induced increase in iNOS expression in the lung (Szabó *et al.*, 1997d).

Thus, from these experimental data it appears that PARS, via a not yet characterized mechanism, appears to regulate the process of iNOS expression. Inhibition of this process may represent an additional mode of beneficial action of PARS inhibition in various forms of inflammation. However, caution should be exercised when interpreting the above data. For instance, in the studies quoted above, extremely high concentrations (10–30 mM) of the PARS inhibitors 3-aminobenzamide and nicotinamide were required in order to see suppression of iNOS induction, and these high concentrations of these agents may have had additional pharmacological actions, such as inhibition of total protein and RNA synthesis, and/or free radical scavenging actions (Hauschildt *et al.*, 1992; Pellat-Seceunyk *et al.*, 1994; Zingarelli *et al.*, 1996a). On the other hand, INH$_2$BP effectively suppressed the expression of iNOS at lower concentrations (100–300 μM). However, in the case of INH$_2$BP several additional modes of action should be considered, since this agent is a known inducer of alkaline phosphatases, with a secondary, pleiotropic modulation of cellular responses (Bauer *et al.*, 1996; Szabó *et al.*, 1997e): these effects may well include inhibition of MAP kinase activity, which, in its own right, may suppress the process of iNOS induction. In this respect it is also noteworthy that 3-aminobenzamide at 1 mM concentration already has significant protective effects as a PARS inhibitor in protecting cells against peroxynitrite-induced injury, whereas a marked inhibition of iNOS mRNA expression can only be seen with 30-times higher concentrations of the agent (Zingarelli *et al.*, 1996a). Clearly, experiments in cells or animals with ablation of the PARS gene are required to definitely address the question as to whether inhibition of PARS *per se* sup-presses the process of iNOS induction. In the first of such experiments, PARS deficient pulmonary fibroblasts stimulated with interferon-gamma and endotoxin produced less nitric oxide when compared to the wild-type counterparts (Szabó *et al.*, 1998).

In *in vivo* experiments, PARS inhibitors have beneficial effects in concentrations which do not affect the expression of iNOS (Zingarelli *et al.*, 1996a, 1996b; Szabó *et al.*, 1997c), or even in conditions where iNOS is not even expressed, such as the early phases of myocardial and splanchnic reperfusion (Cuzzocrea *et al.*, 1997; Zingarelli *et al.*, 1997c; Thiemermann *et al.*, 1997). Therefore, it appears that inhibition of iNOS expression is not obligatory for the protective effect of PARS inhibitors in inflammation and reperfusion injury. Nevertheless, it is possible that, during more chronic administration, pharmacological inhibitors of PARS may suppress the expression of iNOS, and thereby reduce the generation of NO and peroxynitrite.

CONCLUSIONS — FUTURE DIRECTIONS

Current strategies aimed at limiting NO-mediated cell/organ injury include agents which inhibit the induction of iNOS, NOS enzyme inhibitors, preferably with selectivity for iNOS, agents that scavenge or inactivate NO, as well as agents that limit substrate or co-factor availability for iNOS. Less attention has been directed to strategies that interfere with intracellular cytotoxic pathways initiated by NO or its toxic derivatives. Direct and indirect experimental evidence reviewed in this paper supports the view that peroxynitrite-induced DNA strand breakage and PARS activation importantly contribute to the pathophysiology of shock, inflammation, and reperfusion injury. Some preliminary data also implicate PARS in the pathophysiology of rheumatoid arthritis (Miesel *et al.*, 1995; Ehrlich *et al.*, 1995; Kroger *et al.*, 1996; Szabó *et al.*, unpublished observations) and drug-related cytotoxicity (see: Szabó, 1996b). Although the enhanced formation of peroxynitrite has been demonstrated in a variety of pathophysiological conditions not discussed in this chapter (transplant rejection, chronic neurodegenerative diseases, inflammatory bowel disease, adult respiratory distress syndrome, atherosclerosis, carbon monoxide poisoning, etc.), the role of PARS in the associated cytotoxicity remains to be investigated.

Based on the data reviewed in the current chapter, we conclude that pharmacologic inactivation of PARS represents a novel, therapeutically viable strategy to limit cellular injury and improve the outcome of a variety of pathophysiological conditions associated with peroxynitrite production. The viability of this potential therapeutic strategy is strengthened by recent observations demonstrating that the absence of PARS does not compromise DNA repair (Wang *et al.*, 1995). PARS inhibition is unlikely to interfere with the important antimicrobial effects of NO, since invading bacteria do not contain PARS. (On the other hand, as demonstrated by Redegeld and colleagues [1992] and Monti and colleagues [1994], inhibition of PARS may suppress target cell death elicited by cytotoxic activated T-lymphocytes. This process may be, in fact, related to peroxynitrite formation [Filep *et al.*, 1996]). Since inhibition of iNOS has immunosuppressive effects and may facilitate the reoccurrence of latent infections in the treated organism, especially during a chronic treatment regime (Stenger *et al.*, 1996; Augusto *et al.*, 1996), inhibition of PARS may have less side effects in this respect, when compared to inhibition of iNOS. On the other hand, PARS inhibition is only expected to inhibit *part of* the oxidant-induced cytotoxicity, and thus may be inferior in its protective efficacy to the therapeutic efficacy of a highly potent, cell permeable oxidant scavenger molecule. Considering the antibacterial and anti-parasitic roles of peroxynitrite (Zhu *et al.*, 1992; Brunelli *et al.*, 1995; Gatti *et al.*, 1995; Augusto *et al.*, 1996), however, chronic treatment with peroxynitrite scavengers (similar to chronic treatment with iNOS inhibitors) would be expected to increase the risk of microbial infections.

In the above sections, we presented evidence that suggests that peroxynitrite is the key trigger of PARS activation in various pathophysiological conditions. In the *in vivo* experiments, this proposal was, by necessity, based on indirect, rather than direct evidence. It is clear, that hydroxyl radical (produced by superoxide and hydrogen peroxide via the Fenton reaction) is also produced in various forms of inflammation and reperfusion injury, and that hydroxyl radical is an important cytotoxic species, which causes cytotoxicity, in part, via activation of PARS. Although theoretical considerations favor the role of peroxynitrite, as opposed to hydroxyl radical, as a pathophysiologically relevant oxidant species (see: Beckman and Koppenol, 1996; Squadrito *et al.*, 1996; Squadrito and Pryor, 1995), the clear distinction between the role of hydroxyl radicals and peroxynitrite would require further investigations, using specific pharmacological tools. Many of the hydroxyl radical scavengers are also peroxynitrite scavengers (such as glutathione or desferoxamine, but not mannitol), which does not allow clear distinction. Superoxide dismutase mimetics are expected to reduce the formation of both peroxynitrite and hydroxyl radical. Potent and specific peroxynitrite scavengers are not available for experimental purposes, although it is clear that the development of such compounds is underway (Stern *et al.*, 1996; Masumoto *et al.*, 1996; Bribiva *et al.*, 1996; Fici *et al.*, 1996; Fici *et al.*, 1997; Szabó *et al.*, 1997e). Clearly, careful future studies (using specific and potent scavengers) must be performed in order to dissect the contribution of peroxynitrite and hydroxyl radical in the oxidant processes and cell and organ injury in various pathophysiological conditions. Such experiments may be also complicated by the fact that there are multiple interactions between oxyradicals and peroxynitrite. For instance, excess superoxide (not unlike excess NO), limits the oxidant potency of peroxynitrite (see above). Moreover, hydrogen peroxide prolongs the half-life or peroxynitrite (Alvarez *et al.*, 1995) and enhances the cytotoxic potential of NO donors (Volk *et al.*, 1995; Pacelli *et al.*, 1995) and of authentic peroxynitrite (Szabó, 1997). The reaction of peroxynitrite with hydrogen peroxide can also result in the production of singlet molecular oxygen, which is a highly cytotoxic species (DiMascio *et al.*, 1994).

It is noteworthy that the vast majority of the evidence implicating the role of PARS in peroxynitrite-induced toxicity and in other forms of cellular oxidant injury was obtained using pharmacological inhibitors, such as 3-aminobenzamide. This agent is a prototypical PARS inhibitor which has been used in a large number of investigations to inhibit the catalytic activity of PARS, when DNA single strand breakage was triggered by oxyradicals or by peroxynitrite. In these studies, 3-aminobenzamide provided cytoprotective effects but did not interfere with the development of DNA strand breakage, supporting the view that the agent, at 1–3 mM, does not scavenge oxyradicals or peroxynitrite (Zhang *et al.*, 1994; Zingarelli *et al.*, 1996a). There is also ample of prior evidence that 3-aminobenzamide, in low millimolar concentrations, does not directly oxyradical-triggered oxidative processes (Berger *et al.*, 1986; Schraufstatter *et al.*, 1988; Schraufstatter *et al.*, 1986; Thies and Autor, 1991; Berger, 1991; Cochrane, 1991). Moreover 3-aminobenzamide is not an inhibitor of NO synthase and does not scavenge NO (Zhang *et al.*, 1994; Zingarelli *et al.*, 1996). Thus, it appears that 3-aminobenzamide, indeed, exerts its protective actions by inhibition of the catalytic activity of PARS, rather than by interfering with proximal processes (such as the direct actions of peroxynitrite or oxyradicals). Nevertheless, it has been repeatedly put forward that 3-aminobenzamide may act as a hydroxyl radical scavenger (Wilson *et al.*, 1984; Farber *et al.*, 1990), thereby contributing to the cytoprotection. Clearly, follow-up studies with cells and animals with ablated PARS gene are awaited to confirm the conclusions of the studies with pharmacological inhibitors of PARS. In the first such study, Heller and co-workers have, indeed, observed (1995) that islets of the PARS$^{-/-}$ mice are resistant to NO and oxidant-related injury, when compared to the response in islets of the wild-type mice. Similarly, we observed that lung fibroblasts from the PARS$^{-/-}$ mice are protected from peroxynitrite-induced cell injury when compared to the fibroblasts of the corresponding wild-type animals (Szabó *et al.*, 1998).

It is clear that PARS is not the single factor involved in the pathogenesis of cell and organ injury in response to oxidant stress. As discussed above, the relative importance of PARS in mediating oxidant injury is dependent on cell type. Furthermore, as mentioned above, inhibition of PARS activity results in only partial restoration of mitochondrial respiration and cytotoxicity in macrophages or pancreatic islet cells, when exposed to extremely high concentrations of NO donors or authentic peroxynitrite. These high concentrations of oxidants (peroxynitrite, superoxide, hydroxyl radical, etc.) may, therefore, trigger PARS-independent cytotoxic effects, such as direct inhibitory effects on the mitochondrial respiratory chain or inhibitory effects on other intracellular energetic or redox processes. Furthermore, direct interactions of the oxidants with proteins, lipids, arachidonic acid, and other molecules may also play a significant role in the development of cellular injury. It is likely that there are important synergistic interactions between PARS activation and these other cellular processes of cytotoxicity. Nevertheless, from the evidence presented above it appears that inhibition of PARS alone can "tip the balance" and significantly influences the outcome of the various pathophysiological conditions. It is possible, on the other hand, that combination of PARS inhibitors with additional experimental therapeutic approaches (NOS inhibition, oxyradical scavenging, steroidal or non-steroidal anti-inflammatory compounds, etc.) may increase the therapeutic potential of pharmacological inhibitors of PARS.

ACKNOWLEDGEMENTS

This work was supported by a Grant from the National Institutes of Health (R29GM54773) to C.S.

REFERENCES

Althaus, F.R., Hofferer, L., Kleczkowska, H.E., Malanga, M., Naegeli, H., Panzeter, P.L. and Realini, CA. (1994) Histone shuttling by poly ADP-ribosylation. *Mol. Cell. Biochem.*, **138**, 53–9.

Alvarez, B., Denicola, A. and Radi, R. (1995) Reaction between peroxynitrite and hydrogen peroxide: formation of oxygen and slowing of peroxynitrite decomposition. *Chem. Res. Toxicol.*, **8**, 849–864.

Andreoli, S.P. (1989) Mechanisms of endothelial cell ATP depletion after oxidant injury. *Pediatr. Res.*, **25**, 97–101.

Augusto, O., Gatti, R.M. and Radi, R. (1994) Spin-trapping studies of peroxynitrite decomposition and of 3-morpholinosydnonimine N-ethylcarbamide autooxidation: direct evidence for metal-independent formation of free radical intermediates. *Arch. Biochem. Biophys.*, **310**, 118–125.

Augusto, O., Linares, E. and Giorgio, S. (1996) Possible roles of nitric oxide and peroxynitrite in murine leishmaniasis. *Brazilian J. Med. Biol. Res.*, **29**, 853–862.

Bartlett, D., Church, D.F., Bounds, P.L. and Koppenol, W.H. (1995) The kinetics of the oxidation of L-ascorbic acid by peroxynitrite. *Free Rad. Biol. Med.*, **18**, 85–92.

Bauer, M.L., Beckman, J.S., Bridges, R.J., Fuller, C.M. and Matalon, S. (1992) Peroxynitrite inhibits sodium uptake in rat colonic membrane vesicles. *Biochim. Biophys. Acta*, **1104**, 87–94.

Bauer, P.I., Kirsten, E., Young, L.J.T., Varadi, G., Csonka, E., Buki, K.G., Mikala, G., Hu, R., Comstock, J.A., Mendeleyev, J., Hakam, A. and Kun, E. (1996) Modification of growth related enzymatic pathways and apparent loss of tumorigenicity of a ras-transformed bovine endothelial cell line by treatment with 5-iodo-6-amino-1,2-benzopyrone. *Int. J. Oncol.*, **8**, 239–252.

Beckman, J.S. (1991) The double-edged role of nitric oxide in brain function and superoxide-mediated injury. *J. Develop. Physiol.*, **15**, 53–9.

Beckman, J.S. and Koppenol, W.H. (1996) Nitric oxide, superoxide, and peroxynitrite: the good, the bad, and ugly. *Am. J. Physiol.*, **271**, C1424–37.

Beckman, J.S., Beckman, T.W., Chen, J., Marshall, P.A. and Freeman, B.A. (1990) Apparent hydroxyl radical production by peroxynitrite: implication for endothelial injury from nitric oxide and superoxide. *Proc. Natl. Acad. Sci. USA*, **87**, 1620–1624.

Bellmann, K., Jaattela, M., Wissing, D., Burkart, V. and Kolb, H. (1996) Heat shock protein hsp70 overexpression confers resistance against nitric oxide. *FEBS Lett.*, **391**, 185–188.

Bellmann, K., Wenz, A., Radons, J., Burkart, V., Kleemann, R. and Kolb, H. (1995) Heat shock induces resistance in rat pancreatic islet cells against nitric oxide, oxygen radicals and streptozotocin toxicity *in vitro*. *J. Clin. Invest.*, **95**, 2840–2845.

Berger, N.A. (1991) Oxidant-induced cytotoxicity: a challenge for metabolic modulation. *Am. J. Respir. Cell Mol. Biol.*, **4**, 1–3.

Berger, S.J., Sudar, D.C. and Berger, N.A. (1986) Metabolic consequences of DNA damage: DNA damage induces alterations in glucose metabolism by activation of poly (ADP-ribose) polymerase. *Biochem. Biophys. Res. Comm.*, **134**, 227–32.

Bingisser, R., Stey, C., Weller, M., Groscurth, P., Russi, E. and Frei, K. (1996) Apoptosis in human alveolar macrophages is induced by endotoxin and is modulated by cytokines. *Am. J. Resp. Cell. Mol. Biol.*, **15**, 64–70.

Bolanos, J.P., Heales, S.J., Land, J.M. and Clark, J.B. (1995) Effect of peroxynitrite on the mitochondrial respiratory chain: differential susceptibility of neurones and astrocytes in primary culture. *J. Neurochem.*, **64**, 1965–72.

Bonfoco, E., Krainc, D., Ankarcrona, M., Nicotera, P. and Lipton, S. (1995) Apoptosis and necrosis: two distinct events induced, respectively, by mild and intense insults with NMDA or nitric oxide/superoxide in cortical cell cultures. *Proc. Natl. Acad. Sci. USA*, **92**, 7162–7166.

Briviba, K., Roussyn, I., Sharov, V.S. and Sies H. (1996) Attenuation of oxidation and nitration reactions of peroxynitrite by selenomethionine, selenocysteine and ebselen. *Biochem. J.*, **319**, 13–15.

Brunelli, L., Crow, J. and Beckman, J. (1995) The comparative toxicity of nitric oxide and peroxynitrite to Escherichia coli. *Arch. Biochem. Biophys.*, **316**, 327–334.

Burkart, V., Gross-Eick, A., Bellmann, K., Radons, K.J. and Kolb, H. (1995) Suppression of nitric oxide toxicity in islet cells by alpha-tocopherol. *FEBS Lett.*, **364**, 259–263.

Campo, G.M., Squadrito, F., Ioculano, M., Altavilla, D., Zingarelli, B., Pollicino, A.M., Rizzo, A., Calapai, G., Calandra, S. and Scuri, R. (1994) Protective effects of IRFI-016, a new antioxidant agent, in myocardial damage, following coronary artery occlusion and reperfusion in the rat. *Pharmacology*, **48**, 157–66.

Castro, L., Rodriguez, M. and Radi, R. (1994) Aconitase is readily inactivated by peroxynitrite, but not by its precursor, nitric oxide. *J. Biol. Chem.*, **269**, 29409–29415.

Cazevieille, C.. Muller, A.. Meynier, F. and Bonne, C. (1993) Superoxide and nitric oxide cooperation in hypoxia/reoxygenation-induced neuron injury. *Free Radical Biol. Med.*, **14**, 389–95.

Cochrane, C.G. (1991) Mechanisms of oxidant injury of cells. *Molec. Aspects Med.*, **12**, 137–147.

Corbett, J.A., Mikhael, A., Shimizu, J., Frederick, K., Misko, T.P., McDaniel, M.L., Kanagawa, O. and Unanue, E.R. (1993) Nitric oxide production in islet cells from nonobese mice: aminoguanidine-sensitive and -resistant stages in the immunological diabetic process. *Proc. Natl. Acad. Sci. USA*, **90**, 8992–8995.

Cosi, C., Suzuki, H., Milani, D., Facci, L., Menegazzi, M., Vantini, G., Kanai, Y. and Skaper, S.D. (1994) Poly(ADP-ribose) polymerase: early involvement in glutamate-induced neurotoxicity in cultured cerebellar granule cells. *J. Neurosci. Res.*, **39**, 38–46.

Crow, J.P. and Beckman, J.S. (1995) The role of peroxynitrite in nitric oxide-mediated toxicity. *Current Top. Microbiol. Immunol.*, **196**, 57–73.

Crow, J.P., Beckman, J.S. and McCord, J.M. (1995) Sensitivity of the essential zinc-thiolate moiety of yeast alcohol dehydrogenase to hypochlorite and peroxynitrite. *Biochemistry*, **34**, 3544–52.

Cuzzocrea, S., Zingarelli, B., Costantino, G., Szabó, A., Salzman, A.L., Caputi, A.P. and Szabó, C. (1997): Beneficial effects of 3-aminobenzamide, an inhibitor of poly (ADP-ribose) synthetase in a rat model of splanchnic artery occlusion and reperfusion. *Br. J. Pharmacol.*, **121**, 1065–74.

Darley-Usmar, V., Wiseman, H. and Halliwell, B. (1995) Nitric oxide and oxygen radicals: a question of balance. *FEBS Lett.*, **369**, 131–135.

Dawson, V.L. (1995) Nitric oxide: role in neurotoxicity. *Clin. Exp. Pharmacol. Physiol.*, **22**, 305–8.

Dawson, V.L., Dawson, T.M., Bartley, D.A., Uhl, G.R. and Snyder, S.H. (1993) Mechanisms of nitric oxide-mediated neurotoxicity in primary brain cultures. *J. Neurosci.*, **13**, 2651–61.

de Groot, H., Hegi, U. and Sies, H. (1993) Loss of alpha-tocopherol upon exposure to nitric oxide or the sydnonimine SIN-1. *FEBS Lett.*, **315**, 139–42.

Delaney, C.A., Tyrberg, B., Bouwens, L., Vaghef, H., Hellmann, B. and Eizirik, D.L. (1996) Sensitivity of human pancreatic islets to peroxynitrite-induced cell dysfunction and death. *FEBS Lett.*, **394**, 300–306.

DeMurcia, G., Huletsky, A. and Poirier, G.G. (1988) Modulation of chromatin structure by polyADP ribosylation. *Biochem. Cell Biol.*, **66**, 626–635.

Denicola, A., Rubbo, H., Rodriguez, D. and Radi, R. (1993) Peroxynitrite-mediated cytotoxicity to Trypanosoma cruzi. *Arch. Biochem. Biophys.*, **304**, 279–86.

Denicola, A., Souza, J.M., Gatti, R.M., Augusto, O. and Radi, R. (1995) Desferrioxamine inhibition of the hydroxyl radical-like reactivity of peroxynitrite: role of the hydroxamic groups. *Free Rad. Biol. Med.*, **19**, 11–9.

Didier, M., Bursztajn, S., Adamec, E., Passani, L., Nixon, R.A., Coyle JT, Wei, J.Y. and Berman, S.A. (1996) DNA strand breaks induced by sustained glutamate excitotoxicity in primary neuronal cultures. *J. Neurosci.*, **16**, 2238–2250.

DiMascio, P., Bechara, E.J.H., Medairos, M.H.G., Briviba, K. and Sies, H. (1994) Singlet molecular oxygen production in the reaction of peroxynitrite with hydrogen peroxide. *FEBS Lett.*, **355**, 287–289.

Ehrlich, W., Huser, H. and Kroger, H. (1995) Inhibition of the induction of collagenase by interleukin-1 beta in cultured rabbit synovial fibroblasts after treatment with the poly (ADP-ribose)-polymerase inhibitor 3-aminobenzamide. *Rheumat. Int.*, **15**, 171–2.

Eizirik, D.L., Delaney, C.A., Green, M.H.L., Cunningham, J.M., Thorpe, J.R, Pipeleers, D.G., Hellerstrom, C. and Green, I.C. (1996) Nitric oxide donors decrease the function and survival of human pancreatic islets. *Mol Cell Endocrinol.*, **118**, 71–83.

Eliasson, M.J.L., Sampei, K., Mandir, A.S., Hurn, P.D., Traystman, R.J., Bao, J., Pieper, A., Wang, Z.Q., Dawson, T.M., Snyder, S.H. and Dawson, V.L. (1997) Poly (ADP-ribose) polymerase gene disruption renders mice resistant to cerebral ischemia. *Nature Med.*, **3**, 1089–1095.

Elliott, S.J. (1996) Peroxynitrite modulates receptor-activated Ca^{2+} signaling in vascular endothelial cells. *Am. J. Physiol.*, **270**, L954–61.

Endres, M., Wang, Z-Q., Namura, S., Waeber, C. and Moskowitz, M.A. (1997) Ischemic brain injury is mediated by the activation of poly(ADP-ribose)polymerase. *J. Cerebr. Blood Flow Metab.*, **17**, 1143–1151.

Epe, B., Ballmaier, D., Roussyn, I., Briviba, K. and Sies, H. (1996) DNA damage by peroxynitrite characterized with DNA repair enzymes. *Nucleic Acid Res.*, **24**, 4105–4110.

Estevez, A.G., Radi, R., Barbeito, L., Shin, J.T., Thompson, J.A. and Beckman, J.S. (1995) Peroxynitrite-induced cytotoxicity in PC12 cells: evidence for an apoptotic mechanism differentially modulated by neurotrophic factors. *J. Neurochem.*, **65**, 1543–50.

Fagni, L., Lafon-Cazal, M., Roundouin, G., Manzoni, O., Lerner-Natoli, M. and Bockaert, J. (1994) The role of free radicals in NMDA-dependent neurotoxicity. *Progr. Brain Res.*, **103**, 381–390.

Farber, L.J., Kyle, M.E. and Coleman, J.B. (1990) Biology of disease. mechanisms of cell injury by activated oxygen species. *Lab. Invest.*, **62**, 670–679.

Fatehi-Hassanabad, Z., Burns, H., Aughey, E.A., Paul, A., Plevin, R., Parratt, J.R. and Furman, B.L. (1996) Effects of L-canavenine, an inhibitor of inducible nitric oxide synthase on endotoxin mediated shock in rats. *Shock*, **6**, 194–200.

Faulkner, K.M., Liochev, S.I., Fridowich, I. (1994) Stable Mn(III) prophyrins mimic cuperoxide dismutase *in vitro* and substitute for it *in vivo*. *J. Biol. Chem.*, **269**, 23471–23476.

Ferrari, F. (1994) Oxygen-free radicals at myocardial level: effects of ischaemia and reperfusion. *Adv Exp Med Biol*, **366**, 99–111.

Ferrari, R., Ceconi, S., Curello, S., Cargnoni, A., De Giuli, F. and Visioli, O. (1992) Occurrence of oxidative sress during myocardial reperfusion. *Mol. Cell Biochem.*, **111**, 61–69.

Fici, G.J., Althaus, J.S. and VonVoigtlander, P.F. (1997) Effects of lazaroids and a peroxynitrite scavenger in a cell model of peroxynitrite toxicity. *Free Rad. Biol. Med.*, **22**, 223–228.

Fici, G.J., Althaus, J.S., Hall, E.D. and VonVoigtlander, P.F. (1996) Protective efects of tirilazad mesylate in a cellular model of peroxynitrite toxicity. *Res. Comm. Chem. Path. Pharmacol.*, **91**, 327–371.

Filep., J.G., Baron, C., Lachance, S., Perreault, C. and Chan J.S.D. (1996) Involvement of nitric oxide in target-cell lysis and DNA fragmentation induced by murine natural killer cells. *Blood*, **87**, 4136–5143.

Gatti, R.M., Augusto, O., Kwee, J.K. and Giorgio, S. (1996) Leishmanicidal activity of peroxynitrite. *Rediox Report*, **1**, 261–265.

Gilad, E., Cuzzocrea, S., Zingarelli, B., Salzman, A.L. and Szabó, C. (1997) Melatonin is a scavenger of peroxynitrite. *Life Sci.*, **60**, PL169–174.

Gryglewski, R.J., Palmer, R.M. and Moncada, S. (1986) Superoxide anion is involved in the breakdown of endothelium-derived vascular relaxing factor. *Nature*, **320**, 454–6.

Gunasekar, P.G., Kantasamy, A.G., Borowitz, J.L. and Isom, G.E. (1995) NMDA receptor activation produces cncurrent generation of nitric oxide and reactive oxygen species: implication for cell death. *J. Neurochem.*, **65**, 2016–2021.

Hammer, B., Parker, W.D. Jr. and Bennett, J.P. Jr. (1993) NMDA receptors increase OH radicals *in vivo* by using nitric oxide synthase and protein kinase C. *Neuroreport*, **5**, 72–4.

Hauschildt, S., Scheipers, P., Bessler, W.G. and Mulsch, A. (1992) Induction of nitric oxide synthase in L929 cells by tumour-necrosis factor alpha is prevented by inhibitors of poly (ADP-ribose) polymerase. *Biochem. J.*, **288**, 255–260.

Hausladen, A. and Fridovich, I. (1994) Superoxide and peroxynitrite inactivate aconitases, but nitric oxide does not. *J. Biol. Chem.*, **269**, 29405–29408.

Heller, B., Wang, Z.Q., Wagner, E.F., Radons, J., Burkle, A., Fehsel, K., Burkart, V. and Kolb, H. (1995) Inactivation of the poly(ADP-ribose) polymerase gene affects oxygen radical and nitric oxide toxicity in islet cells. *J. Biol. Chem.*, **270**, 11176–80.

Hiromatsu, Y., Sato, M., Yamada, K. and Nonaka, K. (1992) Nicotinamide and 3-aminobenzamide inhibit recombinant human interferon-gamma-induced HLA-DR antigen expression, but not HLA-A, B, C antigen expression, on cultured human thyroid cells. *Clin. Endocrinol.*, **36**, 91–5.

Hogg, N., Joseph, J. and Kalyanaraman, B. (1994) The oxidation of alpha-tocopherol and trolox by peroxynitrite. *Arch. Biochem. Biophys.*, **314**, 153–8.

Hoshida, S., Kuzuya, T., Nishida, M., Yamashita, N., Hori, M., Kamada T. and Tada, M. (1994) Ebselen protects against ischemia-reperfusion injury in a canine model of myocardial infarction. *Am. J. Physiol.*, **267**, H2342–7.

Hough, C.J. and Smulson, M.E. (1994) Association of poly(adenosine diphosphate ribosylated) nucleosomes with transcriptionally active and inactive regions of chromatin. *Biochemistry*, **23**, 5016–23.

Hu, P., Ischiropoulos, H., Beckman, J.S. and Matalon, S. (1994) Peroxynitrite inhibition of oxygen consumption and sodium transport in alveolar type II cells. *Am. J. Physiol.*, **266**, L628–34.

Inada, C., Yamada, K., Takane, N. and Nonaka, K. (1995) Poly(ADP-ribose) synthesis induced by nitric oxide in a mouse beta-cell line. *Life Sci.*, **56**, 1467–74.

Inoue, S. and Kawanishi, S. (1995) Oxidative DNA damage induced by simultaneous generation of nitric oxide and superoxide. *FEBS Lett.*, **371**, 86–88.

Ischiropoulos, H., al-Mehdi, A.B. and Fisher, A.B. (1995) Reactive species in ischemic rat lung injury: contribution of peroxynitrite. *Am. J. Physiol.*, **269**, L158–64.

Ischiropoulos, H., al-Mehdi, A.B. and Fisher, A.B. (1995) Reactive species in ischemic rat lung injury: contribution of peroxynitrite. *Am. J. Physiol.*, **269**, L158–64.

Ischiropoulos, H., Duran, D. and Horwitz, J. (1995) Peroxynitrite-mediated inhibition of DOPA synthesis in PC12 cells. *J. Neurochem.*, **65**, 2366–2372.

Ischiropoulos, H., Zhu, L., Chen, J., Tsai, M., Martin, J.C., Smith, C.D. and Beckman, J.S. (1992) Peroxynitrite-mediated tyrosine nitration catalyzed by superoxide dismutase. *Arch. Biochem. Biophys.*, **298**, 431–7.

Juedes, M.J. and Wogan, G.N. (1966) Peroxynitrite-induced mutation spectra of pSP189 following replication in bacteria and in human cells. *Mutation Res.*, **349**, 51–61.

Junod, A.F., Jornot, L. and Petersen, H. (1989) Differential effects of hyperoxia and hydrogen peroxide on DNA damage, polyadenosine diphosphate-ribose polymerase activity, and nicotinamide adenine dinucleotide and adenosine triphosphate contents in cultured endothelial cells and fibroblasts. *J. Cell Physiol.*, **140**, 177–85.

Kennedy, M., Szabó, C. and Salzman, A.L. (1997) Activation of poly (ADP-ribose) synthetase mediates hyperpermeability induced by peroxynitrite in human intestinal epithelial cells. *Crit. Care Medicine*, **5** (Suppl), A68.

King, P.A., Anderson, V.E., Edwards, J.O., Gustav, G., Plumb, R.C. and Suggs, J.W. (1992) A stable solid that generates hydroxyl radical dissolutions in aqueous solutions: reaction with proteins and nucleic acid. *J. Am. Chem. Soc.*, **114**, 5430–5432.

Kirkland, J.B. (1991) Lipid peroxidation, protein thiol oxidation and DNA damage in hydrogen peroxide-induced injury to endothelial cells: role of activation of poly(ADP-ribose)polymerase. *Biochim. Biophys. Acta*, **1092**, 319–25.

Kooy, N., Royall, J., Ischiropoulos, H. and Beckman, J. (1994) Peroxynitrite-mediated oxidation of dihydrorhodamine 123. *Free Rad. Biol. Med.*, **16**, 149–155.

Kroger, H., Miesel, R., Diethrich, A., Ohde, M., Rajnavolgyi, E. and Ockenfels, H. Synergistic effects of thalidomide and poly (ADP-ribose) polymerase inhibition of type II collagen-induced arthritis in mice. *Inflammation*, **20**, 203–215.

Kröncke, K.D., Fehsel, K., Sommer, A., Rodriguez, M.L. and Kolb-Bachofen, V. (1995) Nitric oxide generation during cellular metabolization of the diabetogenic N-methyl-N-nitroso-urea streptozotocin contributes to islet cell DNA damage. *Biol. Chem. Hoppe-Seyler*, **376**, 179–185.

Kuo, M.L., Chau, Y.P., Wang, J.H. and Shiah, S.G. (1996) Inhibitors of poly (ADP-ribose) polymerase block nitric oxide-induced apoptosis but not differentiation in human leukemia HL-60 cells. *Biochem. Biophys. Res. Comm.*, **219**, 502–508.

Kwon, N.S., Lee, S.H., Choi, C.S., Kho, T. and Lee, H.S. (1994) Nitric oxide generation from streptozotocin. *FASEB J.*, **8**, 529–533.

Lafon-Cazal, M., Culcasi, M., Gaven, F., Pietri, S. and Bockaert, J. (1993) Nitric oxide, superoxide and peroxynitrite: putative mediators of NMDA-induced cell death in cerebellar granule cells. *Neuropharmacology*, **32**, 1259–66.

Lautier, D., Lagueux, J., Thiboldeau, J., Menard, L. and Poirier, G.G. (1993) Molecular and biochemical features of poly (ADP-ribose) metabolism. *Mol. Cell. Biochem.*, **122**, 171–193.

Laval, F. and Wink, D.A. (1994) Inhibition by nitric oxide of the repair protein, O6-methylguanine-DNA-methyltransferase. *Carcinogenesis*, **15**, 443–7.

Lefer, D.J., Scalia, R., Campbell, B., Nossuli, T., Hayward, R., Salamon, M., Grayson, J. and Lefer, A.M. (1997) Peroxynitrite inhibits leukocyte-endothelial interactions and protects against ischemia-reperfusion injury in rats. *J. Clin. Invest.*, **99**, 684–691.

Lin, K.T., Xue, J.Y., Nomen, M., Spur, B. and Wong, P.Y. (1995) Peroxynitrite-induced apoptosis in HL-60 cells. *J. Biol. Chem.*, **270**, 16487–90.

Lucchesi, B.R. (1990) Modulation of leukocyte-mediated myocardial reperfusion injury. *Ann. Rev. Physiol.*, **52**, 561–576.

Lymar, S.V. and Hurst, J.K. (1995) Rapid reaction between peroxynitrite ion and carbon dioxide: implications for biological activity. *J. Am. Chem. Soc.*, **117**, 8867–8868.

Ma, T.T., Ischiropoulos, H. and Brass, C.A. (1995) Endotoxin-stimulated nitric oxide production increases injury and reduces rat liver chemiluminescence during reperfusion. *Gastroenterology*, **108**, 463–9.

MacKenzie, L., Szele, F.G. and Chesselet, M.F. (1993) In situ translation: a novel approach to identify neuronal injury after ischemic lesions. *Soc. Neurosci. Abstr.*, **19**. 1656.

Masiello, P., Cubeddu, T.L., Frosina, G. and Bergamini, E. (1985) Protective effect of 3–aminobenzamide, an inhibitor of poly (ADP-ribose) synthetase, against streptozotocin-induced diabetes. *Diabetologia*, **28**, 683–6.

Masiello, P., Novelli, M., Fierabbrachi, V. and Bergamini, E. (1990) Protection by 3–aminobenzamide and nicotinamide afainst streptozotocin-induced beta-cell toxicity *in vivo* and *in vitro*. *Res. Comm. Chem. Path. Pharmacol.*, **69**, 17–32.

Masumoto, H. and Sies, H. (1996) The reaction of 2–methylseleno benzanilide with peroxynitrite. *Chem. Res. Toxicol.*, **9l** 262–267.

Matheis, G., Sherman, M.P., Buckberg, G.D., Haybron, D.M., Young, H.N. and Ignarro, L.J. (1992) Role of L-arginine – nitric oxide pathway in myocardial reoxygenation injury. *Am. J. Physiol.*, **262**, H616–620.

Mayer, B., Schrammel, A., Klatt, P., Koesling, D. and Schmidt, K. (1995) Peroxynitrite-induced accumulation of cyclic GMP in endothelial cells and stimulation of purified soluble guanylyl cyclase. Dependence on glutathione and possible role of S-nitrosation. *J. Biol. Chem.*, **270**, 17355–60.

Mehta, J.L., Nichols, W.W., Donnelly, W.H., Lawson, D.L, Thompson, L., Riet, M. and Saldeen, T.G. (1989) Protection by superoxide dismutase from myocardial dysfunction and attenuation of vasodilator reserve after coronary occlusion and reperfusion in dog. *Circ. Res.*, **65**, 1283–95.

Meisterernst, M., Stelzer, G. and Roeder, R.G. (1997) Poly (ADP-ribose) polymerase enhances activator-dependent transcription *in vitro*. *Proc. Natl. Acad. Sci. USA*, **94**, 2261–2265.

Messmer, U.K. and Brune, B. (1996) Nitric oxide in apoptotic cersus necrotic RAW 264.7 macrophage cell death: the role of NO-donor exposure, NAD content and p53 accumulation. *Arch. Biochem. Biophys.*, **327**, 1–10.

Messmer, U.K., Reimer, D.M., Reed, J.C. and Brune, B. (1996) Nitric oxide induced poly (Adp-ribose) polymerase cleavage in RAS 264.7 macrophage apoptosis is blocked by Bcl-2. *FEBS Lett.*, **384**, 162–166.

Miesel, R., Kurpisz, M. and Kroger H. (1995) Modulation of inflammatory arthritis by inhibition of poly(ADP ribose) polymerase. *Inflammation*, **19**, 379–87.

Mohr, S., Stamler, J.S. and Brune, B. (1994) Mechanism of covalent modification of glyceraldehyde-3-phosphate dehydrogenase at its active site thiol by nitric oxide, peroxynitrite and related nitrosating agents. *FEBS Lett.*, **348**, 223–7.

Monti, D., Cossarizza, A., Salvioli, S., Franceschi, C. Rainaldi, G., Straface, E., Rivabene, R. and Malorni, W. (1994) Cell death protection by 3–aminobenzamide and othert poly (ADP-ribose) polymerase inhibitors: different effects on human natural killer and lymphokine activated killer cell activities. *Biochem. Biophys. Res. Comm.*, **199**, 525–30.

Monti, D., Troiano, L., Tropea, F., Grassilli, E., Cossarizza, A., Barozzi, D., Pelloni, M.C., Tamassia, M.G., Bellomo, G. and Francheschi, C. (1995) Apoptosis – programmed cell death: a role in the aging process? *Am. J. Clin. Nutr.*, **55**, 1208S–1214S.

Moro, M.A., Darley-Usmar, V.M., Goodwin, D.A., Read, N.G., Zamora-Pino, R., Feelisch, M., Radomski, M.W. and Moncada, S. (1994) Paradoxical fate and biological action of peroxynitrite on human platelets. *Proc. Natl. Acad. Sci. USA*, **91**, 6702–6.

Nagao, M., Nakayasu, M., Aonuma, S., Shima, H. and Sugimura, T. (1991) Loss of amplified genes by poly(ADP-ribose) polymerase inhibitors. *Environ. Health Persp.*, **93**, 169–74.

Naseem, S.A., Kontos, M.C., Rao, P.S., Jesse, R.L., Hess, M.L. and Kukreja, R.C. (1995) Sustained inhibition of nitric oxide by NG-nitro-L-arginine improves myocardial function following ischemia/reperfusion in isolated perfused rat heart. *J. Mol. Cell. Cardiol.*, **27**, 419–26.

Nathan, C. (1992) Nitric oxide as a secretory product of mammalian cells. *FASEB J.*, **6**, 3051–3064.

Nicholson, D.W., All, A., Thornberry, N.A., Vaillancourt, J.P., Ding, C.K., Gallant, M., Gareau, Y., Griffin, P.R., Labelle, M., Lazebnik, Y.A., Munday, N.A., Raju, S.M., Smulson, M.E., Yamin, T.T., Yu, V.L. and Miller, D.K. (1995) Identification and inhibition of the ICE/CED-3 protease necessary for mammalian apoptosis. *Nature*, **376**, 37–43.

Pacelli, R., Wink, C.A., Cook, J.A., Krishna, M.C., deGraff, W., Friedman,.N., Tsokos, M., Samuni, A. and Mitchell, J.B. (1995) Nitric oxide potentiates hydrogen peroxide-induced killing of Eschericic coli. *J. Exp. Med.*, **182**, 1469–1479.

Packer, L. (1994) Oxygen radicals in biological systems. *Methods Enzymol.*, **234**, 1–642.

Pellat-Seceunyk, K., Wietzerbin, J. and Drapier, J.C. (1994) Nicotinamide inhibits nitric oxide synthase mRNA induction in activated macrophages. *Biochem. J.*, **297**, 53–58.

Pellat-Seceunyk, K., Wietzerbin, J. and Drapier, J.C. (1994) Nicotinamide inhibits nitric oxide synthase mRNA induction in activated macrophages. *Biochem. J.*, **297**, 53–58.

Phelphs, D.T., Ferro, T.J., Higgins, P.J., Shankar, R., Parkler, D.M. and Johnson, A. (1995) TNF-alpha induces peroxynitrite-mediated depletion of lung endothelial glutathione via protein kinase C. *Am. J. Physiol.*, **269**, L551–559.

Phelps, D.T., Ferro, T.J., Higgins, P.J., Shankar, R., Parker, D.M. and Johnson. (1995) TNF-alpha induces peroxynitrite-mediated depletion of lung endothelial glutathione via protein kinase C. *Am. J. Physiol.* **269**, L551–559.

Pryor, W. and Squadrito, G. (1995) The chemistry of peroxynitrite: a product from the reaction of nitric oxide with superoxide. *Am. J. Physiol.*, **268**, L699–L722.

Pryor, W.A., Jin, X. and Squadrito, G.L. (1994) One- and two-electron oxidations of methionine by peroxynitrite. *Proc. Natl. Acad. Sci. USA*, **91**, 11173–7.

Qu, Z., Fujimoto, S. and Taniguchi, T. (1994) Enhancement of interferon-gamma induced major histocompatibility complex class II gene expression by expressing an antisense RNA of poly(ADP-ribose) synthetase. *J. Biol. Chem.*, **269**, 5543–5547.

Rabinovitch, A., Suarez-Pinzon, W.L., Sorensen, O. and Bleackley, R.C. (1996) Inducible nitric oxide synthase (iNOS) in pancreatic iclets of nonobese diabetic mice: identification of iNOS expressing cells and relationships to cytokines expressed in the islet. *Endocrinology*, in press.

Radi, R., Beckman, J.S., Bush, K.M. and Freeman, B.A. (1991) Peroxynitrite-induced membrane lipid peroxidation: the cytotoxic potential of superoxide and nitric oxide. *Arch. Biochem. Biophys.*, **288**, 481–7.

Radi, R., Rodriguez, M., Castro, L. and Telleri, R. (1994) Inhibition of mitochondrial electron transport by peroxynitrite. *Arch. Biochem. Biophys.*, **308**, 89–95.

Radons, J., Heller, B., Burkle, A., Hartmann, B., Rodriguez, M.L., Kroncke, K.D., Burkart, V. and Kolb, H. (1994) Nitric oxide toxicity in islet cells involves poly (ADP-ribose) polymerase activation and concomitant NAD depletion. *Biochem. Biophys. Res. Comm.*, **199**, 1270–1277.

Redegeld, F.A., Chatterjee, S., Berger, N.A. and Sitkovsky, M.V. (1992) Poly (ADP-ribose) polymerase partially contributes to target cell death triggered by cytolytic T lymphocytes. *J. Immunol.*, **149**, 3509–3516.

Rees, D.D. (1995) Role of nitric oxide in the vascular dysfunction of septic shock. *Biochem. Soc. Transact.*, **23**, 1025–9.

Richter, C., Schweizer, M., Cossariza, A. and Francheschi, C. (1996) Control of apoptosis by the cellular ATP level. *FEBS Lett.*, **378**, 107–110.

Rojas-Walker, T., Tamir, S., Ji, H., Wishnok, J.S. and Tannenbaum, S.R. (1995) Nitric oxide induces oxidative damage in addition to deamination in macrophage DNA. *Chem. Res. Toxicol.*, **8**, 473–477.

Rubanyi, G.M. and Vanhoutte, P.M. (1986) Superoxide anions and hyperoxia inactivate endothelium-derived relaxing factor. *Am. J. Physiol.*, **250**, H822–7.

Rubbo, H., Radi, R., Trujillo, M., Telleri, R., Kalyanaraman, B., Barnes, S., Kirk, M. and Freeman, B.A. (1994) Nitric oxide regulation of superoxide and peroxynitrite-dependent lipid peroxidation. Formation of novel nitrogen-containing oxidized lipid derivatives. *J. Biol. Chem.*, **269**, 26066–75.

Said, S.I., Berisha, H.I. and Pakbaz, H. (1996) Excitotoxicity in the lung: N-methyl-D-aspartate-induced, nitric oxide-dependent, pulmonary edema is attenuated by vasoactive intestinal peptide and by inhibitors of poly(ADP-ribose) polymerase. *Proc. Natl. Acad. Sci. USA*, **93**, 4688–92.

Salgo, M.G., Bermudez, E., Squadrito, G. and Pryor, W. (1995a) DNA damage and oxidation of thiols peroxynitrite causes in rat thymocytes. *Arch. Biochem. Biophys.*, **322**, 500–505.

Salgo, M.G., Stone, K., Squadrito, G.L., Battista, J.R. and Pryor, W.A. (1995b) Peroxynitrite causes DNA nicks in plasmid pBR322. *Biochem. Biophys. Res. Comm.*, **210**, 1025–30.

Salgo, M.G., Squadrito, G.L. and Pryor, W.A. (1995c) Peroxynitrite causes apoptosis in rat thymocytes. *Biochem. Biophys. Res. Comm.*, **215**, 1111–1118.

Sastry, S.S., Buki, K. and Kun, E. (1989) Binding of adenosine diphosphoribosyltransferase to the termini and internal regions of linear DNA. *Biochemistry*, **28**, 5680–5687.

Schraufstatter, I., Hinshaw, D., Hyslop, P., Spragg, R. and Cochrane, C. (1986) Oxidant injury of cells. DNA strand breaks activate polyadenosine diphosphate-ribose polymerase and lead to depletion of nicotinamide adenine dinucleotide. *J. Clin. Invest.*, **77**, 1312–1330.

Schraufstatter, I.U., Hyslop, P.A., Jackson, J.H. and Cochrane, C.G. (1988) Oxidant-induced DNA damage of target cells. *J. Clin. Invest.*, **82**, 1040–1050.

Schulz, J., Mathews, R.T., Jenkins, B.G., Ferrante, R.J., Siwek, D., Ross Henshaw, D., Cipolloni, P.B., Mecocci, P., Kowall, N.W., Rosen, B.R. and Beal, F.B. (1995) Blockade of neuronal nitric oxide synthase protects against excitotoxicity *in vivo*. *J. Neurosci.*, **15**, 8419–8429.

Schulz, J.B., Matthews, R.T., Muqit, M.M., Browne, S.E. and Beal, M.F. (1995) Inhibition of neuronal nitric oxide synthase by 7–nitroindazole protects against MPTP-induced neurotoxicity in mice. *J. Neurochem.*, **64**, 936–9.

Schulz, R. and Wambolt, R. (1995) Inhibition of nitric oxide synthesis protects the isolated working rabbit heart from ischaemia-reperfusion injury. *Cardiovasc. Res.*, **30**, 432–439.

Selden, L.A., Gersham, L.C., Estes, J.E., Ferro, T.J. and Johnson, A. (1995) Peroxynitrite-induced tyrosine nitration mediates decreased polymerization of actin. *Mol. Biol. of the Cell*, **6**, 20a.

Shi, X., Rojanasakul, Y., Gannett, P., Liu, K., Mao, Y.; Daniel, L.N., Ahmed, N. and Saffiotti, U. (1994) Generation of thiyl and ascorbyl radicals in the reaction of peroxynitrite with thiols and ascorbate at physiological pH. *J. Inorg. Biochem.*, **56**, 77–86.

Shoji, Y., Uedono, Y., Ishikura ,H., Takeyama, N. and Tanaka, T. (1995) DNA damage induced by tumour necrosis factor-alpha in L929 cells is mediated by mitochondrial oxygen radical formation. *Immunology*, **84**, 543–8.

Smulson, M.E., Kang,V.H., Ntambi, J.M., Rosenthal, D.S., Ding, R. and Simbulan, C.M. (1995) Requirement for the expression of poly(ADP-ribose) polymerase during the early stages of differentiation of 3T3–L1 preadipocytes, as studied by antisense RNA induction. *J. Biol. Chem.*, **270**, 119–27.

Snyder, S.H. (1996) No NO prevents parkinsonism. *Nature Med.*, **2**, 65–6.

Southan, G.J. and Szabó, C. (1995) Selective pharmacological inhibition of distinct nitric oxide synthase isoforms. *Biochem. Pharmacol.*, **51**, 383–394.

Spragg, R.G. (1991) DNA strand break formation following exposure of bovine pulmonary artery and aortic endothelial cells to reactive oxygen products. *Am. J. Respir. Cell. Mol. Biol.*, **4**, 4–10.

Squadrito, G.L. and Pryor, W.A. (1995) The formation of peroxynitrite *in vivo* from nitric oxide and superoxide. *Chemico-Biological Interactions*, **96**, 203–206.

Squadrito, G.L., Jin, X., Uppu, R.M. and Pryor, W.A. (1996) Distinguishing reactivities of peroxynitrite and hydroxyl radical. *Methods Enzymol.*, **269**, 366–74.

Stenger, S., Donhauser, N., Thuring, H., Rollinghf, M. and Bogdan, C. (1996) Reactivation of latent leishmaniasis by inhibition of inducible nitric oxide synthase. *J. Exp. Med.*, **183**, 1501–1514.

Stern, M.K., Jensen, M.P. and Kramer, K. (1996) Peroxynitrite decomposition catalysts. *J. Am. Chem. Soc.*, **118**, 8735–8736.

Suarez-Pinzon, W.L, Szabó, C. and Rabinovitch, A. (1997) Development of autoimmune diabetes in NOD mice is associated with the formation of peroxynitrite in pancreatic islet beta-cells. *Diabetes*, **46**, 907–911.

Suarez-Pinzon, W.L., Strynadka, K., Schulz, R. and Rabinovitch, A. (1994) Mechanism of cytokine-induced destruction of rat insulinoma cells: the role of nitric oxide. *Endocrinology*, **134**, 1006–1010.

Szabó, C. (1995) Alterations in nitric oxide production in various forms of circulatory shock. *New Horizons*, **3**, 2–32.

Szabó, C. (1996) DNA strand breakage and activation of poly-ADP ribosyltransferase: a cytotoxic pathway triggered by peroxynitrite. *Free Rad. Biol. Med.*, **21**, 855–869.

Szabó, C. (1996) The role of peroxynitrite in the pathophysiology of shock, inflammation and ischemia-reperfusion injury. *Shock*, **6**, 79–88.

Szabó, C. (1997) Activation of poly (ADP ribose) synthetase (PARS) by peroxynitrite: a cytotoxic pathway in the pathophysiology of circulatory shock In: *Yearbook of Intensive Care and Emergency Medicine*, (Ed: JL Vincent), Springer Verlag, 217–229.

Szabó, C. and Thiemermann, C. (1994) Invited opinion: role of nitric oxide in hemorrhagic, traumatic, and anaphylactic shock and thermal injury. *Shock*, **2**, 145–55

Szabó, C. and Salzman, A.L. (1995) Endogenous peroxynitrite is involved in the inhibition of cellular respiration in immuno-stimulated J774.2 macrophages. *Biochem. Biophys. Res. Comm.*, **209**, 739–743.

Szabó, C., Salzman, A.L. and Ischiropoulos, H. (1995a) Endotoxin triggers the expression of an inducible isoform of NO synthase and the formation of peroxynitrite in the rat aorta *in vivo*. *FEBS Lett.*, **363**, 235–238.

Szabó, C., Salzman, A.L. and Ischiropoulos, H. (1995b) Peroxynitrite-mediated oxidation of dihydrorhodamine 123 occurs in early stages of endotoxic and hemorrhagic shock and ischemia-reperfusion injury. *FEBS Lett.*, **372**, 229–232.

Szabó, C., Zingarelli, B. and Salzman, A.L. (1996a) Role of poly-ADP ribosyltransferase activation in the nitric oxide- and peroxynitrite-induced vascular failure. *Circ. Res.*, **78**, 1051–1063.

Szabó, C., Zingarelli, B., O'Connor, M., and Salzman, A.L. (1996b) DNA strand breakage, activation of poly-ADP ribosyl synthetase, and cellular energy depletion are involved in the cytotoxicity in macrophages and smooth muscle cells exposed to peroxynitrite. *Proc. Natl. Acad. Sci. USA*, **93**, 1753–1758.

Szabó, C., Wong, H.R. and Salzman, A.L. (1996c) Heat shock phenotype inhibits peroxynitrite-induced activation of poly(ADP) ribosyl synthetase and protects against peroxynitrite cytotoxicity in J774 macrophages. *Eur. J. Pharmacol.*, **315**, 221–226.

Szabó, C., Bryk, R., Zingarelli, B., Southan, G.J., Gahman, T.C., Bhat, V., Salzman, A.L., and Wolff, D.J. (1996) Pharmacological characterization of guanidinoethyldisulphide (GED), a novel inhibitor of nitric oxide synthase with selectivity towards the inducible isoform. *Br. J. Pharmacol.*, **118**, 1659–1668.

Szabó, C., Day, B.J. and Salzman, A.L. (1996e) Evaluation of the relative contribution of nitric oxide and peroxynitrite to the suppression of mitochondrial respiration in immunostimulated macrophages, using a novel mesoporphyrin superoxide dismutase analog and peroxynitrite scavenger. *FEBS Lett.*, **381**, 82–86.

Szabó, C., Zingarelli, B., Cuzzocrea, S. and Salzman, A.L. (1997a) Peroxynitrite and endotoxic shock cause vascular hyporeactivity and endothelial dysfunction: role of poly (ADP-ribose) synthetase activation and protection by endogenous glutathione. *Shock*, **7**(Suppl. 2), 20–21.

Szabó, C., Saunders, C., O'Connor, M. and Salzman, A.L. (1997b) Peroxynitrite causes energy depletion and increases permeability via activation of poly-ADP ribosyl synthetase in pulmonary epithelial cells. *Am. J. Mol. Cell. Respir. Biol.*, **16**, 105–109.

Szabó, C., Zingarelli, B., Cuzzocrea, S. and Salzman, A.L. (1997c) Activation of poly (ADP-ribose) synthetase is involved in the endothelial dysfunction in response to peroxynitrite and endotoxin shock. *Crit Care Medicine*, **5** (Suppl): A25.

Szabó, C., Wong, H.R., Bauer, P.I., Kirsten, E., O'Connor, M., Zingarelli, B., Mendeleyev, J., Hasko, G., Vizi, E.S., Salzman, A.L. and Kun, E. (1997d). Regulation of components of the inflammatory response by 5-iodo-6-amino-1,2-benzopyrone, and inhibitor of poly (ADP-ribose) synthetase and pleiotropic modifier of cellular signal pathways. *Int. J. Oncol.*, **10**, 1093–1101.

Szabó, C., Ferrer-Sueta, G., Zingarelli, B., Southan, G.J., Salzman, A.L. and Radi, R. (1997e) Mercaptoethylguanidine and related guanidine nitric oxide synthase inhibitors react with peroxynitrite and protect against peroxynitrite-induced oxidative damage. *J. Biol. Chem.*, in press.

Szabó, C., Virág, L., Cuzzocrea, S., Scott, G.J., Hake, P., O'Connor, M.P., Zingarelli, B., Salzman, A.L. and Kun, E. (1998) Protection against peroxynitrite-induced fibroblast injury and arthritis development by inhibition of poly (ADP-ribose) synthetase. *Proc. Natl. Acad. Sci. USA*, **95**, 3867–3872.

Tamir, S., Burney, S. and Tannenbaum, S.R. (1996) DNA damage by nitric oxide. *Chem. Res. Toxicol.*, **9**, 821–827.

Taniguchi, T., Nemoto, Y., Ota, K., Imai, T. and Tobari, J. (1993) Participation of poly(ADP-ribose) synthetase in the process of norepinephrine-induced inhibition of major histocompatibility complex class II antigen expression in human astrocytoma cells. *Biochem. Biophys. Res. Comm.*, **193**, 886–9.

Tannenbaum, S.R.. Tamir, S., DeRojas-Walker, T. and Wishnok, J.S. (1994) DNA damage and cytotoxicity by nitric oxide. In: *ACS Symposium Series* Nitrosamines and related N-nitroso compounds. (Loeppky, R.A. and Michejda, C.J, eds). **553**, 120–135.

Tarpey, M.M., Beckman, J.S., Ischiropoulos, H., Gore, J.Z. and Brock, T.A. (1995) Peroxynitrite stimulates vascular smooth muscle cell cyclic GMP synthesis. *FEBS Lett.*, **364**, 314–8.

Thiemermann, C., Bowes, J., Myint, F.P. and Vane, J.R. (1997) Inhibition of the activity of poly (ADP ribose) synthetase reduces ischemia-reperfusion injury in the heart and skeletal muscle. *Proc. Natl. Acad. Sci. USA*, **94**, 679–683.

Thies, R.L. and Autor, A.P. (1991) Reactive oxygen injury to cultured pulmonary artery endothelial cells: mediation by polyADP-ribose polymerase activation causing NAD depletion and altered energy balance. *Arch. Biochem. Biophys.*, **286**, 353–363.

Troy, C.M., Derossi, D., Prochiantz, A., Greene, L.A. and Shelanski, M.L. (1996) Downregulation of Cu/Zn superoxide dismutase leads to cell death via the nitric oxide-peroxynitrite pathway. *J. Neurosci.*, **16**, 253–261.

Turk, J., Corbett, J.A., Ramanadham, S., Bohrer, A. and McDaniel, M.L. (1993) Biochemical evidence for nitric oxide formation from streptozotocin in isolated pancreatic islets. *Biochem. Biophys. Res. Comm.*, **197**, 1458–1464.

Uchigata, Y., Yamamoto, H., Kawamura, A. and Okamoto, H. (1982) Protection by superoxide dismutase, catalase, and poly(ADP-ribose) synthetase inhibitors against alloxan- and streptozotocin-induced islet DNA strand breaks and against the inhibition of proinsulin synthesis. *J. Biol. Chem.*, **257**, 6084–8.

Uchigata, Y., Yamamoto, H., Nagai, H. and Okamoto, H. (1983) Effect of poly(ADP-ribose) synthetase inhibitor administration to rats before and after injection of alloxan and streptozotocin on islet proinsulin synthesis. *Diabetes*, **32**, 316–8.

Ueta, K. and Hayashi, O. (1985) ADP-ribosylation. *Ann. Rev. Biochem.*, **54**, 73–100.

Van der Vliet, A., Smith, D., O'Neill, C.A., Kaur, H., Darley-Usmar, V., Cross, C.E. and Halliwell, B. (1994) Interactions of peroxynitrite with human plasma and its constituents: oxidative damage and antioxidant depletion. *Biochem. J.*, **303**, 295–301.

Vasquez-Vivar, J., Santos, A.M., Junqueira, V.B. and Augusto, O. (1996) Peroxynitrite-mediated formation of free radicals in human plasma: EPR detection of ascorbyl, albumin-thyil and uric acid-derived free radicals. *Biochem. J.*, **314**, 869–876.

Vatassery, G.T. (1996) Oxidation of vitamin E, vitamin C, and thiols in rat brain synaptosomes by peroxynitrite. *Biochem. Pharm.*, **52**, 579–586.

Villa, L.M., Salas, E., Darley-Usmar, M., Radomski, M.E.W. and Moncada, S. (1994) Peroxynitrite induces both vasodilatation and impaired vascular relaxation in the isolated perfused rat heart. *Proc. Natl. Acad. Sci., USA*, **91**, 12383–12387.

Viner, R.I., Huhmer, A.F., Bigelow, D.J. and Schoneich, C. (1996) The oxidative inactivation of sarcoplasmic reticulum Ca^{2+}-ATPase by peroxynitrite. *Free Rad. Res.*, **24**, 243–259.

Virág, L., Scott, G.S., Salzman, A.L. and Szabó, C. (1998) Peroxynitrite-induced thymocyte apoptosis: the role of caspases and poly-(ADP-ribose) synthetase (PARS) activation. *Immunology*, **94**, 345–355.

Volk, T., Ioannidis, I., Hensel, M., deGroot, H. and Kox, W.J. (1995) Endothelial damage induced by nitric oxide: synergizm with reactive oxygen species. *Biochem. Biophys. Res. Comm.*, **213**, 196–203.

Wallis, R.A., Panizzon, K.L. and Girard, J.M. (1996) Traumatic neuroprotection with inhibitors of nitric oxide and ADP-ribosylation. *Brain Res.*, **710**, 169–177.

Wallis, R.A., Panizzon, K.L., Hanry, D. and Wasterlain, C.G. (1993) Neuroprotection against nitric oxide injury with inhibitors of ADP-ribosylation. *Neuroreport.*, **5**, 245–248.

Wang, P. and Zweier, J.L. (1996) Measurement of nitric oxide and peroxynitrite generation in the postischemic heart. *J. Biol. Chem.*, **271**, 29223–29230.

Wang, Z.Q., Auer, B., Stingl, L., Berghammer, H., Haidacher, D., Schweiger, M. and Wagner, E.F. (1995) Mice lacking ADPRT and poly(ADP-ribosyl)ation develop normally but are susceptible to skin disease. *Genes Develop.*, **9**, 510–520.

Watson, A.J., Askew, J.N. and Benson, R.S. (1995) Poly(adenosine diphosphate ribose) polymerase inhibition prevents necrosis induced by H_2O_2 but not apoptosis. *Gastroenterology*, **109**, 472–82.

White, C.R., Brock, T.A., Chang, L.Y., Crapo, J., Briscoe, P., Ku, D., Bradley, W.A., Gianturco, S.H., Gore, J. and Freeman, B.A. (1994) Superoxide and peroxynitrite in atherosclerosis.*Proc. Natl. Acad. Sci. USA*, **9**, 1044–8.

White, C.R., Darley-Usmar, V., Berrington, W.R., McAdams, M., Gore, J.Z., Thompson, J.A., Parks, D.A., Tarpey, M.M. and Freeman, B.A. (1996) Circulating plasma xanthine oxidase contributes to vascular dysfunction in hypercholesterolemic rabbits. *Proc. Natl. Acad. Sci. USA*, **93**, 8745–9.

Wilson, G.L., Patton, N.J., McCord, J.M., Mullins, D.W., and Mossman, B.T. (1984) Mechanisms of streptozotocin- and alloxan-induced damage in rat B cells. *Diabetologia*, **27**, 587–91.

Wink, D.A. and Laval, J. (1994) The Fpg protein, a DNA repair enzyme, is inhibited by the biomediator nitric oxide *in vitro* and *in vivo*. *Carcinogenesis*, **15**, 2125–2129.

Wizemann, T., Gardner, C., Laskin, J., Quinones, S., Durham, S., Goller, N., Ohnishi, T. and Laskin D. (1994) Production of nitric oxide and peroxynitrite in the lung during acute endotoxemia. *J. Leukoc. Biol.*, **56**, 759–768.

Xia, Y., Dawson, V.L., Dawson, T.M., Snyder, S.H., and Zweier, J.L. (1996) Nitric oxide synthase generates superoxide abnd nitric oxide in arginine-depleted cells leading to peroxynitrite-mediated cellular injury. *Proc. Natl. Acad. Sci. USA*, **93**, 6770–6774.

Xie, Y.W. and Wolin, M.S. (1996) Role of nitric oxide and its interaction with superoxide in the suppression of cardiac muscle mitochondrial respiration. *Circulation*, **94**, 2580–2586.

Yamamoto, H., Uchigata, Y. and Okamoto, H. (1981) Streptozotocin and alloxan induce DNA strand breaks and poly(ADP-ribose) synthetase in pancreatic islets. *Nature*, **294**, 284–6.

Yamamoto, K., Tsukidate, K. and Farber, J.L. (1993) Differing effects of the inhibition of polyADP-ribose polymerase onthe course of oxidative cell injury in hepatocytes and fibroblasts. *Biochem. Pharmacol.*, **46**, 483–491.

Yermilow, V., Rubio, J., Becchi, M., Friesen, M.D., Pignatelli, B. and Ohshima, H. (1995) Formation of 8-nitroguanine by the reaction of guanine with peroxynitrite *in vitro*. *Carcinogenesis*, **16**, 2045–2050

Yu, L., Gengaro, P.E., Niederberger, M., Burke, T.J. and Schrier, R.W. (1994) Nitric oxide: a mediator in rat tubular hypoxia/reoxygenation injury. *Proc. Natl. Acad. Sci. USA*, **91**, 1691–5.

Zhang, J. and Steiner, J.P. (1995) Nitric oxide synthase, immunophyllins and poly (ADP-ribose) synthetase: novel targets for the development of neuroprotective drugs. *Neurol. Res.*, **17**, 285–288.

Zhang, J., Dawson, V.L., Dawson, T.M. and Snyder, S.H. (1994) Nitric oxide activation of poly (ADP-ribose) synthetase in neurotoxicity. *Science*, **263**, 687–689.

Zhang, J., Pieper, A. and Snyder, S.H. (1995) Poly(ADP-ribose) synthetase activation: an early indicator of neurotoxic DNA damage. *J. Neurochem.*, **65**, 1411–4.

Zhu, L., Gunn, C. and Beckman, J.S. (1992) Bactericidal activity of peroxynitrite. *Arch. Biochem. Biophys.*, **298**, 452–7.

Zingarelli, B., O'Connor, M., Wong, H., Salzman, A.L. and Szabó, C. (1996a) Peroxynitrite-mediated DNA strand breakage activates poly-ADP ribosyl synthetase and causes cellular energy depletion in macrophages stimulated with bacterial lipopolysaccharide. *J. Immunol.*, **156**, 350–358.

Zingarelli, B., Salzman, A.L. and Szabó, C. (1996b) Protective effects of nicotinamide against nitric oxide-mediated vascular failure in endotoxic shock: potential involvement of polyADP ribosyl synthetase. *Shock*, **4**, 258–264.

Zingarelli, B., Day, B.J., Crapo, J., Salzman, A.L. and Szabó, C. (1997a) The potential involvement of peroxynitrite in the pathogenesis of endotoxic shock. *Br. J. Pharmacol.*, **120**, 259–267.

Zingarelli, B., Salzman, A.L. and Szabó, C. (1997b) Mercaptoethylguanidine, a selective inhibitor of the inducible isoform of nitric oxide synthase has beneficial effects in a rat model of delayed hemorrhagic shock. *Crit. Care Med.*, **5** (Suppl): A138.

Zingarelli, B., Cuzzocrea, S., O'Connor, M., Zsengeller, Z., Salzman, A.L. and Szabó, C. (1997c) Poly (ADP-ribose) synthetase activation induced by peroxynitrite mediates myocardial ischemia and reperfusion injury. In: *Immune Consequences of Trauma, Shock and Sepsis, Mechanisms and Therapeutic Approaches, Monduzzi Editione* (Bologna) (Faist, E., ed.), pp. 681–685.

Zingarelli, B., Salzman, A.L. and Szabó, C. (1998) Genetic disruption of poly (ADP ribose) synthetase inhibits the expression of P-selectin and intercellular adhesion molecule-1 in myocardial ischemia-reperfusion injury. *Circ. Res.*, **83**, 85–94.

7 Nitric Oxide and Gene Transcription

James K. Liao and Peter Libby

Atherosclerosis & Vascular Medicine Unit, Cardiovascular Division, Department of Medicine, Brigham & Women's Hospital and Harvard Medical School, Boston, Masschusetts 02115, USA

Some twenty years ago, Furchgott and Zawadzki first identified endothelium-dependent relaxing factor (EDRF) as an important regulator of vascular tone (Furchgott and Zawadzki 1980). Subsequent work established that nitric oxide ($\cdot$NO) or closely related molecules accounts for most of the activity of EDRF (Ignarro *et al.*, 1987; Palmer *et al.*, 1987; Myers *et al.*, 1990). At least three distinct $\cdot$NO synthases have been cloned and found to produce various amounts of $\cdot$NO under different conditions (Bredt *et al.*, 1991; Lyons *et al.*, 1992; Xie *et al.*, 1992; Lowenstein *et al.*, 1992; Janssens *et al.*, 1992; Lamas *et al.*, 1992) (Table 7-1). For example, both Type I and Type III $\cdot$NO synthase produce low levels of basal $\cdot$NO; whereas, sustained, high levels of $\cdot$NO are produced by Type II $\cdot$NO synthase only after induction by certain cytokines. Such differences in the regulation of these specific $\cdot$NO synthases may underlie their importance and contributions in homeostatic and inflammatory processes. Furthermore, the localization of these various $\cdot$NO synthases in specific cell types other than vascular wall cells suggest that $\cdot$NO may mediate other signaling pathways in different organ systems (Geller *et al.*, 1993; Mannick *et al.*, 1994; Chen *et al.*, 1996). As a result, the functional roles of $\cdot$NO in various biological processes have expanded beyond those of a vasodilator.

The diversity of $\cdot$NO-mediated processes results not only from the biological actions of $\cdot$NO itself, but also, from the complex biochemical interactions between $\cdot$NO and its surrounding environment (Stamler *et al.*, 1992a; Stamler 1994). Such environmental factors including the presence of reactive oxygen species and antioxidants can determine the ultimate biological effects and biochemical status of $\cdot$NO and its derivatives (Beckman *et al.*, 1990; Stamler *et al.*, 1992b, 1992c). For example, $\cdot$NO can form a potent oxidant, peroxynitrite, through its interaction with superoxide anion (Beckman *et al.*, 1990; Radi *et al.*, 1991). Indeed, some of the neurotoxicity of $\cdot$NO derives from the formation of peroxynitrite (Lipton *et al.*, 1993; Zhang *et al.*, 1994). Furthermore, interaction of $\cdot$NO

Table 7-1. NO Synthases

Isoforms	*Cell Types*	*Basal $\cdot$NO Levels*	*Stimulated $\cdot$NO Levels*
Type I (nNOS)	Neurons, Skeletal Muscle, ? Smooth Muscle	Low	Transient, Low
Type II (iNOS)	Macrophages, Myocytes, Smooth Muscle, Hepatocytes,	None	Sustained, High
Type III (ecNOS)	Endothelial Cells, Platelets	Low	Transient, High

with thiol derivatives produces nitrosothiols which can modify the activities of signaling molecules such as mitogen-activated protein (MAP) kinases and p21*ras* via S-nitrosylation (Lander *et al.*, 1996a, 1996b). Thus, it is not surprising to find that ·NO can mediate disparate and sometimes contradictory effects in different cell types and conditions.

Recent studies from many laboratories have shown that ·NO modulates gene expression via its effects on multiple signaling pathways (Lander *et al.*, 1993a, 1993b; Miyamoto *et al.*, 1997; Hausladen *et al.*, 1996). For example, ·NO activates protein tyrosine phosphatases in peripheral blood mononuclear cells (Lander *et al.*, 1993a) leading to the possiblity that ·NO regulates various transcription factors which are tyrosine phosphorylated such as signal transducers and activators of transcription (STATs) and interferon regulatory factor (IRF)-1. Oxidants such as hydrogen peroxide and pervanadate stimulate protein tyrosine phosphorylation and activate transcription factors downstream of MAP kinase pathways such as p38 c-*jun* kinase (JNK) and nuclear factor-kappa B (NF-κB) (Adler *et al.*, 1995; Schreck *et al.*, 1991). Since ·NO can avidly scavenge superoxide anion, it can prevent superoxide anion from forming its dismutation product, hydrogen peroxide (Huie and Padmaja, 1993). Furthermore, under certain conditions, ·NO may act as an electron donor or antioxidant (Stamler *et al.*, 1992a). Indeed, antioxidants such as N-acetylcysteine or pyrrolidine dithiocarbamate (PDTC) inhibit the activation of the transcription factors, AP-1 and NF-κB (Devary *et al.*, 1993; Marui *et al.*, 1993; Legrand-Poels *et al.*, 1995; Pinkus *et al.*, 1996). Finally, S-nitrosylation of p21*ras* by nitrosothiols may also activate NF-κB and JNK which are involved in endothelial cell activation and proliferation (Lander *et al.*, 1996a, 1996b).

Several oxidant-sensitive transcription factors such as *egr*-1, activator protein (AP)-1, and NF-κB may modulate endothelial cell gene expression and activation (Collins, 1993; Ohba *et al.*, 1994; Bandyopadhyay *et al.*, 1995; Khachigian *et al.*, 1996). In particular, the activation of NF-κB is required for the transcriptional induction of many pro-inflammatory mediators involved in vascular inflammation such as adhesion molecules, cytokines, and growth factors (Pober *et al.*, 1983; Collins, 1993; Peng *et al.*, 1995a). Hence, factors which modulate NF-κB activation could regulate the inflammatory process. This chapter will focus on how endogenous and exogenous ·NO can attenuate endothelial cell activation and vascular inflammation via its inhibitory effects on NF-κB.

·NO AND ATHEROGENESIS

One of the proximal events of atherogenesis and vascular inflammation is the activation of endothelial cells (Pober *et al.*, 1983; Cybulsky *et al.*, 1991). The activated endothelium expresses surface adhesion molecules such as vascular cell adhesion molecule (VCAM)-1, intercellular adhesion molecule (ICAM)-1, and E-selectin which facilitate the attachment of peripheral blood leukocytes to the endothelial surface (Collins, 1993). Indeed, monocyte adhesion to the vessel wall and subsequent differentiation of these cells into macrophages is crucial for the development of macrophage-derived foam cells in atherosclerotic vessels (Libby and Clinton, 1993; Ross and Harker, 1976; Berliner *et al.*, 1995).

Vessels affected by atherosclerosis develop endothelial dysfunction as indicated by impaired vasomotor function due to the decreased synthesis, release, and/or activity of endothelial-derived ·NO (Ignarro *et al.*, 1987; Tanner *et al.*, 1991). It remains uncertain whether this aspect of endothelial dysfunction serves merely as a marker for atherosclerosis or an important contributor to the atherogenic process. However, growing evidence indicates that ·NO can exert anti-atherogenic actions. For example, reduced endothelial-derived ·NO activity contributes to impaired vascular relaxation (Berliner *et al.*, 1995; Tanner *et al.*, 1991), enhanced platelet aggregation (Radomski *et al.*, 1990; Wolf *et al.*, 1997), and increased vascular smooth muscle proliferation (Garg and Hassid, 1989; Nakaki *et al.*, 1990), and endothelial-leukocyte interactions (Gauthier *et al.*, 1994; Kurose *et al.*, 1993). Inactivation of endothelial-derived ·NO by ·O_2^- derived from angiotensin II-stimulated NADH oxidase can lead to nitrate tolerance, vasoconstriction, and hypertension (Munzel *et al.*, 1995; Rajagopaian *et al.*, 1996). Mutant mice lacking endothelial Type III NO synthase have hypertension and develop greater proliferative and inflammatory vascular response to cuff injury (Huang *et al.*, 1995; Moroi *et al.*, 1996). *In vivo* transfer of endothelial cell ·NO synthase gene into balloon-injured vessels decreases intimal smooth muscle cell proliferation in rat carotid arteries (von der Leyen *et al.*, 1995). Enriching the diets of cholesterol-fed rabbits with L-arginine, the precursor of ·NO, improves endothelial-dependent relaxation, reduces leukocyte attachment to the endothelial surface, and limits the extent of atherosclerotic lesions (Cooke *et al.*, 1992; Tsao *et al.*, 1994). These findings agree with a study showing that a single bolus treatment with an NO-generating compound, S-nitrosothiol, inhibits intimal smooth muscle proliferation of the rabbit carotid artery following injury with a balloon-tip catheter (Marks *et al.*, 1995).

·NO AND ENDOTHELIAL-LEUKOCYTE INTERACTION

Although stimulation of soluble guanylyl cyclase mediates many effects of ·NO such as vasodilation and inhibition of vascular smooth muscle proliferation and platelet aggregation, the ability of ·NO to attenuate leukocyte attachment to the endothelial surface does not depend on cGMP formation (De Caterina *et al.*, 1995; Khan *et al.*, 1996). For example, NO donors, but not cGMP analogues decrease cytokine-induced endothelial expression of VCAM-1, E-selectin, and to a lesser extent, ICAM-1. Furthermore, NO inhibits the expression of other pro-inflammatory mediators such as macrophage-colony stimulating factor (M-CSF or CSF-1), IL-6 and IL-8 via cGMP-independent pathways (Peng *et al.*, 1995a; De Caterina *et al.*, 1995).

Recent studies have shown that ·NO inhibits monocyte adhesion to the endothelial surface (Gauthier *et al.*, 1994; Kurose *et al.*, 1993; Tsao *et al.*, 1994; Kubes *et al.*, 1991). *In vivo* studies demonstrate that inhibition of endogenous ·NO production by N^ω-nitro-L-arginine methylester (L-NAME) promotes vasoconstriction and leukocyte adhesion to the endothelial during ischemia reperfusion injury in splanchnic vessels (Gauthier *et al.*, 1994; Davenpeck *et al.*, 1994). Bath and co-workers reported that ·NO inhibits monocyte adhesion to endothelium *in vitro*, without altering expression of CD11b/CD18, one of the cognate ligands on monocytes for certain endothelial-leukocyte adhesion molecules (Bath *et al.*, 1991). In experiments superfusing a cat mesenteric preparation with inhibitors of ·NO production, Kubes and collaborators also showed that such inhibitors increase leukocyte (mostly neutrophil) adhesion to mesenteric venules (Kubes *et al.*, 1991). They also showed that increased expression of CD11b/CD18 on the leukocyte surface could not account for L-NAME's effect.

FIGURE 7-1. Monocyte adhesion (mean ± SEM) to endothelial cell monolayers stimulated with interleukin (IL)-1α (10 ng/ml) in the presence and absence of ·NO donor, S-nitrosoglutathione (GSNO, 0.2 mM) for 24 h. (De Caterina *et al.*, 1995; reproduced with permission)

Studies by Kurose and colleagues, however, have documented increased monocyte adhesion to mesenteric endothelial cells *in vivo* by L-NAME and reversal by cGMP analogues (Kurose *et al.*, 1993). These studies contrast with finding from others who found that cGMP analogues do not affect cytokine-induced expression of adhesion molecules (De Caterina *et al.*, 1995). Several possibilities could account for this difference. First, these *in vivo* studies showed induction of endothelial-leukocyte adhesion by L-NAME rather than by cytokines. Perhaps induction of VCAM-1 by L-NAME, but not by cytokines, is amenable to reversal by cGMP analogues. Second, these studies cannot separate cGMP effects on hemodynamics and/or leukocytes from direct alterations in endothelial functions. Furthermore, induction of other adhesion molecules (i.e. ICAM-1 or E-Selectin) by L-NAME treatment may depend partially on cGMP-dependent pathways.

We have found that inhibition of endogenous endothelial ·NO production by N^G-monomethyl-L-arginine (LNMA) mildly induces VCAM-1 expression in cultured saphenous vein endothelial cells (De Caterina *et al.*, 1995). Furthermore, we find that exposure of cytokine-activated endothelial cells to an ·NO donor, S-nitrosoglutathione (GSNO), greatly diminishes leukocyte adhesion to the endothelial surface (Figure 7-1). Thus, endothelial cells serve as a potential target of ·NO's effect in limiting leukocyte adhesion to the vessel wall.

NF-κB AND TRANSCRIPTIONAL INDUCTION OF PRO-INFLAMMATORY GENES

The genes that encode a number of cellular adhesion molecules and cytokines implicated in atherogenesis share specific DNA binding motifs in their promoters which interact with the transcription factor, NF-κB (Collins *et al.*, 1993, 1995) (Table 7-2). NF-κB was originally described as a heterodimeric cytosolic protein in B-cells, which upon activation, translocates into the nucleus where it binds to specific decameric sequences in the IgG κ light chain enhancer (Sen and Baltimore, 1986a, 1986b; Wirth and Baltimore, 1988).

Table 7-2. Selected Genes Requiring NF-κB for Transactivation.

Cellular Adhesion Molecules
 — Intercellular Adhesion Molecule-1 (ICAM-1)
 — Vascular Cell Adhesion Molecule-1 (VCAM-1)
 — Endothelial Leukocyte Adhesion Molecule-1 (ELAM-1 or E-Selectin)

Inflammatory Cytokines
 — Interleukin (IL)-2, -6 and -8
 — Tumor Necrosis Factor (TNF)-α and -β
 — Macrophage-Colony Stimulating Factor (M-CSF or CSF-1)
 — Granulocyte Colony Stimulating Factor (G-CSF)
 — Granulocyte/Macrophage-Colony Stimulating Factor (GM-CSF)
 — Macrophage Chemotactic Protein-1 (MCP-1)

Immunologic Mediators
 — Immunoglobulin (IgG) κ light chain
 — T Cell Receptor α and β chain
 — Major Histocompatability Complex (MHC) Class I
 — Major Histocompatability Complex (MHC) Class II Invariant Chain (Ii)
 — β₂-Microglobulin
 — Type II Inducible Nitric Oxide Synthase (iNOS)

Viral Enhancers
 — Human Immunodeficiency Virus-1 (HIV-1)
 — Cytomegalovirus (CMV)
 — Adenovirus
 — Simian Virus 40 (SV40)

Transcription Factors
 — IκB-α
 — c-Rel
 — NF-κB p105
 — c-myc
 — Interferon Regulatory Factor-1 (IRF-1)

Subsequent studies have shown that this nuclear binding protein can also activate viral enhancer elements (Perkins *et al.*, 1993; Speir *et al.*, 1996). Members of the mammalian NF-κB family possess *Rel* homology domains which are necessary for dimerization, nuclear translocation, and DNA binding (Thanos and Maniatis, 1995; Baeuerle and Baltimore, 1996). The *Rel*-related proteins initially described factors which determine the relative spatial orientation during differentiation of the Drosophila body (i.e. dorsal vs. ventral) (Lemaitre *et al.*, 1996).

The NF-κB family can be further divided into two groups based upon their structure and function (Table 7-3). The first group consists of p65 (*Rel* A), c-*Rel*, and *Rel* B which contain transcriptional activation domains necessary for gene induction (Ballard *et al.*, 1992). The second group consists of p105 and p100 which upon proteolytic processing give rise to p50 (NF-κB1) and p52 (NF-κB2), respectively (Rice *et al.*, 1992; Scheinman *et al.*, 1993). The carboxy terminal regions of p105 and p100 share structural features with the endogenous NF-κB inhibitor, IκB, and thus, functions to retain NF-κB in the cytoplasm. The mature proteins, p50 and p52, however, can form functional dimers with members of

Table 7-3. NF-κB/Rel Family

NF-kB	Species
Rel A (p65)	Mammalian
c-Rel	Mammalian
p50 (NF-κB1)	Mammalian
p105 (IκB-γ)	Mammalian
p52 (NF-κB2)	Mammalian
p100 (IκB-δ)	Mammalian
Rel B	Mammalian
dorsal	Drosophila
v-Rel	Viral

both groups (Thanos and Baltimore, 1995). With the exception of *Rel* B which cannot form homodimers, members of both groups can bind in a tissue-specific manner as homo- or heterodimers to enhancer elements of target genes. In vascular endothelial cells, the transcriptional induction of pro-inflammatory genes depends predominantly upon the activation of the *Rel* A/p50 heterodimer (Ahmad *et al.*, 1995; Osnes *et al.*, 1995).

NF-κB is activated under numerous conditions associated with vascular inflammation (Table 7-4). For example, bacterial or viral products such as lipopolysaccharide (LPS) or double-stranded RNA can trigger the activation of NF-κB (Osnes *et al.*, 1996; Visvanathan and Goodbourn, 1989). In addition, the production of cytokines and growth factors at sites of inflammation can also activate NF-κB (Baeuerle and Henkel, 1994). There is growing evidence that reactive oxygen species (ROS) such as hydrogen peroxide (H_2O_2) and possibly peroxynitrite ($ONOO^-$) contibutes to cellular "oxidative stress" and may be a common pathway for the activation of NF-κB (Collins, 1993; Marui *et al.*, 1993) (Figure 7-2). Indeed, recent studies have shown that the dismutated product of superoxide anion (O_2^-), H_2O_2 is a potent oxidant which not only can activate NF-κB, but also other transcription factors such as AP-1 and *egr*-1 (Abate *et al.*, 1990; Schreck *et al.*, 1991; Nose *et al.*, 1991; Ohba *et al.*, 1994). These findings suggest that activation of NF-κB and endothelial cells occur through generation of ROS and that antioxidants would be effective inhibitors of NF-κB and endothelial cell activation.

The activation of NF-κB involves the degradation of its cytoplasmic inhibitor, IκB and activation of bound protein kinase A (Sen and Baltimore, 1986b; Henkel *et al.*, 1993; Zhong *et al.*, 1997) (Figure 7-2). Presently, five distinct IκB proteins have been described and function to retain NF-κB in the cytoplasm and render it inactive (Beg and Baldwin 1993, Baeuerle and Baltimore, 1996) (Table 7-3). These IκB proteins contain ankyrin repeat motifs which mask the nuclear localization sequence of NF-κB subunits (Thanos and Maniatis, 1995). Of the different IκB proteins, the most well-characterized is IκB-α, a 37 kD protein (Haskill *et al.*, 1991). Following cytokine stimulation or oxidative stress, IκB-α is phosphorylated by a novel ubiquitinated serine kinase called IκB kinase (Palombella *et al.*, 1994; Brown *et al.*, 1995) (Figure 7-2). Phosphorylation of IκB-α targets the IκB-α for ubiquitination and rapid degradation by 26S proteasomes (Chen *et al.*, 1996). The proteasomes are large multimeric assemblies of intracellular enzymes which digest proteins marked for degradation by ubiquitination. Serine phosphorylation of p105 and p100 targets these proteins for limited proteolytic cleavage rather than complete degradation through

Table 7-4. Certain Factors Which Activate NF-κB

Inflammatory Mediators
— Tumor Necrosis Factor-α
— Interleukin-1 and -2
— Lymphotoxin
— Leukotriene B4

Growth Factors
— Platelet-Derived Growth Factor (PGDF)
— Transforming Growth Factor-β1 (TGF-β1)

Viral Mediators
— Viral Infection (HIV, EBV, CMV, etc.)
— Double-Stranded RNA
— Epstein-Barr Nuclear Antigen-2

Bacterial Mediators
— Lipopolysaccharide (LPS)
— Muramyl Proteins
— Exotoxin B

Oxidants
— Hydrogen Peroxide
— Ultraviolet light

Drugs
— Phorbolesters
— Okadaic Acid
— Cycloheximide
— Anisomycin
— Pervanadate

Physical Stress
— Laminar Shear Stress
— Stretch
— Cyclic Strain

the ubiquitin-proteasome pathway (Palombella *et al.*, 1994; Chen *et al.*, 1996). The degradation of IκB-α or proteolytic cleavage of p105/p100 then allows the unbound NF-κB to translocate into the nucleus where it can transactivate the κB enhancer elements in the regulator region of pro-inflammatory genes (Collins, 1993; Collins *et al.*, 1995) (Table 7-2).

Since the phosphorylation of IκB-α is a key regulatory step in the activation of NF-κB, factors which influence IκB-α phosphorylation might modulate inflammatory processes. Indeed, recent studies indicate that salicylates and agents with antioxidant activity such as N-acetylcysteine (NAC) or pyrrolidine dithiocarbamate (PDTC) inhibit cytokine-induced NF-κB by preventing IkB-a phosphorylation (Blackwell *et al.*, 1996; Munoz *et al.*, 1996; Kopp and Ghosh, 1994; Pierce *et al.*, 1996). Alternatively, stabilization of phosphorylated IκB-a by inhibitors of 26S proteasomes such as the peptide aldehyde derivatives MG132 or lactacystin attenuates cytokine-induced VCAM-1 expression in endothelial cells (Read *et al.*, 1995, 1996).

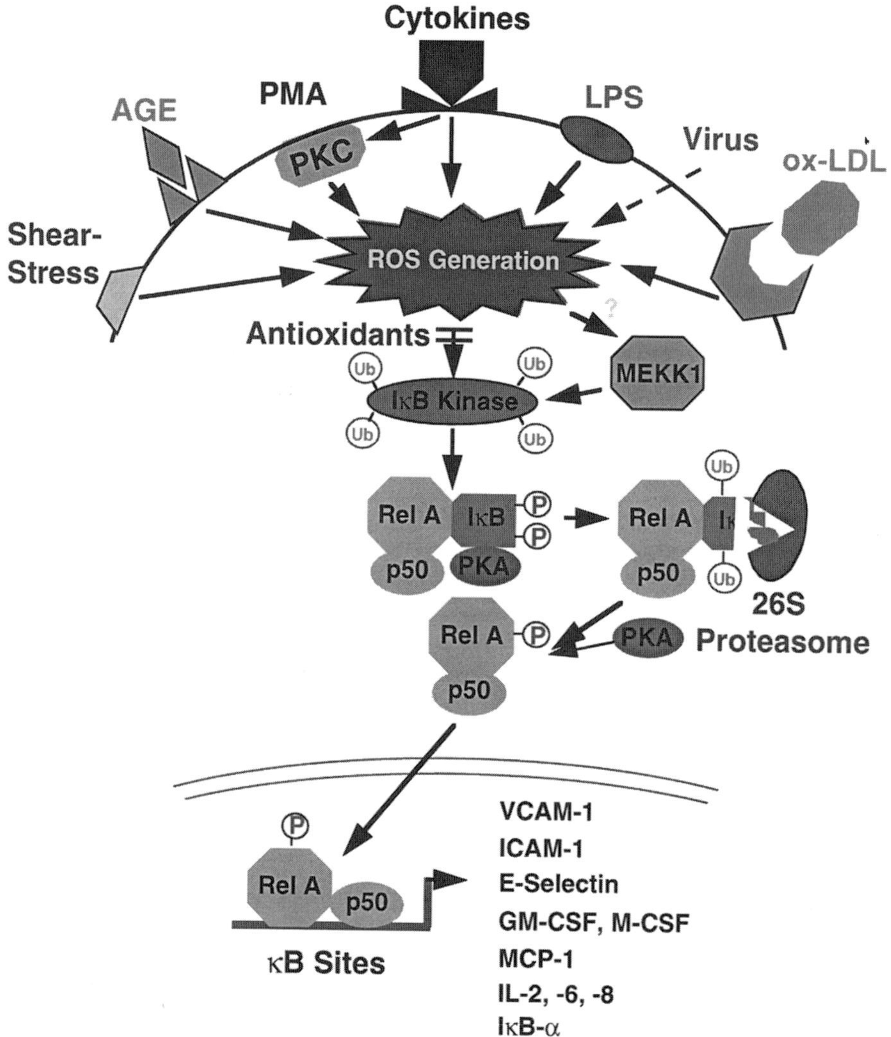

FIGURE 7-2. Schematic diagram of signaling pathway leading to the activation of NF-κB.

Interestingly, one of the main factors which limit NF-κB activity is NF-κB itself. Depending on the species, the IκB-α promoter contains at least three to five functional κB binding sites (LeBail *et al.*, 1993; Ito *et al.*, 1994; de Martin *et al.*, 1993; Chiao *et al.*, 1994). Thus, the induction of IκB-α by NF-κB serves as an inducible autoregulatory feedback mechanism for terminating NF-κB activity (Sun *et al.*, 1993; Read *et al.*, 1994). However, the induction of IκB-α is not limited to NF-κB. Recent studies have demonstrated that the glucocorticoids can modulate NF-κB stability and transcriptionally induce the expression of IκB-α in T-lymphocytes (Jurkat cells) (Scheinman *et al.*, 1995; Auphan *et al.*, 1995). Consequently, the induction of IκB-α by glucocorticoids and the subsequent inhibition of NF-κB activity may underlie some of the anti-inflammatory effects of glucocorticoids.

·NO AND NF-κB ACTIVATION

Vascular endothelial cells produce ·NO constitutively and can express the inducible isoform of ·NO synthase in response to cytokine stimulation (Janssens *et al.*, 1992; Lamas *et al.*, 1992; Gross *et al.*, 1991). Thus, the net activation of NF-κB in vascular wall cells during inflammation probably depends on a complex balance between stimulatory and inhibitory factors. The finding that inhibition of endogenous endothelial ·NO production by L-NMA could activate NF-κB, induce VCAM-1 expression, and stimulate endothelial-leukocyte adhesion suggests that constitutively-produced ·NO may play an important physiologic role in tonically inhibiting endothelial cell activation under basal conditions (De Caterina *et al.*, 1995). This inhibitory effect of ·NO is also supported by the finding *in vivo* that inhibition of endogenous ·NO production by L-NAME promotes endothelial-leukocyte interactions through the expression of NF-κB-dependent adhesion molecules (Tsao *et al.*, 1996).

Although under static conditions, cultured endothelial cells do not produce sufficient ·NO to inhibit cytokine-induced activation of NF-κB, higher levels of ·NO may be produced by endothelial cells *in vivo* in response to fluid shear stress which is known to stimulate endothelial cell Type III ·NO synthase (ecNOS) gene transcription (Uematsu *et al.*, 1995, Nadaud *et al.*, 1996). As ecNOS expression increases with shear stress, regions of normal laminar flow may have sufficient ·NO production to suppress NF-κB activation. However, regions of distrubed flow, lacking the stimulus to augment ·NO production, may lose this potentially protective action of ·NO. This phenomenon could help explain why the regions of physiologic shear stress tend to be less susceptible to the development of nascent atherosclerotic lesions than regions of distrubed flow (i.e. near brach points or flow dividers) which have increased predilection to develop these lesions (Golledge *et al.*, 1997).

In addition to ·NO produced by ecNOS, endothelial cells may encounter higher levels of ·NO generated by inducible Type II ·NO synthase (iNOS) at sites of vascular inflammation (Lyons *et al.*, 1992; Xie *et al.*, 1992; Lowenstein *et al.*, 1992). Exposure of rodent macrophages and human vascular smooth muscle cells to cytokines leads to the induction of iNOS which produces substantially higher levels of ·NO compared to that of ecNOS (Dinerman *et al.*, 1993). Furthermore, compared to *in vitro* studies, higher ·NO concentrations could be achieved locally since endothelial cells are in close proximity to cells expressing iNOS *in vivo*. Thus, the amount of ·NO produced by iNOS can be approximated with ·NO donors which have been shown to inhibit NF-κB activation (Figure 7-3). Additionally, ·NO may form adducts with sulfhydryl compounds forming more stable nitrosothiols (Stamler *et al.*, 1992b, 1992c). Such increased availability of ·NO may be required to suppress cytokine-activated NF-κB in endothelial cells. Interestingly, the higher levels of ·NO produced by activated macrophages may also provide for an endogenous feedback mechanism for limiting ·NO production since the induction of iNOS requires the activation of NF-κB (Xie *et al.*, 1994). The regulation of NF-κB by ·NO, therefore, might serve as a mechanism by which ·NO can modulate the expression of various pro-inflammatory mediators during atherogenesis and vascular inflammation.

STABILIZATION OF NF-κB BY ·NO

Treatment of cells with ·NO appears to stabilize the inactive NF-κB/IκB-α complex following stimulation by cytokines (Figure 7-4). By preventing the degradation of IκB-

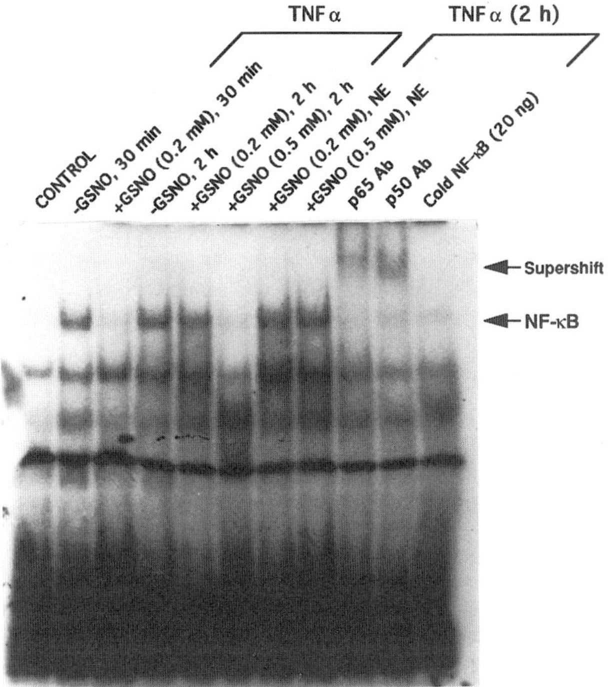

FIGURE 7-3. Electrophoretic mobility shift assay showing the activation of NF-κB following stimulation with TNF-α (10 ng/ml) in the presence and absence of GSNO. NF-κB activation was not affected when GSNO was added directly to nuclear extracts (NE). Specificity of NF-κB band was determined by "supershifting" with p65 and p50 antibodies and in the presence of excess (100×) unlabeled (cold) κB oligonucleotides. (Peng *et al.*, 1995b; reproduced with permission)

α, the NF-κB is retained in the cytoplasm and hence cannot activate transcription of target genes within the nucleus. The mechanism by which ·NO stabilizes IκB-α differs from that of antioxidants such as N-acetylcysteine or pyrrolidine dithiocarbamate (PDTC) which stabilize the NF-κB/IκB-α complex through scavenging reactive oxygen species such as superoxide anion (Schreck *et al.*, 1991; Legrand-Poels *et al.*, 1995; Devary *et al.*, 1993). Although it is not known whether NF-κB activation induced by oxidative stress results from hydrogen peroxide-mediated IκB-α phosphorylation, ·NO can bind superoxide anion with extremely high affinity (Huie and Padmaja, 1993). Thus, one mechanism by which ·NO may stabilize IκB-α and inhibit NF-κB activation is through scavenging superoxide anion thereby decreasing its dismutation product, hydrogen peroxide.

·NO may also stabilize IκB-α via its effects on other signaling molecules which regulate IκB-α phosphorylation. For example, although serine phosphorylation by IκB-α kinase(s) signals the degradation of IκB-α, the activation of IκB-α kinase occurs through the protein tyrosine phosphorylation MAP kinase (MEKK1) pathway (Chen *et al.*, 1996). The local generation of hydrogen peroxide increases protein tyrosine phosphorylation and activates NF-κB in cells (Schreck *et al.*, 1991; Fialkow *et al.*, 1993; Schreck *et al.*, 1992). Perhaps, ·NO inhibits IκB-α phosphorylation its antioxidant properties. In addition, ·NO can directly activate protein tyrosine phosphatases such as p56[lck], thereby potentially inhibiting MEKK1

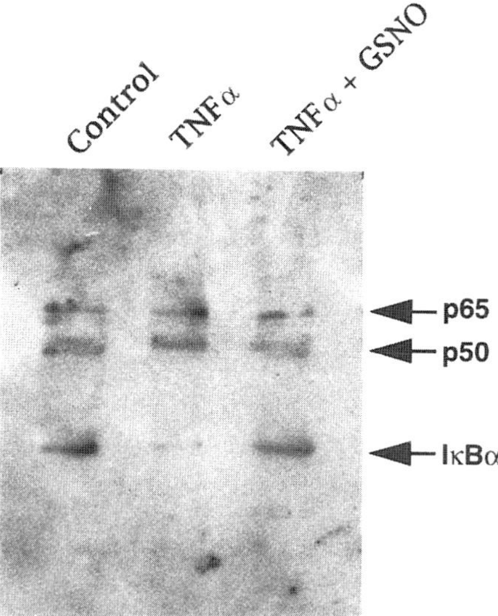

FIGURE 7-4. Immunoprecipitation of endothelial cellular extracts with agarose-conjugated p65 antibody followed by immunoblotting with antibodies to p65, p50 and IκB-α. Endothelial cells were stimulated with TNF-α in the presence and absence of GSNO (0.2 mM). (Peng *et al.*, 1995b; reproduced with permission)

activation and IκB-α phosphorylation (Lander *et al.*, 1993b). Indeed, inhibition of protein tyrosine phosphatases by pervanadate causes serine phosphorylation of IκB-α (Singh and Aggarwal, 1995; Weber *et al.*, 1995). It is not known whether the direct activation of phosphatases by ·NO or the indirect scavenging effects of ·NO with reactive oxygen species is the mechanism by which NO prevents the phosphorylation of IκB-α. Furthermore, some studies suggest that NF-κB can be activated through direct protein tyrosine phosphorylation of IκB-α (Imbert *et al.*, 1996).

Recent studies have shown that ·NO produces some of its effects through its interaction with thiol groups to form nitrosothiols (Stamler *et al.*, 1992b, 1992c). Nitrosothiols such as GSNO, in turn, can modify the activity of signaling molecules via protein S-nitrosylation (Stamler, 1994). For example, S-nitrosylation of the prokaryotic oxygen-sensitive transcription factor, *Oxy*R, leads to its activation via S-nitrosylation of critical cysteine residues (Hausladen *et al.*, 1996). S-nitrosylation of p21*ras* which is involved in mitogenesis and possibly NF-κB activation induces GTP/GDP exchange on p21*ras* (Lander *et al.*, 1996b). Thus, the generation of nitrosothiols by ·NO may be an important pathway for activating NF-κB in unstimulated endothelial cells. However, these effects of nitrosothiols are not consistent with the net effect of ·NO on cytokine-stimulated endothelial cells, an inhibitor of NF-κB activation.

ENHANCED NUCLEAR TRANSLOCATION OF NF-κB BY ·NO

Of particular interest is the ability of IκB-α to inhibit NF-κB-mediated signaling events in the nucleus. Recent studies have demonstrated that IκB-α can translocate into the nucleus and displace NF-κB subunits from their cognate DNA sequences (Zabel *et al.*, 1993; Read *et al.*, 1996). It is somewhat surprising though that IκB-α can enter the nucleus since the carboxy terminus of IκB-α contains multiple ankyrin repeat motifs which may inhibit its entry into the cell nucleus (Beg and Baldwin, 1993). Nevertheless, the nuclear translocation of IκB-α has been documented by several investigators and appears to constitute an effective mechanism which can rapidly terminate NF-κB-mediated transactivation of pro-inflammatory genes.

It is not known for certain, however, whether enhanced nuclear translocation of IκB-α results from an active IκB-α nuclear transporter or the result of passive diffusion of increased amounts of newly-synthesized IκB-α. For example, nuclear translocation of IκB-α occurs following stimulation with TNF-α which also induces IκB-α synthesis indirectly via activation of NF-κB (Chiao *et al.*, 1994; Sun *et al.*, 1993; Read *et al.*, 1994). Thus, the process of nuclear translocation of IκB-α may not be restricted to the actions of ·NO, but rather, occurs through a more general mechanism for terminating NF-κB transactivation.

INDUCTION OF THE NF-κB INHIBITOR, IκB-α, BY ·NO

A particularly novel effect of ·NO is its ability to induce the expression of IκB-α (Figure 7-5). Exposure to ·NO does not affect the expression of NF-κB subunits *Rel* A and p50, or cause the activation of NF-κB in human endothelial cells. Several recent studies, however, noted an increase in NF-κB activation by ·NO-generating compounds in peripheral blood mononuclear cells (Lander *et al.*, 1993a). Perhaps differences in the intracellular redox regulation of thiols in different cell types could account for the observed differences between endothelial cells and peripheral blood mononuclear cells. Furthermore, the actions of ·NO may be quite different depending on whether or not the cells have been activated by cytokines.

Transfection studies using the IκB-α promoter linked to the chloramphenicol acetyltransferase reporter gene suggests that the induction of IκB-α by ·NO occurs at the transcriptional level (Figure 7-6). Previous analyses of the human, porcine, and murine IκB-α promoters have revealed multiple functional κB sites necessary for transcriptional induction by NF-κB (Le Bail *et al.*, 1993; Ito *et al.*, 1994; de Martin *et al.*, 1993; Chiao *et al.*, 1994). All of these sites are located within −350 bp of the transcriptional start site and provide for an inducible autoregulatory pathway for terminating the activation of NF-κB. However, ·NO's effects on IκB-α gene transcription is probably not mediated by NF-κB since ·NO inhibits NF-κB and alone, does not induce VCAM-1 gene transcription. Further analyses of the upstream IκB-α promoter will be necessary to determine which DNA binding domain(s) constitute the ·NO-responsive *cis*-regulatory element(s) in the IκB-α promoter.

The induction of IκB-α gene transcription also occurs following treatment with corticosteroids in Jurkat or T-cells (Scheinman *et al.*, 1995; Auphan *et al.*, 1995). However, we have observed that the induction of IκB-α by glucocorticoids in vascular endothelial cells is modest at best, producing only about a 2-fold increase in IκB-α steady-state mRNA levels compared to greater than 10-fold with ·NO (unpublished observations). It is possible that ·NO and glucocorticoids may share similar or identical pathways for transactivating the IκB-α gene. It remains to be determined whether ·NO production in endothelial cells mediates the induction of IκB-α by glucocorticoids.

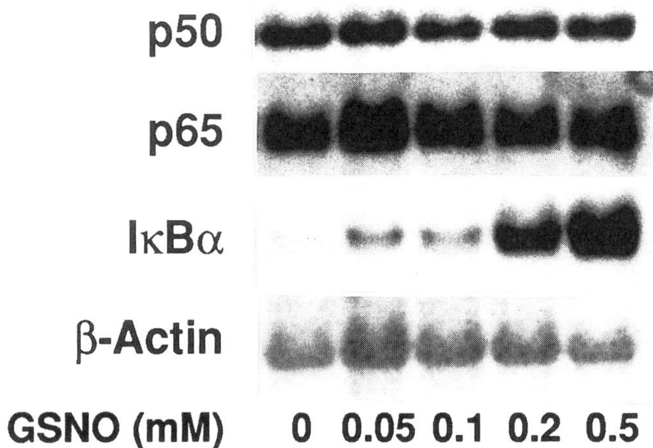

FIGURE 7-5. Northern analyses showing the concentration-dependent induction of IκB-α, but not p65 or p50, mRNA expression by GSNO. (Peng *et al.*, 1995b; reproduced with permission)

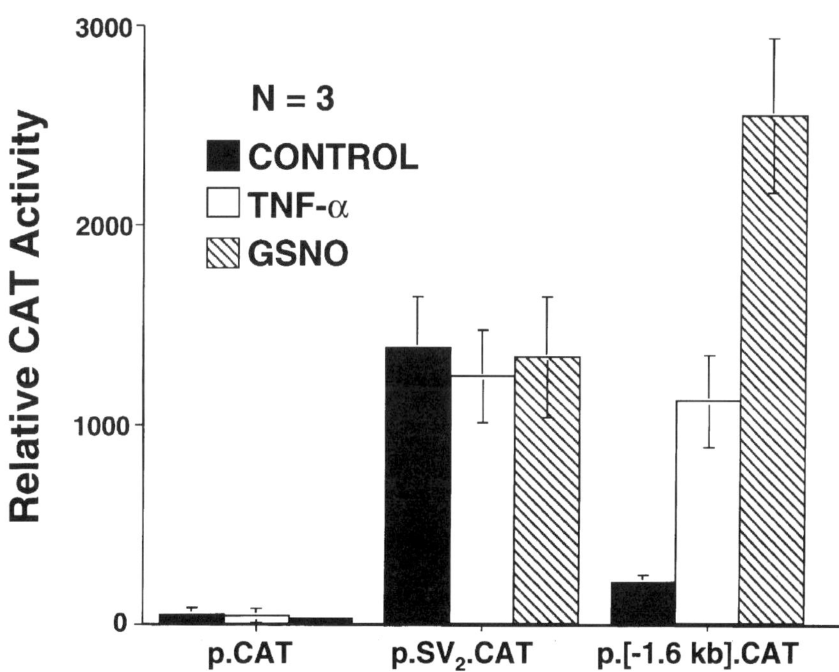

FIGURE 7-6. Induction of an IκB-α promoter activity following stimulation with TNF-α or GSNO (0.2 mM). Bovine aortic endothelial cells were transfected with a –1.6 kb murine IκB-α promoter construct linked to the chloramphenicol acetyltransferase (CAT) reporter gene, promoterless vector (p.CAT) and highly-constitutive SV40 promoter (SV₂.CAT) before stimulation with TNF-α or GSNO. (Peng *et al.*, 1995b; reproduced with permission)

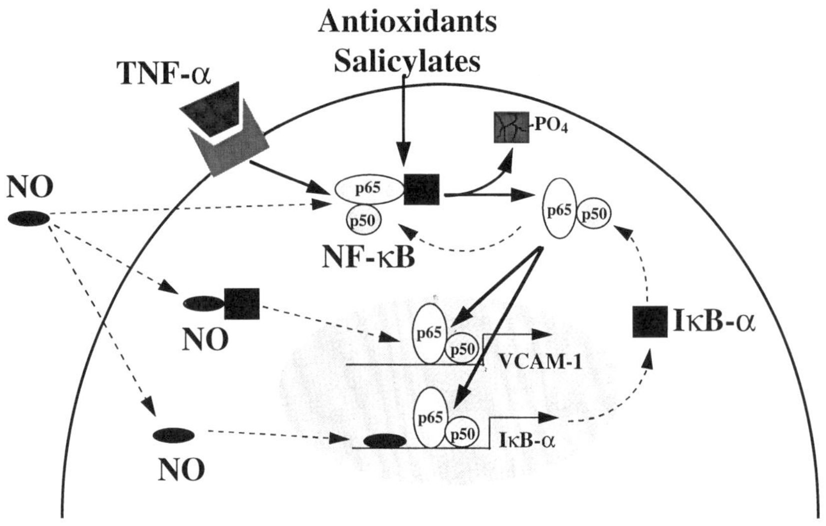

FIGURE 7-7 Schematic diagram showing the potential mechanisms by which ·NO inhibits NF-κB activity.

·NO AND VASCULAR SMOOTH MUSCLE ACTIVATION

Similar to endothelial cells, vascular smooth muscle cells (SMC) when stimulated by appropriate cytokines also express adhesion molecules and pro-inflammatory mediators (Libby and Hansson, 1991). Recent studies indicate that SMC require NF-κB to express certain pro-inflammatory mediators and for proliferation (Lawrence *et al.*, 1994; Bellas *et al.*, 1995). Indeed, activation of NF-κB in SMC by cytomeglovirus infection promotes SMC proliferation *in vitro* (Bellas *et al.*, 1995; Speir *et al.*, 1996). In contrast to cultured SMC which have serum-induced basal NF-κB activity, quiescent SMC *in vivo* do not possess constitutive NF-κB activity (Bourcier *et al.*, 1997). A particular difference between *in vivo* and cultured SMC is the presence of constitutive ·NO produced by the overlying endothelium *in vivo*. Thus, it is conceivable that endothelial-derived ·NO tonically inhibits basal NF-κB activity in subjacent SMC. The paracrine regulation of basal NF-κB activity in SMC by endothelial-derived ·NO would support the notion that factors such as oxidized low-density lipoproteins which cause endothelial dysfunction by inhibiting endothelial ·NO activity might promote SMC proliferation and expression of pro-inflammatory mediators or adhesion molecules.

Several studies support the role of endothelial-derived ·NO in inhibiting SMC proliferation. *In vivo* transfer of the Type III ·NO synthase gene into balloon-injured vessels decreases intimal smooth muscle proliferation in rat carotid arteries (von der Leyen *et al.*, 1995). The mechanism by which ·NO inhibits SMC proliferation may involve cGMP-mediated pathways since high concentrations of cGMP analogues can also inhibit SMC proliferation (Garg and Hassid, 1989; Nakaki *et al.*, 1990).

The expression of ICAM-1 and VCAM-1 in SMC, however, are not affected by two different cGMP analogues, 8-bromo-cGMP and dibutyryl-cGMP (Shin *et al.*, 1996). Indeed, several groups have shown that ·NO can exert non-cGMP-dependent effects on other cell types such as platelets (Brune and Lapetina, 1989), fibroblasts (Garg and Hassid,

1991), and macrophages (Lancaster and Hibbs, 1990). Interestingly, the inhibitory effects of ·NO on basal and stimulated NF-κB activation resemble those of antioxidants such as N-acetylcysteine and pyrrolidine dithiocarbamate (PDTC). Antioxidants have been shown to inhibit SMC proliferation, and at higher concentrations, appear to induce SMC apoptosis (Tsai *et al.*, 1996). Thus, endothelial ·NO production in the vessel wall may influence SMC not only in their transient vasomotor functions, but also perhaps in their more prolonged transcriptional responses to NF-κB activation.

An example of how ·NO can regulate NF-κB activity in SMC is illustrated by SMC response to the lymphokine, interferon-gamma (IFN-γ). In contrast to certain cytokines such as TNF-α and IL-1α, IFN-γ does not activate NF-κB but can potently induce the expression of VCAM-1 and major histocompatibility complex (MHC) Class II antigens in cultured SMC (Warner *et al.*, 1989; Silvennoinen *et al.*, 1993). The signaling pathway for IFN-γ-stimulated responses involves the protein tyrosine phosphorylation of signal transducers and activators of transcription (STATs) and gamma activating factor (GAF) (Silvennoinen *et al.*, 1993; Shuai *et al.*, 1992). Activation of GAF, in turn, induces the expression of another transcription factor, interferon regulatory factor (IRF)-1, which binds to the promoters of target genes containing the interferon-stimulated response element (ISRE) (Miyamoto *et al.*, 1988). The VCAM-1 promoter contains two tandem κB sites located in close promixity to an ISRE site (Neish *et al.*, 1992; Iademarco *et al.*, 1992). Recent studies have shown that IRF-1 synergizes with NF-κB in transactivating the VCAM-1 gene (Neish *et al.*, 1995). Thus, we find that ·NO attenuates IFN-γ-induced VCAM-1 expression SMC cultured under serum conditions by inhibiting basal NF-κB activity rather than IRF-1 (Shin *et al.*, 1996).

INDUCIBLE NITRIC OXIDE: PRO- OR ANTI-INFLAMMTORY MEDIATOR?

During inflammatory responses, vascular wall cells and macrophages can produce sustained high levels of ·NO via the induction of Type II inducible NO synthase (iNOS) (Lyons *et al.*, 1992; Xie *et al.*, 1992; Gross *et al.*, 1991; Buttery *et al.*, 1996). Indeed, the expression of iNOS is increased in lesions of atherosclerosis, experimental transplantation arteriosclerosis, and in balloon-injured vessels (Buttery *et al.*, 1996; Russell *et al.*, 1995; Douglas *et al.*, 1994; Joly *et al.*, 1992). Consequently, growing evidence implicates ·NO via interaction with superoxide anion to form a potent oxidant, peroxynitrite, as a pro-inflammatory mediator (Beckman *et al.*, 1990). Inhibition of inducible ·NO production by N^G-monomethyl-L-arginine (LNMA) reduces transplant rejection (Worrall *et al.*, 1995).

In contrast, recent *in vitro* studies suggest that ·NO may also function as an anti-inflammatory mediator. For example, the ·NO released by exogenous ·NO donors can attenuate vascular inflammation by decreasing cytokine-induced endothelial cell activation (Peng *et al.*, 1995b; Khan *et al.*, 1996) and by inhibiting endothelial-leukocyte interaction (Gauthier *et al.*, 1994; De Caterina *et al.*, 1995; Khan *et al.*, 1996). It is uncertain, however, whether the ·NO donors used in these studies were biologically equivalent to endogenous-derived inducible ·NO produced at sites of vascular inflammation *in vivo* since the ·NO released from ·NO donors may not be physiologic in terms of their high concentrations and relatively short duration of action (Kowaluk and Fung, 1990). Furthermore, controversies exist as to whether some of the ·NO donors used in these studies such as S-nitrosoglutathione (GSNO) and 3-morpholino sydnonomine (SIN-1) actually produce their effects by releasing ·NO (Kowaluk and Fung, 1990; Stamler *et al.*, 1992a; Singh *et al.*, 1996). In particular, the pro-drug SIN-1 generates superoxide anion as well as ·NO, and thus, can effectively also release peroxynitrite (Brunelli *et al.*, 1995). Hence, the relationship between endogenously-derived inducible ·NO and endothelial cell activation remains uncertain.

Most of the studies implicating inducible ·NO as a pro-inflammatory and cytotoxic mediator, however, are based upon association rather than causality (Russell *et al.*, 1995; Worrall *et al.*, 1995; Adamson *et al.*, 1996). Nevertheless, some of the reported inconsistencies of inducible ·NO's vascular effects may relate to where the ·NO is produced, the level of ·NO production, and the environmental milieu into which it is released. Such factors can determine the ultimate biological effects of ·NO by modulating its ability to function as an oxidant or reducing agent (Stamler *et al.*, 1992a). For example, inducible ·NO may function as a cytotoxic mediator through its ability to activate poly(ADP-ribose) synthetase (Zhang *et al.*, 1994). However, other neurotoxic effects of inducible ·NO may result from ·NO's interaction with superoxide anion to form a potent oxidant, peroxynitrite (Beckman *et al.*, 1990; Lipton *et al.*, 1993). Furthermore, the interaction of inducible ·NO with thiol groups can lead to modification of intracellular proteins (Stamler, 1994). Thus, inducible ·NO may mediate either pro- and anti-inflammatory processes depending on its local concentration and interactions with environmental factors such as thiols and reactive oxygen species.

SUMMARY AND CONCLUSIONS

The appreciation of the role of ·NO in various biological processes continues to expand as knowledge of the effects of ·NO increases. This discussion has focused on how this endogenous mediator can regulate NF-κB and endothelial cell activation through its ability to stabilize and induce the NF-κB inhibitor, IκB-α (Figure 7-7). Elucidation of the further molecular details of the mechanisms by which ·NO or its derivatives regulate IκB-α stability and gene transcription will require more work. As we gain familiarity with the myriad effects of this multipotent mediator, we may discover yet other, as yet unrecognized effects of ·NO on gene regulation.

What started as a mysterious and evanescent vasodilator in the pharmacology laboratory, and as a nefarious pollutant by atmospheric chemists, has now emerged as a novel regulator of gene regulation at a transcriptional level scarcely suspected a only a few years ago. No doubt, continued exploration of the roles for ·NO in vascular biology will furnish further surprises in the future.

REFERENCES

Abate, C., Patel, L., Rauscher, F.J. III and Curran, T. (1990) Redox regulation of Fos and Jun DNA-binding activity *in vitro. Science*, **249**, 1157–1160.

Adamson, D.C., Wildemann, B., Sasaki, M., Glass, J.D., McArthur, J.C., Christov, V.I.. Dawson, T.M. and Dawson, V.L. (1996) Immunologic NO synthase: elevation in severe AIDS dementia and induction by HIV-1 gp41. *Science*, **274**, 1917–1921.

Adler, V., Schaffer, A., Kim, J., Dolan, L. and Ronai, Z. (1995) UV irradiation and heat shock mediate JNK activation via alternate pathways. *J. Biol. Chem.*, **270**, 26071–26077.

Ahmad, M., Marui, N., Alexander, R.W. and Medford, R.M. (1995) Cell type-specific transactivation of the VCAM-1 promoter through an NF-κB enhancer motif. *J. Biol. Chem.*, **270**, 8976–8983.

Auphan, N., DiDonato, J.A., Rosette, C., Helmberg, A. and Karin, M. (1995) Immunosuppression by glucocorticoids: inhibition of NFκB activity through induction of IκB synthesis. *Science*, **270**, 286–290.

Bandyopadhyay, R.S., Phelan, M. and Faller, D.V. (1995) Hypoxia induces AP-1-regulated genes and AP-1 transcription factor binding in human endothelial and other cell types. *Biochim. Biophys. Acta.*, **1264**, 72–78.

Baeuerle, P.A. and Henkel, T. (1994) Function and activation of NF-κB in the immune system. *Ann. Rev. Immunol.*, **12**, 141–179.

Baeuerle, P.A. and Baltimore, D. (1996) NF-κB: ten years after. *Cell*, **87**, 13–20.

Ballard, D.W., Dixon, E.P., Peffer, N.J., Bogerd, H., Doerre, S., Stein, B. and Greene, W.C. (1992) The 65-kDa subunit of human NF-κB functions as a potent transcriptional activator and a target for v-Rel-mediated repression. *Proc. Natl. Acad. Sci. USA*, **89**, 1875–1879.

Bath, P.M.W., Hassal, D.G., Gladwin, A.-M., Palmer, R.M.J. and Martin, J.F. (1991) Nitric oxide and prostacyclin: divergence of inhibitory effects on monocyte chemotaxis and adhesion to endothelium in vitro. *Arterioscl Thromb.*, **11**, 254–260.

Beckman, J.S., Beckman, T.W., Chen, J., Marshall, P.A. and Freeman, B.A. (1990) Apparent hydroxyl radical production by peroxynitrite: implications for endothelial injury from nitric oxide and superoxide. *Proc. Natl. Acad. Sci. USA*, **87**, 1620–1624.

Beg, A.A. and Baldwin, A.S. (1993) The IκB proteins: multifunctional regulators of Rel/NF-κB transcription factors. *Genes Dev.*, **7**, 2064–2070.

Bellas, R.E., Lee, J.S. and Sonenshein, G.E. (1995) Expression of a constitutive NF-κB-like activity is essential for proliferation of cultured bovine vascular smooth muscle cells. *J. Clin. Invest.*, **96**, 2521–2527.

Berliner, J.A., Navab, M., Fogelman, A.M., Frank, J.S., Demer, L.L., Edwards, P.A., Watson, A.D. and Lusis, A.J. (1995) Atherosclerosis: basic mechanisms. Oxidation, inflammation and genetics. *Circulation*, **91**, 2488–2496.

Blackwell, T.S., Blackwell, T.R., Holden, E.P., Christman, B.W. and Christman, J.W. (1996) In vivo antioxidant treatment suppresses nuclear factor-kappa B activation and neutrophilic lung inflammation. *J. Immunol.*, **157**, 1630–1637.

Bourcier, T., Sukhova, G. and Libby, P. (1997) The nuclear factor κ-B signaling pathway participates in dyregulation of vascular smooth muscle cells in vitro and in human atherosclerosis. *J. Biol. Chem.*, **272**, 15817–15824.

Bredt, D.S., Hwang, P.H., Glatt, C., Lowenstein, C., Reed, R.R. and Snyder, S.H. (1991) Cloned and expressed nitric oxide synthase structurally resembles cytochrome P-450 reductase. *Nature*, **351**, 714–718.

Brown K., Gerstberger, S., Carlson, L., Franzoso, G. and Siebenlist, U. (1995) Control of IkBa proteolysis by site-specific signal-induced phosphorylation. *Science*, **267**, 1485–1488.

Brune, B. and Lapetina, E.G. (1989) Activation of a cytosolic ADP-ribosyltransferase by nitric oxide-generating agents. *J. Biol. Chem.*, **264**, 8455–8461.

Brunelli, L., Crow, J.P. and Beckman, J.S. (1995) The comparative toxicity of nitric oxide and peroxynitrite to Escherichia coli. *Arch. Biochem. Biophys.*, **316**, 327–334.

Buttery, L.D., Springall, D.R., Chester, A.H., Evans, T.J., Standfield, E.M., Parums, D.V., Yacoub, M.H. and Polak, J.M. (1996) Inducible nitric oxide synthase is present within human atherosclerotic lesions and promotes the formation and activity of peroxynitrite. *Lab. Invest.*, **75**, 77–85.

Chen, L.Y., Mehta, P. and Mehta, J.L. (1996) Oxidized LDL decreases L-arginine uptake and nitric oxide synthase protein expression in human platelets: relevance of the effect of oxidized LDL on platelet function. *Circulation*, **93**, 1740–1746.

Chen, Z.J., Parent, L. and Maniatis, T. (1996) Site-specific phosphorylation of IκBα by a novel ubiquitination-dependent protein kinase activity. *Cell*, **84**, 853–862

Chiao, P.J., Miyamoto, S. and Verma, I. (1994) Autoregulation of IκBα activity. *Proc. Natl. Acad. Sci. USA*, **91**, 28–32

Collins T. (1993) Endothelial nuclear factor-κB and the initiation of the atherosclerotic lesion. *Lab. Invest.*, **68**, 499–508.

Collins, T., Read, M.A., Neish, A.S., Whitley, M.Z., Thanos, D. and Maniatis, T. (1995) Transcriptional regulation of endothelial cell adehsion molecules: NFκB and cytokine-inducible enhancers. *FASEB J.*, **9**, 1–7.

Cooke, J.P., Singer, A.H., Tsao, P., Zera, P., Rowan, R.A. and Billingham, M.E. (1992) Antiatherogenic effects of L-arginine in the hypercholesterolemic rabbit. *J. Clin. Invest.*, **90**, 1168–1172.

Cybulsky, M.I. and Gimbrone, M.A. Jr. (1991) Endothelial expression of a mononuclear leukocyte adhesion molecule during atherogenesis. *Science*, **251**, 788–791.

Davenpeck, K.L., Gauthier, T.W., Albertine, K.H. and Lefer, A.M. (1994) Role of P-selectin in microvascular leukocyte-endothelial interaction in splanchnic ischemia-reperfusion. *Am. J. Physiol.*, **267**, H622–H630.

De Caterina, R., Libby, P., Peng, H.B., Thannickal, V.J., Rajavashisth, T.B., Gimbrone, M.A. Jr., Shin, W.S. and Liao, J.K. (1995) Nitric oxide decreases cytokine-induced endothelial activation. Nitric oxide selectively reduces endothelial expression of adhesion molecules and proinflammatory cytokines. *J. Clin. Invest.*, **96**, 60–68.

de Martin, R., Vanhove, B., Cheng, Q., Hofer, E., Csizmadia, V., Winkler, H. and Bach, F. (1993) Cytokine-inducible expression in endothelial cells of an IκBα-like gene is regulated by NF-κB. *EMBO J.*, **12**, 2773–2778.

Devary, Y., Rosette, C., DiDonato, J.A. and Karin, M. (1993) NF-κB activation by ultrviolet light not dependent on a nuclear signal. *Science*, **261**, 1442–1445.

Dinerman, J.L., Lowenstein, C.J. and Snyder, S.H. (1993) Molecular mechanisms of nitric oxide regulation. *Circ. Res.*, **73**, 217–222.

Douglas, S.A., Vickery-Clark, L.M. and Ohlstein, E.H. (1994) Functional evidence that balloon angioplasty results in transient nitric oxide synthase induction. *Eur. J. Pharmacol.*, **225**, 81–89.

Fialkow, L., Chan, C.K., Grinstein, S. and Downey, G.P. (1993) Regulation of tyrosine phosphorylation in neutrophil by the NADPH oxidase. Role of reactive oxygen intermediates. *J. Biol. Chem.*, **268**, 17131–17137.

Furchgott, R.F. and Zawadzki, J.V. (1980) The obligatory role of endothelial cells in the relaxation of arterial smooth muscle muscle by acetylcholine. *Nature*, **288**, 373–376.

Garg, U.C. and Hassid, A. (1989) Nitric oxide-generating vasodilators and 8–bromo-cyclic guanosine monophosphate inhibit mitogenesis and proliferation of cultured rat vascular smooth muscle cells. *J. Clin. Invest.*, **83**, 1774–1777.

Garg, U.C. and Hassid, A. 1991. Nitric oxide decreases cytosolic free calcium in Balb/c 3T3 fibroblasts by a cyclic GMP-independent mechanism. *J. Biol. Chem.*, **266**, 9–12.

Gauthier, T.W., Davenpeck, K.L. and Lefer, A.M. (1994) Nitric oxide attenuates leukocyte-endothelial interaction via P-selectin in splanchnic ischemia-reperfusion. *Am. J. Physiol.*, **267**, G562–G568.

Geller, D.A., Lowenstein, C.J., Shapiro, R.A., Nussler, A.K., Di Silvio, M., Wang, S.C. *et al.*, (1993) Molecular cloning and expression of inducible nitric oxide synthase from human hepatocytes. *Proc. Natl. Acad. Sci. USA*, **90**, 3491–3495.

Golledge, J., Turner, R.J., Harley, S.L., Springall, D.R. and Powell, J.T. (1997) Circumferential deformation and shear stress induce differential responses in saphenous vein endothelium exposed to arterial flow. *J. Clin. Invest.*, **99**, 2719–2726.

Gross, S.S., Jaffe, E.A., Levi, R. and Kilbourn, R.G. (1991) Cytokine-activated endothelial cells express an isotype of nitric oxide synthase which is tetrahydrobiopterin-dependent, calmodulin-independent and inhibited by arginine analogs with a rank-order of potency characteristic of activated macrophages. *Biochem. Biophys. Res. Commun.*, **178**, 823–829.

Haskill, S., Beg, A.A., Tompkins, S.M., Morris, J.S., Yurochko, A.D., Sampson-Johannes, A., Mondal, K., Ralph, P. and Baldwin, A.S. (1991) Characterization of an immediate-early gene induced in adherent monocytes that encodes IκB-like activity. *Cell*, **65**, 1281–1289.

Hausladen, A., Privalle, C.T., Keng, T., DeAngelo, J. and Stamler, J.S. (1996) Nitrosative stress: Activation of the transcription factor OxyR. *Cell*, **86**, 719–729.

Henkel, T., Machleidt, T., Alkalay, I., Krönke, M., Ben-Neriah, Y. and Baeuerle, P.A. (1993) Rapid proteolysis of IκB-α is necessary for activation of transcription factor NF-κB. *Nature*, **365**, 182–185.

Huang, P.L., Huang, Z., Mashimo, H., Bloch, K.D., Moskowitz, M.A., Bevan, J.A. and Fishman, M.C. (1995) Hypertension in mice lacking the gene for endothelial nitric oxide synthase. *Nature*, **377**, 239–242.

Huie, R.E. and Padmaja, S. (1993) Reaction of nitric oxide with superoxide anion. *Free Rad. Res. Commun.*, **18**, 195–200.

Iademarco, M.F., McQuillan, J.J., Rosen, G.D. and Dean, D.C. (1992) Characterization of the promoter for vascular cell adhesion molecule-1 (VCAM-1). *J. Biol. Chem.*, **267**, 16323–16329.

Ignarro, L.J., Buga, G.M., Wood, K.S., Byrns, R.E. and Chaudhuri, G. (1987) Endothelium-derived relaxing factor produced and released from artery and vein is nitric oxide. *Proc. Natl. Acad. Sci. USA*, **84**, 9265–9269.

Imbert, V., Rupec, R.A., Livolsi, A., Pahl, H.L., Traenckner, E.B.-M., Mueller-Dieckmann, C., Farahifar, D., Rossi, B., Auberger, P., Baeuerle, P.A. and Peyron, J.F. (1996) Tyrosine phorphorylation of IκB-α activates NF-κB without proteolytic degradation of IκB-α. *Cell*, **86**, 787–798.

Ito, C.Y., Kazantsev, A.G. and Baldwin, A.S. Jr. (1994) Three NF-kappa B sites in the I kappa B-alpha promoter are required for induction of gene expression by TNF alpha. *Nucleic Acids Res.*, **22**, 3787–3792.

Janssens, S.P., Shimouchi, A., Quertermous, T., Bloch, D.B. and Bloch, K.D. (1992) Cloning and expression of a cDNA encoding human endothelium-derived relaxing factor/nitric oxide synthase. *J. Biol. Chem.*, **267**, 14519–14522.

Joly, G.A., Schini, V.B. and Vanhoutte, P.M. (1992) Balloon injury and interleukin-1 beta induce nitric oxide synthase activity in rat carotid arteries. *Circ. Res.*, **71**, 331–338.

Khachigian, L.M., Lindner, V., Williams, A.J. and Collins, T. (1996) Egr-1-induced endothelial gene expression: a common theme in vascular injury. *Science*, **271**, 1427–1431.

Khan, B.V., Harrison, D.G., Olbrych, M.T., Alexander, R.W. and Medford, R.M. (1996) Nitric oxide regulates vascular cell adhesion molecule 1 gene expression and redox-sensitive transcriptional events in human vascular endothelial cells. *Proc. Natl. Acad. Sci. USA*, **93**, 9114–9119.

Kopp, E. and Ghosh, S. (1994) Inhibition of NF-κB by sodium salicylate and aspirin. *Science*, **265**, 956–958.

Kowaluk, E.A. and Fung, H.L. (1990) Spontaneous liberation of nitric oxide cannot account for in vitro vascular relaxation by S-nitrosothiols. *J. Pharmacol. Exp. Ther.*, **255**, 1256–1264.

Kubes, P., Suzuki, M. and Granger, D.N. (1991) Nitric oxide: an endogenous modulator of leukocyte adhesion. *Proc. Natl. Acad. Sci. USA*, **88**, 4651–4655.

Kurose, I., Kubes, P., Wolf, R., Anderson, D.C., Paulson, J., Miyasaka, M. and Granger, D.N. (1993) Inhibition of nitric oxide production. Mechanisms of vascular albumin leakage. *Circ. Res.*, **73**, 164–171.

Lamas, S., Marsden, P.A., Li, G.K., Tempst, P. and Michel, T. (1992) Endothelial nitric oxide synthase: molecular cloning and characterization of a distinct constitutive isoform. *Proc. Natl. Acad. Sci. USA*, **89**, 6348–6352.

Lancaster, L.R. and Hibbs, J.B. (1990) EPR demonstration of iron-nitrosyl complex formation by cytotoxic activated macrophages. *Proc. Natl. Acad. Sci. USA*, **87**, 1223–1227.

Lander, H.M., Jacovina, A.T., Davis, R.J. and Tauras, J.M. (1996a) Differential activation of mitogen-activated protein kinases by nitric oxide-related species. *J. Biol. Chem.*, **271**, 19705–19709.

Lander, H.M., Milbank, A.J., Tauras, J.M., Hajjar, D.P., Hempstead, B.L., Schwartz, G.D., Draemer, R.T., Mirza, V.A., Chait, B.T., Burk, S.C. and Quilliam, L.A. (1996b) Redox regulation of cell signalling. *Nature*, **381**, 380–381.

Lander, H.M., Sehajpal, P., Levine, D.M. and Novogrodsky, A. (1993a) Activation of human peripheral blood mononuclear cells by nitric oxide-generating compounds. *J. Immunol.*, **150**, 1509–1516.

Lander, H.M., Sehajpal, P.K. and Novogrodsky, A. (1993b) Nitric oxide signaling: a possible role for G proteins. *J. Immunol.*, **151**, 7182–7187.

Lawrence, R., Chang, L.-J., Siebenlist, U., Bressler, P. and Sonenshein, G.E. (1994) Vascular smooth muscle cells express a constitutive NF-κB-like activity. *J. Biol. Chem.*, **269**, 28913–28918.

Le Bail, O., Schmidt-Ullrich, R. and Israel, A. (1993) Promoter analysis of the gene encoding the IκBα/MAD3 inhibitor of NF-κB: positive regulation by members of the rel/NF-κB family. *EMBO J.*, **12**, 5043–5048.

Legrand-Poels, S., Bours, V., Piret, B., Pflaum, M., Epe, B., Rentier, B. and Piette, J. (1995) Transcription factor NF-κB is activated by photosensitization generating oxidative DNA damages. *J. Biol. Chem.*, **270**, 6925–6934.

Lemaitre, B., Nicolas, E., Michaut, L., Reichhart, J.M. and Hoffmann, J.A. (1996) The dorsoventral regulatory gene cassette spatzle/Toll/cactus controls the potent antifungal response in Drosophila adults. *Cell*, **86**, 973–983.

Libby, P. and Hansson, G.K. (1991) Involvement of the immune system in human atherogenesis: current knowledge and unanswered questions. *Lab Invest.*, **64**, 5–15.

Libby, P. and Clinton, S.K. (1993) The role of macrophages in atherogenesis, *Curr. Opin. Lipidol.*, **4**, 355–363.

Lipton, S.A., Choi, Y.B., Pan, Z.H., Lei, S.Z., Chen, H.S., Sucher, N.J., Loscalzo, J., Singel, D.J. and Stamler, J.S. (1993) A redox-based mechanism for the neuroprotective and neurodestructive effects of nitric oxide and related nitroso-compounds. *Nature*, **364**, 626–632.

Lowenstein, C.J., Glatt, C.S., Bredt, D.S. and Snyder, S.H. (1992) Cloned and expressed macrophage nitric oxide synthase contrasts with brain enzyme. *Proc. Natl. Acad. Sci. USA*, **89**, 6711–6715.

Lyons, C.R., Orloff, G.J. and Cunningham, J.M. (1992) Molecular cloning and functional expression of an inducible nitric oxide synthase from a murine macrophage cell line. *J. Biol. Chem.*, **267**, 6370–6374.

Mannick, J.B., Asano, K. Izumi, K., Kieff, E. and Stamler, J.S. (1994) Nitric oxide produced by human B lymphocytes inhibits apoptosis and Epstein-Barr virus reactivation. *Cell*, **79**, 1137–1146,

Marks, D.S., Vita, J.A., Folts, J.D., Keaney, J.F. Jr., Welch, G.N. and Loscalzo, J. (1995) Inhibition of neointimal proliferation in rabbits after vascular injury by a single treatment with a protein adduct of nitric oxide. *J. Clin. Invest.*, **96**, 2630–2638.

Marui, N., Offermann, M.K., Swerlick, R., Kunsch, C., Rosen, C.A., Ahmad, M., Alexander, R.W. and Medford, R.M. (1993) Vascular cell adhesion molecule-1 (VCAM-1) gene transcription and expression are regulated through an antioxidant-sensitive mechanism in human vascular endothelial cells. *J. Clin. Invest.*, **92**, 1866–1872.

Miyamoto, M., Fujita, T., Kimura, Y., Maruyama, M., Harada, H., Sudo, Y., Miyata, T. and Taniguchi, T. (1988) Regulated expression of agene encoding a nuclear factor, IRF-1, that specifically binds to IFN-β gene regulatory elements. *Cell*, **54**, 903–913.

Miyamoto, A., Laufs, R., Pardo, C. and Liao, J.K. (1997) Modulation of bradykinin receptor ligand binding affinity and its coupled G-proteins by nitric oxide. *J. Biol. Chem.*, **272**, 19601–19608.

Moroi, M., Gold, H.K., Yasuda, T., Fishman, M.C. and Huang, P.L. (1996) Mice mutant in endothelial nitric oxide synthase: vessel growth and response to injury. *Circulation*, **94**, I154 (abstract).

Munoz, C., Pascual-Salcedo, D., Castellanos, M.C., Alfranca, A., Aragones, J., Vara, A., Redondo, M.J. and de Landazuri, M.O. (1996) Pyrrolidine dithiocarbamate inhibits the production of interleukin-6, interleukin-8 and granulocyte-macrophage colony-stimulating factor by human endothelial cells in reponse to inflammatory mediators: modulation of NF-κB and AP-1 transcription factors activity. *Blood*, **88**, 3482–3490.

Munzel, T., Sayegh, H., Freeman, B.A., Tarpey, M.M. and Harrison, D.G. (1995) Evidence for enhance vascular superoxide anion production in nitrate tolerance. A novel mechanism underlying tolerance and cross-tolerance. *J. Clin. Invest.*, **95**, 187–194.

Myers, R.P., Minor, R.L., Jr., Fuerra, R., Jr., Bates, J.N. and Harrison, D.G. (1990) Vasorelaxant properties of the endothlium-derived relaxing factor more closely resemble S-nitrosocysteine than nitric oxide. *Nature*, **345**, 161–163.

Nadaud, S., Philippe, M., Arnal, J.F., Michel, J.B. and Soubrier, R. (1996) Sustained increase in aortic endothelial nitric oxide synthase expression *in vivo* in a model of chronic high blood flow. *Circ. Res.*, **79**, 857–863.

Nakaki, T., Nakayama, M. and Kato, R. (1990) Inhibition by nitric oxide and nitric oxide-producing vasodilators of DNA synthesis in vascular smooth muscle cells. *Eur. J. Pharmacol.*, **189**, 347–353.

Neish, A.S., Williams, A.J., Palmer, H.J., Whitley, M.Z. and Collins, T. (1992) Functional analysis of the human vascular cell adhesion molecule-1 promoter. *J. Exp. Med.*, **176**, 1583–1593.

Neish, A.S., Read, M.A., Thanos, D., Pine, R., Maniatis, T. and Collins, T. (1995) Endothelial interferon regulatory factor-1 cooperates with NF-κB as a transcriptional activator of vascular cell adhesion molecule-1. *Mol. Cell. Biol.*, **15**, 2558–2569.

Nose, K., Shibanuma, M., Kikuchi, K., Kageyama, H., Sakiyama, S. and Kuroki, T. (1991) Transcriptional activation of early-response genes by hydrogen peroxide in a mouse osteoblastic cell line. *Eur. J. Biochem.*, **201**, 99–106.

Ohba, M., Shibanuma, M., Kuroki, T. and Nose, K. (1994) Production of hydrogen peroxide by transforming growth factor-β1 and its involvement in induction of *egr-1* in mouse osteoblastic cells. *J. Cell Biol.*, **126**, 1079–1088.

Osnes, L.T., Foss, K.B., Joo, G.B., Okkenhaug, C., Westvik, A.B., Ovstebo, R., Kierulf, P., Jo, G.B. and Ovsteb, R. (1996) Acetylsalicylic acid and sodium salicylate inhibit LPS-induced NF-κB/c-Rel nuclear translocation and synthesis of tissue factor (TF) and tumor necrosis factor alfa (TNF-α) in human monocytes. *Thromb. Haemost.*, **76**, 970–976.

Palombella, V.J., Rando, O.J., Goldberg, A.L. and Maniatis, T. (1994) The ubiquitin-proteasome pathway is required for processing the NF-κB1 precursor protein and the activation of NF-κB. *Cell*, **78**, 773–785.

Palmer, R.M.J., Ferrige, A.G. and Moncada, S. (1987) Nitric oxide release accounts for the biological activity of endothelium-derived relaxing factor. *Nature*, **327**, 524–526.

Peng, H.B., Rajavashisth, T.B., Libby, P. and Liao, J.K. (1995a) Nitric oxide inhibits macrophage-colony stimulating factor gene transcription in vascular endothelial cells. *J. Biol. Chem.*, **270**, 17050–17055.

Peng, H.B., Libby, P. and Liao, J.K. (1995b) Induction and stabilization of IκB-α by nitric oxide mediate inhibition of NF-κB. *J. Biol. Chem.*, **270**, 14214–14219.

Perkins, N.D., Edwards, N.L., Duckett, C.S., Agranoff, A.B., Schmid, R.M. and Nabel, G.J. (1993) A cooperative interaction between NF-κB and Sp1 is required for HIV-1 enhancer activation. *EMBO J.*, **12**, 3551–3558.

Pierce, J.W., Read, M.A., Ding, H., Luscinskas, F.W. and Collins, T. (1996) Salicylates inhibit IκB-α phosphorylation, endothelial-leukocyte adhesion molecule expression and neutrophil transmigration. *J. Immunol.*, **156**, 3961–3969.

Pinkus, R., Weiner, L.M. and Daniel, V. (1996) Role of oxidants and antioxidants in the induction of AP-1, NF-κB and glutathione S-trnsferase gene expression. *J. Biol. Chem.*, **271**, 13422–13429.

Pober, J.S., Collins, T., Gimbrone, M.A., Jr., Cotran, R.S., Gitlin, J.D., Fiers, W., Clayberger, C., Krensky, A.M., Burakoff, S.J. and Reiss, C.S. (1983) Lymphocytes recognize human vascular endothelial and dermal fibroblast Ia antigens induced by recombinant immune interferon. *Nature*, **305**, 726–729.

Radi, R., Beckman, J.S., Bush, K.M. and Freeman, B.A. (1991) Peroxynitrite oxidation of sulfhydryls. *J. Biol. Chem.*, **266**, 4244–4250.

Radomski, M.W., Palmer, R.M. and Moncada, S. (1990) An L-arginine/nitric oxide pathway present in human platelets regulates aggregation. *Proc. Natl. Acad. Sci. USA*, **87**, 5193–5197.

Rajagopaian, S., Kurz, S., Münzel, T., Tarpey, M., Freeman, B.A., Griendling, K.K. and Harrison, D.G. (1996) Angiotensin II-mediated hypertension in the rat increases vascular superoxide production via membrane NADH/NADPH oxidase activation: contribution to alterations of vasomotor tone. *J. Clin. Invest.*, **97**, 1916–1923.

Read, M.A., Whitley, M.Z., Williams, A.J. and Collins, T. (1994) NF-κB and IκBα: an inducible regulatory system in endothelial activation. *J. Exp. Med.*, **179**, 503–512.

Read, M.A., Neish, A.S., Luscinskas, F.W., Palombella, V.J., Maniatis, T. and Collins, T. (1995) The proteasome pathway is required for cytokine-induced endothelial-leukocyte adhesion molecule expression. *Immunity*, **2**, 493–506.

Read, M.A., Neish, A.S., Gerritsen, M.E. and Collins, T. (1996) Postinduction transcriptional repression of E-selectin and vascular cell adhesion molecule-1. *J. Immunol.*, **157**, 3472–3479

Rice, N.R., MacKichan, M.L. and Israël, A. (1992) The precursor of NF-κB p50 has I kappa B-like functions. *Cell*, **71**, 243–253.

Ross, R. and Harker, L. (1976) Hyperlipidemia and atherosclerosis and chronic hyperlipidemia initiates and maintains lesions by endothelial cell desquamation and lipid accumulation. *Science*, **193**, 1094–1100.

Russell, M.E., Wallace, A.F., Wyner, L.R., Newell, J.B. and Karnovsky, M.F. (1995) Upregulation and modulation of inducible nitric oxide synthase in rat cardiac allografts with chronic rejection and transplant arteriosclerosis. *Circulation*, **92**, 457–464.

Scheinman, R.I., Beg, A.A. and Baldwin, A.S. (1993) NF-κB p100 (Lyt-10) is a component of H2TF1 and can function as an IκB-like molecule. *Mol. Cell. Biol.*, **13**, 6089–6101.

Scheinman R.I., Cogswell, P.C., Lofquist, A.K. and Baldwin Jr., A.S. (1995) Role of transcriptional activation of IκB-α in mediation of immunosuppression by glucocorticoids. *Science*, **270**, 283–286.

Schreck, R., Rieber, P. and Baeuerle, P. A. (1991) Reactive oxygen intermediates as apparently widely used messengers in the activation of the NF-(B transcription factor and HIV-1. *EMBO J.*, **10**, 2247–2251.

Schreck, R., Albermann, K. and Baeuerle, P.A. Nuclear Factor-κB: an oxidative stress-responsive transcription factor of eukaryotic cells. (1992) *Free Rad. Res. Commun.*, **17**, 221–237

Sen, R. and Baltimore, D. (1986a) Multiple nuclear factors interact with the immunoglobulin enhancer sequences. *Cell*, **46**, 705–716.

Sen, R. and Baltimore, D. (1986b) Inducibility of kappa immunoglobulin enhancer-binding protein NF-kB by a posttranslational mechanism. *Cell*, **47**, 921–928.

Shin, W.S., Hong, Y.H., Peng, H.B., De Caterina, R., Libby, P. and Liao, J.K. (1996) Nitric oxide attenuates vascular smooth muscle cell activation by interferon-g: the role of constitutive NF-κB activity. *J. Biol. Chem.*, **271**, 11317–11324.

Shuai, K., Schindler, C., Prezioso, V.R. and Darnell Jr., J.E. (1992) Activation of transcription by IFN-γ: Tyrosine phosphorylation of a 91–kD DNA binding protein. *Science*, **258**, 1808–1811.

Silvennoinen, O., Ihle, J.N., Schlessinger, J. and Levy, D.E. (1993) Interferon-induced nuclear signalling by Jak protein tyrosine kinases. *Nature*, **366**, 583–585.

Singh, R.J., Hogg, N., Joseph, J. and Kalyanaraman, B. (1996) Mechanism of nitric oxide release from S-nitrosothiols. *J. Biol. Chem.*, **271**, 18596–18603.

Singh, S. and Aggarwal, B.B. (1995) Protein-tyrosine phosphatase inhibitors block tumor necrosis factor-dependent activation of the nuclear transcription factor NF-kappa B. *J. Biol. Chem.*, **270**, 10631–10639.

Speir, E., Shibutani, T., Yu, Z.X., Ferrans, V. and Epstein, S.E. (1996) Role of reactive oxygen intermediates in cytomegalovirus gene expression and in the response of human smooth muscle cells to viral infection. *Circ. Res.*, **79**, 1143–1152.

Spink, J., Cohen, J. and Evans, T.J. (1995) The cytokine responsive vascular smooth muscle cell enhancer of inducible nitric oxide synthase. Activation by nuclear factor-kappa B. *J. Biol. Chem.*, **270**, 29541–29547.

Stamler, J.S., Singel, D.J. and Loscalzo, J. (1992a) Biochemistry of nitric oxide and its redox-activated forms. *Science*, **258**, 1898–1902.

Stamler, J.S. (1994) Redox signaling: nitrosylation and related target interactions of nitric oxide. *Cell*, **78**, 931–936.

Stamler, J.S., Jaraki, O., Osborne, J., Simon, D.I., Keaney, J., Vita, J., Singel, D., Valeri, C.R. and Loscalzo, J. (1992b) Nitric oxide circulates in mammalian plasma primarily as an S-nitroso adduct of serum albumin. *Proc. Natl. Acad. Sci. USA*, **89**, 7674–7677.

Stamler, J.S., Simon, D.I., Osborne, J.A., Mullins, M.E., Jaraki, O., Michel, T., Singel, D.J. and Loscalzo, J. (1992c) S-nitrosylation of proteins with nitric oxide: synthesis and characterization of biologically active compounds. *Proc. Natl. Acad. Sci. USA*, **89**, 444–448.

Sun, S.C., Ganchi, P.A., Ballard, D.W. and Greene, W.C. (1993) NF-κB controls expression of inhibitor IκB-α-Evidence for an inducible autoregulatory pathway. *Science*, **259**, 1912–1915.

Tanner, F.C., Noll, G., Boulanger, C.M. and Lüscher, T.F. (1991) Oxidized low density lipoproteins inhibit relaxations of porcine coronary arteries: Role of scavenger receptor and endothelium-derived nitric oxide. *Circulation*, **83**, 2012–2020.

Thanos, D. and Maniatis, T. (1995) NF-κB: a lesson in family values. *Cell*, **80**, 529–532.

Tsai, J.C., Jain, M., Hsieh, C.M., Lee, W.S., Yoshizumi, M., Patterson, C., Perrella, M.A., Cooke, C., Wang, H., Haber, E., Schlegel, R. and Lee, M.E. (1996) Induction of apoptosis by pyrrolidinedithiocarbamate and N-acetylcysteine in vascular smooth muscle cells. *J. Biol. Chem.*, **271**, 3667–3670.

Tsao, P.S., McEvoy, L.M., Drexler, H., Butcher, E.C. and Cooke, J.P. (1994) Enhanced endothelial adhesiveness in hypercholesterolemia is attenuated by L-arginine. *Circulation*, **89**, 2176–2182.

Tsao, P.S., Buitrago, R., Chan, J.R. and Cooke, J.P. (1996) Fluid flow inhibits endothelial adhesiveness. Nitric oxide and transcriptional regulation of VCAM-1. *Circulation*, **94**, 1682–1689.

Uematsu, M., Ohara, Y., Navas, J.P., Nishida, K., Murphy, T.J., Alexander, R.W., Nerem, R.M. and Harrison, D.G. (1995) Regulation of endothelial cell nitric oxide synthase mRNA expression by shear stress. *Am. J. Physiol.*, **269**, C1371–C1378.

Visvanathan, K.V. and Goodbourn, S. (1989) Double-stranded RNA activates binding of NF-kappa B to an inducible element in the human beta-interferon promoter. *EMBO J.*, **8**, 1129–1138.

von der Leyen, H.E., Gibbons, G.H., Morishita, R., Lewis, N.P., Zhang, L., Nakajima, M., Kaneda, Y., Cooke, J.P. and Dzau, V.J. (1995) Gene therapy inhibiting neointimal vascular lesion: *in vivo* transfer of endothelial cell nitric oxide synthase gene. *Proc. Natl. Acad. Sci. USA*, **92**, 1137–1141.

Warner, S.J.C., Friedman, G.B. and Libby, P. (1989) Immune interferon inhibits proliferation and induces 2'-5'-oligoadenylate synthetase gene expression in human vascular smooth muscle cells. *J. Clin. Invest.*, **83**, 1174–1182.

Weber, C., Negrescu, E., Erl, W., Pietsch, A., Frankenberger, M., Ziegler-Heitbrock, H.W.L., Siess, W. and Weber, P.C. (1995) Inhibitors of protein tyrosine kinase suppress TNF-stimulated induction of endothelial cell adhesion molecules. *J. Immunol.*, **155**, 445–451.

Wirth, T. and Baltimore, D. (1988) Nuclear factor NF-κB can interact functionally with its cognate binding site to provide lymphoid-specific promoter function. *EMBO J.*, **7**, 3109–3113.

Wolf, A., Zalpour, C., Theilmeier, G., Wang, B.Y., Ma, A., Anderson, B., Tsao, P.S. and Cooke, J.P. (1997) Dietary L-arginine supplementation normalizes platelet aggregation in hypercholesterolemic humans. *J. Am. Coll. Cardiol.*, **29**, 479–485.

Worrall, N.K., Lazenby, W.D., Misko, T.P., Lin, T.S., Rodi, C.P., Manning, P.T., Tilton, R.G., Williamson, J.R. and Ferguson, T.B. Jr. (1995) Modulation of *in vivo* alloreactivity by inhibition of inducible nitric oxide synthase. *J. Exp. Med.*, **181**, 63–70.

Xie, Q.W., Cho, H.J., Calaycay, J., Mumford, R.A., Swiderek, K.M., Lee, T.D., Ding, A., Troso, T. and Nathans, C. (1992) Cloning and characterization of inducible nitric oxide synthase from mouse macrophages. *Science*, **256**, 225–228.

Xie, Q.W., Kashiwabara, Y. and Nathan, C. (1994) Role of transcription factor NF-κB/Rel in induction of nitric oxide synthase. *J. Biol. Chem.*, **269**, 4705–4708.

Zabel, U., Henkel, T., Silva, M.S. and Baeuerle, P.A. (1993) Nuclear uptake control of NF-κB by MAD-3, an IκB protein present in the nucleus. *EMBO J.*, **10**, 4159–4167.

Zhang, J., Dawson, V.L., Dawson, T.M. and Snyder, S.H. (1994) Nitric oxide activation of poly(ADP-ribose) synthetase in neurotoxicity. *Science*, **263**, 687–689.

Zhong, H., SuYang, H., Erdjument-Bromage, H., Tempst, P. and Ghosh, S. (1997) The transcriptional activity of NF-κB is regulated by the IκB-associated PKAc subunit through a cyclic AMP-independent mechanism. *Cell*, **89**, 413–424.

8 Nitric Oxide and the Regulation of Vasoactive Genes

Douglas V. Faller

Cancer Research Center, Professor of Medicine, Biochemistry, Pediatrics, Microbiology, Pathology and Laboratory Medicine, Boston University School of Medicine, Rm K-701, 80 E. Concord St., Boston, MA 02118, USA
Tel: 617-638-4173; Fax: 617-638-4176; E-mail: dfaller@bu.edu

Changes in the environmental oxygen tension to which cells are exposed *in vivo* result in physiological and sometimes pathological consequences which are associated with differential expression of specific genes. Low oxygen tension (hypoxia) affects cellular physiology *in vivo* and *in vitro* in a number of ways, including the synthesis and release of, or inhibition of, vasoactive substances involved in modulating the vascular smooth muscle tone in the body. The concept that genes encoding vasoactive factors could themselves be regulated by the vasoactive agent nitric oxide (NO) stemmed from this initial work characterizing the control of these genes by oxygen tension. Nitric oxide and oxygen transduce similar signals; *i.e.*, their absence results in identical patterns of gene expression in endothelial cells and other cell types. The result is transcriptional induction of genes encoding vasoconstrictors, mitogens, and matrix molecules like *PDGF-B*, *endothelin-1*, *VEGF*, *thrombospondin-1*, and *collagenase IV* (*MMP-9*), and inhibition of endothelial nitric oxide synthase (*eNOS*). The implications of this finding is that nitric oxide can feed-back on, and modulate, a signal induced by hypoxia, and *vice versa*. For example, nitric oxide, which can act directly on smooth muscle cells as a vasodilator, can also facilitate vasodilation indirectly by reversing the production of vasoconstrictors induced by hypoxia. Both oxygen and nitric oxide appear to signal through a novel heme-containing sensor. An understanding of the consequences of the reciprocal interactions between nitric oxide and hypoxia may explain both the acute and chronic pathophysiological sequelae of diseases characterized by regional hypoxia, including atherosclerosis, pulmonary hypertension, sickle cell disease and systemic sclerosis (scleroderma).

Key words: endothelial cell, vasoconstriction, platelet-derived growth factor, endothelin-1, hypoxia, oxygen.

INTRODUCTION

The concept that genes encoding vasoactive factors could themselves be regulated by the vasoactive agent nitric oxide (NO) stemmed from the initial work describing the control of these genes by oxygen tension. Changes in the environmental oxygen tension to which cells are exposed *in vivo* result in physiological and sometimes pathological consequences, which are associated with differential expression of genes (Gerritsen and Bloor, 1993; Halliwell and Gutteridge, 1990). Both hyperoxia and hypoxia affect cellular physiology in various ways (Smith, 1994; Meyrick and Reid, 1980; Furchgott and Zawadzki, 1980; Holden and McCall, 1984; Nelin *et al.*, 1994), including the synthesis and release of vasoactive substances involved in modulating the vascular smooth muscle tone in the body (Rubanyi and Vanhoutte, 1985; Vender *et al.*, 1987; Rakugi *et al.*, 1990; Shirakami *et al.*, 1991).

LOCAL REGULATION OF VASCULAR TONE

The overall tone of the vasculature is the result of a dynamic and shifting balance between stimuli provoking contraction and relaxation (see (Faller, 1992, 1994, 1997) for review). Local control of vascular tone is mediated by paracrine, and sometimes autocrine, release of vasoconstrictor or vasodilator substances. The vascular endothelium is the major source of paracrine effects on the contractile elements of the blood vessel (Vane *et al.*, 1990). The endothelial cell layer forms the permeability barrier between circulating blood cells and the underlying vascular tissue, which is composed of fibroblasts and smooth muscle cells. As such, it is in a unique position to respond to circulating factors or blood elements and modulate events in the vasculature *via* paracrine effects. Both vasoconstricting and vasodilating factors are released from the endothelium. Endothelin-1 is the most powerful vasoconstrictive agent released by endothelial cells (Yanagisawa *et al.*, 1988), and works in a paracrine fashion. This 21-amino acid peptide has regions of homology to a group of neurotoxins and is distinct from other known mammalian bioactive peptides. Endothelin-1 represents one member of a family of structurally-related peptides (Endothelin-1, -2, and -3). Human endothelin-1 (ET-1) is derived from a 212-amino acid precursor, preproET-1, via a 38-amino acid intermediate. Endothelin-1 is rapidly cleared from the circulation, but has a long local effect on vascular tone. Physiological responses to endothelin are reported for blood vessels derived from a large variety of tissues including the heart, kidney, and brain (MacCumber *et al.*, 1989). *In vivo*, endothelin-1 elicits a dose-dependent increase in lobar pulmonary artery pressure, and intralobar bolus injections of endothelin-1 evoke pulmonary vasoconstriction (Minkes *et al.*, 1990). In addition, endothelin-1 decreases the glomerular filtration rate by increasing vascular resistance and mesangial contraction. Factors shown to induce endothelin production by endothelial cells include thrombin, transforming growth factor (TGF$_\beta$), shear stress, and hypoxia (Kourembanas *et al.*, 1993; Kourembanas and Faller, 1989; Kourembanas *et al.*, 1990, 1991). Endothelin-1 is thought to act locally, but elevated circulating levels of this peptide were reported in patients undergoing hemodialysis or renal transplant, in patients with myocardial infarction, and more recently in the plasma of patients with pulmonary hypertension (Stewart *et al.*, 1991). In the latter study, nearly all patients diagnosed as having primary pulmonary hypertension had elevated endothelin-1 levels in the arterial compared with venous plasma, suggesting local pulmonary production of endothelin-1. Platelet-derived growth factor (PDGF), a mitogenic peptide produced by endothelial cells (Hannan *et al.*, 1988) known to cause proliferation of vascular smooth muscle cells, is also a powerful paracrine vasoconstrictor for these cells (Berk *et al.*, 1986), more potent than angiotensin II. The specific isoforms of the PDGF dimer cause different spectra of effects on vascular smooth muscle (Sachinidis *et al.*, 1990; Majack *et al.*, 1990). The opposing actions of acetylcholine- and bradykinin-induced vasorelaxation are also dependent on an intact endothelium (Furchgott and Zawadzki, 1980). The relaxing mediator released by endothelial cells, initially termed Endothelial-derived Relaxing Factor(s), was recently identified as nitric oxide (NO) (Ignarro *et al.*, 1987; Palmer *et al.*, 1987). This potent inorganic vasodilator is synthesized by endothelial cells from arginine, and acts on the neighboring smooth muscle cell through guanylate cyclase to promote relaxation. In addition, nitric oxide is a potent inhibitor of platelet aggregation and adhesion, and an inhibitor of smooth muscle and fibroblast mitogenesis (Garg and Hassid, 1989, 1990; Nakaki *et al.*, 1990).

THE VASCULAR RESPONSE TO HYPOXIA

The vascular response to hypoxia has multiple components which contribute to both the initial responses and the longer term sequelae. The acute responses include the induction of reactive vasomotor changes and local margination and extravasation of inflammatory cells. In both the systemic and pulmonary circulations changes in pO_2 induce a change in vessel caliber. The opposing effects of hypoxia on systemic and pulmonary blood flow, however, and the relevance of hypoxic vasodilation/constriction to tissue oxygenation and homeostasis have been recognized for five decades (Von Euler and Lilestrand, 1946). A decrease in pO_2 causes vasodilation in coronary, skeletal muscle, cerebral and gastrointestinal circulations. In certain vascular beds, however, including the pulmonary, renal and umbilical vein circulation, hypoxia results in vasoconstriction. In the pulmonary circulation in particular, airway hypoxia or pulmonary arterial hypoxemia cause vasoconstriction in a process known as hypoxic pulmonary vasoconstriction. Systemic hypoxic vasodilation provides a feedback mechanism of increasing blood flow to tissues with increased metabolic demands. In the pulmonary circulation, in contrast, acute regional hypoxia leads to hypoxic pulmonary vasoconstriction, providing a rapid mechanism for optimizing tissue oxygenation, matching perfusion to ventilation, and therefore improving gas exchange. Chronic hypoxia, however, can lead to remodeling of the vasculature, including excess smooth muscle formation as well as peripheral extension of muscle into the distal arterioles. This can be a maladaptive response, causing significant injury.

Hypoxia in the pulmonary vasculature has been strongly implicated in the pulmonary hypertension resulting from a number of different insults, including high altitude exposure, cystic fibrosis, chronic bronchitis, scleroderma (systemic sclerosis) and alveolar hypoventilation (Vender *et al.*, 1987). The resulting increase in pulmonary vascular resistance is marked not only by vasoconstriction, but also by structural remodeling. Medial hypertrophy of the muscular pulmonary arteries is observed, along with neomuscularization of intra-acinar vessels and a rapid eight-fold increase in proliferation of adventitial fibroblasts and pulmonary interstitial connective tissue cells (Lockhart and Saiag, 1981; Orton *et al.*, 1988; Rabinovitch *et al.*, 1983; Stenmark *et al.*, 1987).

The endothelium is known to be necessary for the vasoconstriction of pulmonary and umbilical vessels induced by hypoxia. Hypoxia mediates its effects by regulating the release of vasoactive substances from endothelial cells (Holden and McCall, 1984; Kourembanas *et al.*, 1990, 1991). Hypoxia increases the secretion and the gene expression of the potent vasoconstrictors and growth factors *endothelin-1* and *PDGF-B* by cultured human endothelial cells (Hannan *et al.*, 1988; Kourembanas *et al.*, 1993; Kourembanas and Faller, 1989; Kourembanas *et al.*, 1990, 1991). The transcription and release of endothelin-1 and PDGF-B are therefore a function of ambient oxygen tension. The behavior of vascular smooth muscle cells in response to changes in oxygen tension may thus be regulated in large part by their interaction with adjacent endothelial cells, which in turn can detect changes in environmental oxygen through a biochemical sensor and respond by regulating the transcription of genes encoding paracrine vasoactive and growth factors.

REGULATION OF VASOACTIVE GENES BY HYPOXIA

A number of studies had previously suggested that a variety of cells in the body could respond to hypoxia by releasing vasoactive factors, including endothelin-1 (Yoshimoto *et al.*, 1991; Hieda and Gomez-Sanchez, 1990; Vanhoutte *et al.*, 1989; Rakugi *et al.*, 1990) and platelet-derived growth factor (Joseph *et al.*, 1990). Further investigation disclosed that this oxygen-regulated production of vasoactive substances occurred at the level of endothelial cell gene regulation (Kourembanas *et al.*, 1990, 1991). Around the same time, hypoxia was shown to induce expression of specific genes in certain other cell types, including renal and hepatoma cell lines (Goldberg *et al.*, 1988; Lacombe *et al.*, 1988; Goldberg *et al.*, 1987).

Hypoxia Induces Transcription and Production of PDGF-B by Endothelial Cells

Because hypoxic states are associated with abnormal proliferation and constriction of the smooth muscle cells and fibroblasts surrounding the distal vessels of the lung (Rubler and Fleischer, 1967; Oppenheimer and Esterly, 1971), we initially hypothesized that in hypoxic as well as in normal states, the endothelial cell layer may play a key role in controlling smooth muscle tone by secreting a number of vasoactive agents. PDGF-BB is both a major growth factor for vascular smooth muscle cells and a powerful vasoconstrictor. We found that hypoxic conditions (0–3% oxygen environments) significantly increased *PDGF-B* mRNA in cultured human umbilical vein endothelial cells by enhancing the transcriptional rate of this gene (Kourembanas *et al.*, 1990). This increase was inversely proportional to oxygen tension and was reversible upon re-exposure of cells to a 21% oxygen atmosphere. *PDGF-A* mRNA levels were not affected nor was the overall rate of cellular gene transcription increased in response to hypoxia. These studies were the first to demonstrate that endothelial cells are not only capable of sensing oxygen tension, but are also able to discriminate and respond to even small differences in oxygen tension, resulting in dramatic upregulation of the *PDGF-B* chain gene. Other growth factors made by endothelial cells can also influence cellular PDGF-B production. Basic fibroblast growth factor (bFGF), a growth factor for endothelial cells in culture, significantly decreases the amount of PDGF-like protein secreted by these cells (Kourembanas and Faller, 1989). The VEGF receptors Flt-1 and Flk-1 (KDR/Flk), which can also serve as PDGF receptors, may be upregulated on endothelial cells and other cell types by hypoxia (Tuder *et al.*, 1995; Waltenberger *et al.*, 1996; Plate *et al.*, 1993), making those cells exposed to low oxygen tensions even more sensitive to autocrine or paracrine PDGF-BB production.

Hypoxia Induces Transcription and Production of Endothelin-1 by Endothelial Cells

Endothelin-1 is a potent vasoconstrictor released by endothelial cells that can also function as a paracrine regulator of vascular tone. Physiologic low oxygen tension ($pO_2 = 30$ torr) increases endothelin-1 secretion from cultured human endothelial cells 4- to 8-fold above the secretion rate at ambient oxygen tension (Kourembanas *et al.*, 1991). This increase in secretion is accompanied by a corresponding increase in the transcriptional rate of the *preproET-1* gene resulting in increased steady-state mRNA levels of *preproET-1*. In contrast, the transcription of a number of other growth-factor-encoding genes, including *TGF-β* and basic *FGF* is unaffected by hypoxia. Interestingly, increases in bFGF release from endothelial cells induced by hypoxia have been reported (Michiels *et al.*, 1994), despite the lack of changes in transcript levels, suggesting release from a intracellular stored pool. *Endothelin-1* transcript production increases within 1 hr of hypoxia and persists for at least 48 hr. In addition, the stimulatory effects of low oxygen tension on *endothelin-1* mRNA levels are reversible upon re-exposure to 21% oxygen environments. These findings suggest a role for endothelin-1 in the control of regional blood flow in the vasculature in response to changes in oxygen tension. Since the concentration of endothelin-1 is likely to be much higher at the interface of endothelium and smooth muscle than in the larger volume of distribution of the blood stream, the local production of endothelin-1 in the microvasculature may actually be significantly higher than the values obtained from plasma. Based on these *in vitro* observations and predictions, other investigators have subsequently demonstrated that transcript and protein levels of PDGF, endothelin-1 and endothelin-1 receptor subtypes are indeed increased in rat or human lung under acute and chronic conditions of hypoxia *in vivo*, and may play a role in the acute and chronic pulmonary responses to hypoxia induced experimentally (Li *et al.*, 1994a, 1994b; Elton *et al.*, 1992; Katayose *et al.*, 1993), or secondary to pulmonary fibrosis (Martinet *et al.*, 1987) or systemic sclerosis (Gay *et al.*, 1989; Ferris *et al.*, 1990). Similarly, circulating endothelin-1 levels are significantly elevated by approximately 2-fold in patients with chronic hypoxia due to pulmonary disease (Ferri *et al.*, 1995), systemic sclerosis (Yamane *et al.*, 1992; Yamane, 1994; Zachariae *et al.*, 1994; Kupper, 1995), and pulmonary hypertension (Giaid *et al.*, 1993).

Regulation of Endothelial Nitric Oxide Synthase and Nitric Oxide Production by Oxygen Tension

Endothelial cell-generated nitric oxide (NO) accounts in large part for the labile vasodilator termed endothelium-derived relaxing factor (Furchgott and Zawadzki, 1980; Palmer *et al.*, 1987; Ignarro *et al.*, 1987). Synthesis of the potent vasodilator substance nitric oxide by the nitric oxide synthases requires one of the two chemically equivalent guanidino nitrogens of L-arginine and molecular oxygen. Two distinct types of nitric oxide synthase have been characterized and molecularly cloned (see Nathan and Xie, 1994, for review). A constitutive type (cNOS), dependent on Ca^{++}-calmodulin, is found in neural tissue, platelets and endothelial cells. An inducible, "Ca^{++}-independent" enzyme (iNOS) has been found in many tissue types (Nathan, 1992). Both enzymes are flavoproteins, are dependent on NADPH as a cofactor, and tetrahydrobiopterin enhances enzyme activity. The product, NO, activates soluble guanylate cyclase, resulting in the generation of cyclic guanosine monophosphate (cGMP) and relaxation of vascular smooth muscle (Ignarro *et al.*, 1984; Fiscus, 1988). In endothelial cells, NOS enzymatic activity is constitutively-expressed, but activation of the calcium-calmodulin pathway is required for maximal activity. Thus, control and regulation of the cNOS enzyme at the cellular level has been proposed to be chiefly through activation of specific cell surface receptors by calcium-mobilizing agonists.

A number of studies have suggested a relationship between oxygen tension and nitric oxide production. Endothelium-dependent vasodilation mediated by NOS is inhibited by low oxygen tension, with resulting reductions in levels of cGMP produced (De Mey and Vanhoutte, 1983; Furchgott and Zawadzki, 1980; Graser and Vanhoutte, 1991; Johns *et al.*, 1989; Peach *et al.*, 1989; Crawley *et al.*, 1992). Furthermore, endothelium-dependent relaxation to agonists such as acetylcholine and thrombin are abolished reversibly by anoxic conditions (De Mey and Vanhoutte, 1983). We recently reported that exposure of human or bovine endothelial cells to low (but physiological) oxygen tensions ($pO_2 = 20$–40 torr) results in a profound decrease in the transcript for *cNOS* and a corresponding fall in cNOS protein levels (Phelan and Faller, 1996). The ability of endothelial cells exposed to hypoxia to produce nitric oxide in response to bradykinin, a stimulator of cNOS activity, was coordinately impaired. In the adult, oxygen tensions in tissues have been measured at 40 torr, with the levels being substantially lower ($pO_2 = 15$–30) under conditions of hypoxemia. The level of suppression of the *cNOS* transcript is dependent on the oxygen tension, and cobalt inhibited the expression of *cNOS* transcripts, suggesting a mechanism comparable to that by which oxygen tension regulates expression of other vasoregulatory genes. In the presence of Actinomycin D, hypoxia had no effect on *cNOS* transcripts, suggesting that new gene transcription is required for cNOS suppression by hypoxia. The reducing agents PDTC and N-Ac did not mimic *cNOS* gene suppression by hypoxia, suggesting that this suppression is not related to the redox state of the intracellular environment. Thus, regulation of cNOS function in response to environmental factors can occur at the level of gene expression as well as at the level of enzyme activation.

The decreases in endothelial cNOS protein observed in response to chronic hypoxia may contribute to the inability of the vasculature to respond normally after exposure to prolonged or repeated hypoxic environments *in vitro* (Adnot *et al.*, 1991; Carville *et al.*, 1993; Yang and Mehta, 1994; Dinh-Xuan *et al.*, 1991; Crawley *et al.*, 1992). Reduced expression of endothelial nitric oxide synthase has been reported in the lungs of patients with pulmonary hypertension of various etiologies (Giaid and Saleh, 1995). Prolonged hypoxia *in vivo* reversibly inhibits the ability of vessels to respond to activators of cNOS, such as acetylcholine or ionophore, suggesting a reversible loss of enzyme activity (Shaul *et al.*, 1993). The decreases in cNOS enzyme levels we observe thus may contribute both to the decreases in nitric oxide production found during prolonged hypoxia and also to the reversible, but acetylcholine-refractory, constricted state of hypoxic vasculature.

REGULATION OF INDUCIBLE NITRIC OXIDE SYNTHASE BY OXYGEN TENSION

An inducible, calcium-independent form (iNOS) has been identified in a number of cell types including vascular smooth muscle cells and perhaps endothelial cells (see Nathan and Xie, 1994). Nitric oxide is synthesized by vascular smooth muscle cells in response to endotoxin or inflammatory mediators such as tumor necrosis factor-α (TNFα) and interleukin-1β (IL-1β) (see (Faller *et al.*, 1996)). We investigated the molecular basis for the induction of nitric oxide synthase (NOS) in response to lipopolysaccharide (LPS) or IL-1β using rat vascular smooth muscle cells derived from pulmonary and systemic vasculature (Faller *et al.*, 1996). The regulation of mRNA levels for this enzyme in response to LPS or IL-1β treatment was examined in parallel with changes in levels of cyclic GMP. There is an increase in expression of inducible NOS transcript corresponding to an increase in cyclic GMP levels beginning with 3 hr of exposure to either LPS or IL-1β. In cells derived from the pulmonary vasculature, initial induction of NOS transcript was detectable at 3 hr and the transcript levels continued to increase to a maximal intensity at 24 hr. In contrast, the cells derived from the systemic vasculature showed a maximal induction of NOS transcript at 3 hr, and the intensity decreased from this time point to 24 hr. The full induction of NOS transcripts was dependent on new protein synthesis and on cellular tyrosine protein kinase activity. Thus, the production of nitric oxide in vascular smooth muscle cells in response to either LPS or IL-1β results from increased expression of the gene for (inducible) NOS. Low oxygen tension has been reported to regulate iNOS gene or protein expression by several-fold (Archer *et al.*, 1995; Shaul *et al.*, 1995; Pohl and Busse, 1989; Xue *et al.*, 1994; Le Cras *et al.*, 1996; Hwang *et al.*, 1994).

Autoregulation of NOS Activity

In addition to regulation of NOS at the level of gene expression, nitric oxide appears also capable of reacting with the NOS enzymes themselves, potentially establishing a feedback loop between the product of the enzyme and the enzyme. Because the NOS enzymes are heme-containing proteins (Klatt *et al.*, 1992), the possibility that NO, like the NOS substrate O_2, could interact with the NOS protein active site was predictable. Both inducible and endothelial NOS activities are inhibited by NO and NO donors (Griscavage *et al.*, 1993; Assreuy *et al.*, 1993; Ravichandran *et al.*, 1995), possibly via interaction with allosteric thiols (Patel *et al.*, 1996). Because endothelial NOS is membrane-associated, and because the cell membrane may be storage site for NO (Kiechle and Maliski, 1993), the effective concentrations of NOS seen by the enzyme may be in a range sufficient for autoregulation.

Nitric Oxide Modifies the Induction of Endothelin-1 and PDGF by Hypoxia

An interesting reciprocal feedback regulation between nitric oxide production and the regulated expression of the vasoconstrictors PDGF-B and endothelin-1 has been predicted and confirmed. Based on the observed responses of these genes to oxygen tension, we predicted that these genes would have the same response to nitric oxide as they do to oxygen (Kourembanas *et al.*, 1993). Whereas low oxygen tension ($pO_2 = 30$–20 Torr) increases *endothelin-1* gene and protein expression 4- to 8-fold above that seen at normal oxygen tension ($pO_2 = 150$ Torr), sodium nitroprusside (SNP), which releases nitric oxide, suppresses this effect (Kourembanas *et al.*, 1993). This inhibition of hypoxia-induced *endothelin-1* expression occurs within the first hour of exposure of cells to SNP. Moreover, when the endogenous constitutive levels of nitric oxide made by endothelial cells are suppressed using N-ω-nitro-L-arginine (L-NNA), a potent competitive inhibitor of nitric oxide synthase, the baseline levels of endothelin-1 produced in normoxic environments are increased 3- to 4-fold. The effects of hypoxia and L-NNA on endothelin-1 expression are additive. The regulation of *endothelin-1* production by nitric oxide is predominantly at the level of transcription. Similar effects of nitric oxide were observed on the expression of the PDGF-B chain gene. *PDGF-B* gene expression was suppressed by nitric oxide in a hypoxic environment and induced by L-NNA in both normoxic and hypoxic environments. These findings suggest that in addition to its role as a vasodilator, nitric oxide may also influence vascular tone *via* the regulated reciprocal production of endothelin-1 and PDGF-B in the vasculature.

Vascular Endothelial Growth Factor (VEGF) Regulation by Oxygen Tension and Nitric Oxide

Vascular endothelial growth factor is a potent mitogen for vascular endothelial cells isolated from both small and large vessels. VEGF is a dimeric glycoprotein whose structure resembles that of two other angiogenic growth factors, platelet-derived growth factor and placenta growth factor. It is a soluble molecule and mediates endothelial cell growth and neovascularization (Leung *et al.*, 1989; Ferrara *et al.*, 1992). VEGF is a selective mitogen for endothelial cells, and *VEGF* gene expression can be induced by hypoxia in a number of cell types, including endothelial cells and vascular smooth muscle cells (Namiki *et al.*, 1995, Stavri *et al.*, 1995a), and returns to background levels during reoxygenation. Its strong activity as an angiogenic factor makes it a major contributor to the tumor angiogenic activity elaborated by non-vascularized, growing tumors as the oxygen-tension in the area of the tumor falls (Shweiki *et al.*, 1995, 1992). VEGF is upregulated in response to cobalt, nickel and manganese, is inhibited by carbon monoxide, and requires protein synthesis for hypoxia-responsive induction (Goldberg and Schneider, 1994). PDGF-BB, itself a hypoxia-inducible factor, has been reported to act synergistically with hypoxia in the induction of VEGF (Stavri *et al.*, 1995a). Treatment with phorbol esters induces VEGF production, but hypoxia-induced expression of VEGF appears independent of functional AP-1 activity (Finkenzeller *et al.*, 1995), but dependent upon c-Src and/or Raf-1 activity (Mukhopadhyay *et al.*, 1995; Grugel *et al.*, 1995). Although hypoxic regulation of VEGF production likely has a transcriptional component (Levy *et al.*, 1995), others have reported that the predominant mode of regulation by oxygen tension is at the level of *VEGF* mRNA stability (Shima *et al.*, 1995b; Stein *et al.*, 1995; White *et al.*, 1995). Studies of *cis*-acting elements by transient transfection have given contradictory results. Sequences at both the 5' and 3' ends of the human gene have been reported to be responsive to oxygen tension, and the 3' region contains an HIF-1 binding site (Minchenko *et al.*, 1994), whereas in the rat or human genes other groups have reported a conserved HIF-1 site further 5' which is necessary for induction by hypoxia (Levy *et al.*, 1995; Liu *et al.*, 1995). The tumor suppressor pVHL (von Hippel-Lindau protein) has recently been reported to negatively regulate the expression of VEGF (and other hypoxia-inducible genes like PDGF-B) exclusively at the level of mRNA stability (Vairo *et al.*, 1996; Gnarra *et al.*, 1996).

The regulation of VEGF expression by hypoxia has a number of striking parallels with endothelin-1 regulation. The receptors for both endothelin-1 (designated ET_A-Receptor) and VEGF (designated Flt-1 and Flk-1 [KDR/flk]) are upregulated in vascular tissue in parallel with their respective ligands by low oxygen tension (Takagi *et al.*, 1996; Tuder *et al.*, 1995). Both genes can be modulated by nitric oxide and carbon monoxide, and by heavy metals (Tuder *et al.*, 1995; Kourembanas *et al.*, 1993). Just as NO, like oxygen, suppressed endothelin-1 gene expression, sodium nitroprusside, a NO donor, decreased, and L-NAME (N-nitro-L-arginine methyl ester, an inhibitor of NO synthesis), increased both VEGF and VEGF receptor transcripts in normoxic and hypoxic lung tissue (Tuder *et al.*, 1995).

Thrombospondin-1 Regulation by Oxygen Tension and Nitric Oxide

One striking feature of chronic *in vivo* hypoxic exposure is medial hypertrophy and gross adventitial thickening of pulmonary vessels (Lockhart and Saiag, 1981; Orton *et al.*, 1988; Rabinovitch *et al.*, 1983; Stenmark *et al.*, 1987). If hypoxia persists, these structural changes may become irreversible. Many of the growth factors and matrix proteins which contribute to vascular remodeling under hypoxia are endothelial-derived. Cell proliferation and matrix synthesis reflects a balance between pro-proliferative and anti-proliferative stimuli regulated by oxygen tension.

We and others have demonstrated that hypoxia induces a variety of genes in endothelial cells whose products are mitogenic for endothelial cells or smooth muscle cells, including *PDGF-B* (Kourembanas *et al.*, 1990; Michiels *et al.*, 1994; Katayose *et al.*, 1993; Kourembanas *et al.*, 1991; Zamora *et al.*, 1996, 1993; Rakugi *et al.*, 1990), *insulin-like growth factor (IGF)* (Perkett *et al.*, 1992), and *vascular endothelial derived growth factor (VEGF)* (Shweiki *et al.*, 1995, 1992). Hypoxia also induces endothelial cell release of a number of other mitogenic factors, including bFGF and certain prostanoids, which stimulate smooth muscle cell growth (Michiels *et al.*, 1994). In addition, hypoxia also decreases endothelial cell expression of genes encoding a number of anti-mitogenic or antiproliferative agents, including *nitric oxide synthase* (Kourembanas *et al.*, 1993; Phelan and Faller, 1996; Nakaki *et al.*, 1990). Hypoxia reduces endothelial cell release of prostacyclin (Assender *et al.*, 1992) and heparan sulfates (Benitz *et al.*, 1990; Humphries *et al.*, 1986), which inhibit proliferation. Thus, pro-proliferative mechanisms are upregulated under hypoxia, while a number of inhibitory influences on mitogenesis are diminished in a low pO_2 environment, and the balance of effects under hypoxia are thus clearly shifted to promote proliferation and matrix synthesis.

In addition to increasing release of growth factors, hypoxia stimulates the production extracellular matrix proteins which themselves may promote mitogenesis. Increased endothelial production of the matrix proteins laminin and fibronectin have been observed under hypoxia (Lynch *et al.*, 1988). Thrombospondin-1 is a large 450 kD matrix-associated glycoprotein protein, first described in platelets (Baenziger *et al.*, 1972, 1971). This protein is reported to exist in a wide variety of cells including vascular smooth muscles, endothelial cells, fibroblasts, and epithelial cells (Frazier, 1991; Mosher, 1990; Mumby *et al.*, 1984). Thrombospondin functions in platelet aggregation (Dixit *et al.*, 1985), cell adhesion, cell proliferation and migration (O'Shea and Dixit, 1988) and may inhibit angiogenesis (Good *et al.*, 1990). Thrombospondin-1 appears to stimulate smooth muscle proliferation (Majack *et al.*, 1988). Low oxygen tension environment induces *thrombospondin-1* transcripts and protein production in human umbilical endothelial cells (Phelan *et al.*, 1998). In early passage human umbilical endothelial cells, a pO_2 of 20–30 torr induces thrombospondin-1 transcripts levels by up to 6-fold. Like the *VEGF* gene, the regulation of *thrombospondin-1* gene expression by low oxygen tension appears to be predominantly post-transcriptional. Actinomycin D chase experiments demonstrated that the increase in steady state *thrombospondin-1* transcript level during exposure to hypoxia is to a large degree secondary to increases in *thrombospondin-1* mRNA stability. It is noteworthy that PDGF-B, the product of a hypoxia-responsive gene, itself induces thrombospondin mRNA in rat vascular smooth muscle cells (Majack *et al.*, 1987). The induction of thrombospondin by PDGF is rapid and parallels PDGF-mediated mitogenesis. The regulation of *thrombospondin-1* by PDGF is at the transcriptional level and is "super-induced" by cycloheximide, similar to other early response genes such as c-*myc*, c-*fos* (Majack *et al.*, 1987). There is similar super-induction of *thrombospondin-1* mRNA in a low oxygen tension environment in the presence of cycloheximide. Inhibition of production of a protein with presumptive RNase activity by cyclohexamide has also been demonstrated in the regulation of *VEGF* (Shima *et al.*, 1995a), c-*jun* (Ryseck *et al.*, 1988), and *IL-2* genes (Zubiaga *et al.*, 1991), among others, and implies a negative basal regulation of these genes *via* intrinsic, but regulatable, mRNA instability. We have demonstrated induction of the *PDGF-B* gene in endothelial cells by hypoxia and regulation of PDGF-B production by nitric oxide (Kourembanas *et al.*, 1993, 1990), however, *thrombospondin-1* induction by hypoxia is unlikely to be mediated by autocrine production of PDGF-B for a number of reasons. The induction of PDGF-B in endothelial cells by hypoxia is delayed, relative to the induction of *thrombospondin-1* transcripts. Furthermore, we have been unable to detect PDGF receptors in human umbilical vein endothelial cells using very sensitive autokinase assays (DVF and L. Mundschau, unpublished). Finally, treatment of human umbilical vein endothelial cells with recombinant PDGF-B or hypoxia-conditioned medium does not result in induction of immediate early gene expression (DVF and L. Mundschau, unpublished).

The effect of low oxygen tension on *thrombospondin-1* gene induction is reversible and can be reproduced by cobalt chloride and other metals capable of substituting for iron in a heme ring. *PDGF-B*, *ET-1*, *cNOS* and *VEGF* genes have all been subsequently shown to respond to cobalt chloride which mimics the effect of low oxygen tension (Kourembanas *et al.*, 1993; Phelan and Faller, 1996; Phelan *et al.*, 1997). The induction of *thrombospondin-1* transcript by cobalt chloride is consistent with these previous findings, implicating a heme-containing oxygen sensor in the regulation of *thrombospondin-1* gene expression. Carbon monoxide and nitric oxide (in the form of soluble nitric oxide or nitroprusside) inhibit endothelial cell induction of *PDGF-B*, *endothelin-1* and *thrombospondin-1*, further supporting a role for a heme-containing oxygen sensor in the regulated expression of these genes (Kourembanas *et al.*, 1993; Phelan *et al.*, 1997). These studies also showed a role for endogenous nitric oxide in the regulation of ET-1. Addition of the competitive inhibitor L-NAME at $pO_2 = 155$ induced *endothelin-1* and *thrombospondin-1* mRNA, implying that endogenous nitric oxide was partly responsible for the regulation of basal levels of these genes and that ambient oxygen levels are not sufficient to fully suppress gene expression (Faller, 1994). It is noteworthy, however, that pO_2 levels of up to 600 torr do not further suppress hypoxia-responsive gene expression below that seen at $pO_2 = 155$. Moreover, the induction of *thrombospondin-1* by the nitric oxide synthase inhibitor L-NAME and hypoxia were additive, demonstrating that even under acutely hypoxic conditions, enough nitric oxide was being generated by endothelial cells in an autocrine fashion to interact with an oxygen (or nitric oxide) sensor.

It is intriguing to speculate on possible physiological ramifications of thrombospondin-1 induction by hypoxic environments, in addition to those involved in adaptive vascular remodeling. For example, low oxygen tension is a feature common to solid tumors, and thrombospondin-1 expression has previously been linked to metastasis and tumor invasion. Similarly, hypoxic areas within tumors have been demonstrated to confer a higher metastatic potential. Thrombospondin was recently demonstrated to modulate human breast adenocarcinoma cell adhesion to human endothelial cells in culture (Incardona *et al.*, 1995). Thrombospondin has also been shown to promote squamous carcinoma cell and breast carcinoma cell invasion of collagen in a dose-dependent manner (Wang *et al.*, 1995, 1996). Thrombospondin may play an additional role in the metastatic process by acting as a chemotactic and haptotactic molecule, in addition to promoting tumor invasion by adhesive mechanisms (Yabkowitz *et al.*, 1993). Furthermore, thrombospondin has been demonstrated to upregulate urokinase type activator and plasminogen activator inhibitor-1 (Hosokawa *et al.*, 1993) as well as binding to and prolonging the activity of urokinase (Harpel *et al.*, 1999). Both of these gene products are believed to regulate extracellular matrix degradation and the invasive process of metastasis (Moscatelli and Rifkin, 1994).

A second possible pathological correlation with hypoxia and thrombospondin-1 production might be in association with sickle cell disease. Adhesion of sickled erythrocytes to endothelium is mediated in large part by endothelial thrombospondin-1, binding to CD36 or sulfated proteoglycans on the erythrocyte (Sugihara *et al.*, 1992; Brittain *et al.*, 1993; Hillery *et al.*, 1996). Hypoxia not only facilitates erythrocyte sickling, but induces endothelial cell binding of a number of blood cell types (Ginis *et al.*, 1993a, 1995, 1993b), in addition to promoting vaso-constriction (Phelan and Faller, 1995, 1996; Faller, 1994). Whether the induction of thrombospondin-1 production during hypoxic episodes in sickle cell disease might further facilitate sickle cell-endothelial cell adhesions, thus reinforcing vaso-occlusive events and contributing to the vicious cycle of sickling, sludging, vasoconstriction, and further local hypoxia, is under study.

Regulation of other Growth Factors by Oxygen Tension and Nitric Oxide

Fibroblasts exposed to hypoxia showed enhanced proliferation in response to serum and growth factors like EGF (Wing *et al.*, 1988; Storch and Talley, 1988). The EGF-receptor is also upregulated by hypoxia, at both the mRNA and the protein level (Laderoute *et al.*, 1992a). Basic fibroblast growth factor (bFGF), an angiogenic factor (Hannan *et al.*, 1988), is regulated differentially by oxygen tension depending on the cell type. Hypoxia has no effect on bFGF transcript levels (Kourembanas *et al.*, 1990, 1991) and may inhibit bFGF release from endothelial cells, but induces bFGF production and release from macrophages (Kuwabara *et al.*, 1995). Acidic FGF (aFGF) and PDGF were upregulated in parallel with bFGF in macrophages, and this induction could be mimicked by cobalt and nickel. Hypoxia modulates the synthesis of TGF-α and -β. TGF-α exerts mitogenic activities on endothelial cells, whereas TGF-β has been reported to act as an inhibitor of endothelial mitogenesis *in vitro*, but acts as a angiogenic factor *in vivo* (Bicknell and Harris, 1991). Hypoxia has been found to induce TGF-β in and TGF-β in cultured cells as a function of oxygen concentration (Grugel *et al.*, 1995; Laderoute *et al.*, 1992b). The placenta growth factor (PLGF), an angiogenic factor with significant sequence homology to VEGF, is upregulated by low oxygen tension, as well as by treatment with cobalt or iron chelators (Gleadle *et al.*, 1995a). Interleukin-6 (IL-6, an anti-inflammatory cytokine which promotes vasorelaxation and smooth muscle cell proliferation (Ohkawa *et al.*, 1994), is induced in endothelial cells at the transcript and protein levels, *in vitro* and *in vitro*, by hypoxia (Yan *et al.*, 1995). A nuclear factor IL-6 binding site in the proximal promoter was sufficient for hypoxic induction, and the binding of CCAAT-enhancer-binding protein β (C/CEBP-β), which is a member of the NF-IL-6 family of DNA-binding proteins, showed increased binding to the NF-IL-6 binding site during exposure to hypoxia.

Many of the vasoactive factors have been demonstrated to regulate the production of other factors in either a positive or negative direction, and further compound the effects induced by hypoxia directly. For example, bFGF has been shown to regulate PDGF-BB transcript levels in endothelial cells (Kourembanas and Faller, 1989). Basic fibroblast growth factor (bFGF), a growth factor for endothelial cells in culture, significantly decreases the amount of PDGF-like protein secreted by these cells (Kourembanas and Faller, 1989). Levels of *PDGF-B* mRNA increase more than ten-fold in the absence of bFGF. This effect is specific for the *B* chain gene of *PDGF*, as transcript levels of other growth factors synthesized by endothelial cells, such as transforming growth factor beta (TGF$_\beta$) are not affected by bFGF. *PDGF-B* mRNA levels increase 12 hrs after bFGF removal, as the cells begin to accumulate in G_o stage of the cell cycle, and remain elevated for at least 96 hours. Transcript levels fall again upon re-exposure of the cells to bFGF.

bFGF may regulate *PDGF-B* mRNA levels and PDGF-B protein production by endothelial cells either directly, or indirectly *via* cell cycle arrest at G_0. This effect is not due to nonspecific cessation of cell growth because arresting the cells at S phase or in metaphase does not result in increased *PDGF-B* transcript levels. Platelet-derived growth factor, in turn, via activation of protein kinase-C, induces the transcription of the VEGF gene (Finkenzeller *et al.*, 1992). Treatment of vascular smooth muscle cells with mild hypoxia (2.5% O_2) has been shown to increase the expression of VEGF slightly, whereas the combination of mild hypoxia and either bFGF or TGF-β showed a marked synergistic effect on VEGF production (Stavri *et al.*, 1995a, 1995b). Thus, vascular endothelial proliferation could be induced both by the direct action of these factors and by and indirect synergistic process.

SENSATION AND TRANSDUCTION OF THE HYPOXIC SIGNAL

The Oxygen/Nitric Oxide Sensor

The mechanism(s) involving the initial generation of the hypoxic signal have been speculated upon at length. It was initially proposed that changes in energy production might mediate the response. Early studies in the pulmonary circulation suggested that inhibitors of energy metabolism could be involved in the mechanism of hypoxic pulmonary vasoconstriction. However, several lines of evidence argue against this hypothesis (Stevens and Rodman, 1995). Endothelium appears to rely in large part on glycolysis for energy production (Dobrina and Rossi, 1983). In one study, cellular ATP, ADP and AMP levels remained constant over 2.5 hr exposure to 0.1 mm Hg pO_2 in cultured endothelial cell (Mertens *et al.*, 1990). Similarly, using real-time NMR on living cells, we have observed no change in high energy phosphate bonds in human endothelial cells growing on microspheres as the oxygen tension was dropped from 150 torr to 20 torr. Furthermore, treatment of human or bovine endothelial cells in culture with sodium azide or potassium cyanide does not reproduce the gene regulation induced by hypoxia.

A specific cellular oxygen sensor has thus been hypothesized. In many systems heme proteins participate in O_2 sensing. The O_2 sensing and signal transduction system of the soil bacterium Rhizobium meliloti during nitrogen fixation (the Fix system) has been well characterized. A lowering of environmental oxygen tension is sensed by a membrane-associated heme-containing protein, Fix L, which induces both autophosphorylation and phosphotransfer activities. This protein then autophosphorylates and transfers a phosphate to a second soluble protein, Fix J (Monson *et al.*, 1995; Da Re *et al.*, 1994; Monson *et al.*, 1992). Studies with carbon monoxide in the intact circulation suggest that heme proteins may play a role in vascular O_2 sensing (Marshall *et al.*, 1988; Sylvester and McGowan, 1978). One model of oxygen sensing has been based on the role of oxygen as an electron acceptor in a variety of redox systems involving electron transport. The heme-containing cytochromes P480 or P558 systems, or flavoprotein oxidoreductases, have been suggested as potential sensors in some tissues (Acker *et al.*, 1989; Sylvester and McGowan, 1978; Gleadle *et al.*, 1995b). Other models of heme oxygenases are also potential candidates, including the membrane-associated NADPH oxidases. Under normoxic conditions, these proteins could bind free O_2 and convert it into H_2O_2, which would then be converted to hydroxyl radicals (OH·) and hydroxide (OH^-), through the iron-dependent Fenton reaction. Such reactive oxygen intermediates could then act as chemical messengers which suppress the expression of genes induced by hypoxia or otherwise could regulate the redox state of a variety of important proteins (Fandrey *et al.*, 1994; Acker *et al.*, 1989; Wegner *et al.*, 1996; Acker, 1994).

Alternatively, a sensor consisting of a heme-binding protein could bind O_2-like molecules and attain a "relaxed" form in the bound state or a "tense" configuration in the unbound state. We predicted that if the oxygen sensor in endothelial cells were a heme-containing protein, nitric oxide could evoke potentially similar intracellular events to those induced by oxygen, leading to the transcriptional regulation of the *endothelin-1* and *PDGF-B* genes. Indeed, any molecule bound by heme proteins in this same manner would be predicted to lead to similar regulatory events. We first tested this hypothesis by exposing cells to carbon monoxide, a known high-affinity ligand for heme (Kourembanas *et al.*, 1993). In the presence of carbon monoxide, both the *endothelin-1* and *PDGF-B* genes continued to be expressed at only low levels, as if in a normoxic environment, despite the absence of oxygen. In addition, we have found that inhibitors of heme biosynthesis blunt the hypoxic response in endothelial cells. Furthermore, heavy metals which can substitute for iron in the heme structure (including cobalt and nickel), but which are incapable of binding oxygen, will render the sensing molecule unresponsive to oxygen tension (Phelan *et al.*, 1995; Phelan and Faller, 1996; Phelan *et al.*, 1998).

Since nitric oxide and carbon monoxide were found to suppress the hypoxic increases in vasoconstrictor production and since they both activate guanylate cyclase, we tested the hypothesis that elevated cGMP levels may also down-regulate *endothelin-1* and *PDGF-B* gene expression (Kourembanas *et al.*, 1993). In the presence of methylene blue, an inhibitor of guanylate cyclase, the basal and hypoxic transcript levels of *endothelin-1* and *PDGF-B* were slightly elevated. In contrast, 8-bromo-cGMP addition had no effect on *endothelin-1* and *PDGF-B* gene expression providing more evidence for an alternative mechanism of hypoxic signal transduction (e.g., *via* a heme-containing sensor, rather than through a nitric oxide-cGMP coupled pathway). Whether or not the nitric oxide signal and the oxygen signal are transduced intracellularly *via* common pathways, both nitric oxide and oxygen do have similar vasodilating effects on the pulmonary vascular tone. Nitric oxide thus exerts its vasodilating effect in at least two ways — *via* previously described direct effects on the baseline tone of vascular smooth muscle, and also indirectly by inhibiting endothelial production of the potent vasoconstrictors, endothelin-1 and PDGF-B (Kourembanas *et al.*, 1993).

Second Messenger Systems

The mechanisms whereby low oxygen tension regulates gene expression in endothelial cells is especially biologically relevant. There is much evidence that a true, physiological signal transduction pathway is used by hypoxia in endothelial cells; i.e., that changes in oxygen tension indeed generate a physiological signal: (1) physiological levels of hypoxia (not anoxia), which do not affect cell growth, induce the observed responses; (2) the signal is dose-dependent and reversible; (3) endothelial cells are extremely resistant to hypoxia, showing no loss of high energy phosphate donors; (4) general metabolic inhibitors such as azide, cyanide, or 2-deoxyglucose, and reducing agents or oxidizing agents do not reproduce the effect; (4) the signal is inhibitable by certain serine-threonine protein kinase inhibitors, by iron chelators, by oxygen, by carbon monoxide, and by nitric oxide, but not by PKA inhibitors, etc; (5) the conditions used are not sufficiently stressful to induce "stress genes or proteins" (Kourembanas and Faller, 1991; Kourembanas *et al.*, 1993; Kourembanas and Faller, 1989; Kourembanas *et al.*, 1990, 1991; Phelan and Faller, 1996). It is generally accepted now that regulation of endothelin-1, PDGF-B, VEGF, and erythropoietin by oxygen tension are physiological and regulated responses, mediated by second messengers, protein kinases, and transcription factors (Stevens and Rodman, 1995; Mukhopadhyay *et al.*, 1995). More recently, it has been shown that known second messengers are regulated by physiologic (non-toxic) changes in oxygen tension. Finally, transcription factors such as AP-1 are (reversibly) activated in endothelial cells as a function of the oxygen tension (Bandyopadhyay *et al.*, 1995).

The pathway leading from the O_2 sensor to activation of the transcription factors described below and ultimately to gene regulation is unknown. To begin to explore what signaling mediators may be involved, studies utilizing various inhibitors of known signaling pathways have been employed. PKC inhibitors inhibit hypoxia-induced endothelin-1 induction (Bandyopadhyay *et al.*, 1995), as do inhibitors of the double-stranded RNA-dependent kinase PKR (DVF, unpublished). These results are congruent with the findings that AP-1 induction may be involved in some hypoxia-mediated signaling (Bandyopadhyay *et al.*, 1995). Therefore, serine-threonine kinases like PKC may mediate part of the signaling pathway. Or, rather than being directly involved in hypoxic signaling, the presence of such kinases may be required for gene induction to occur. In either case, this kinase activity appears to be required late in the process. A potential role for tyrosine kinases in transducing the hypoxic signal has also been suggested. General inhibitors of tyrosine kinases have been reported to block activation of HIF-1 by low oxygen environments (Wang *et al.*, 1995b). Hypoxia activates cellular Src kinase activity, perhaps *via* activation of p21[Ras] (Seko *et al.*, 1996). Conversely, both c-Src and p21[Ras] activity have recently been shown to be necessary for induction of VEGF by hypoxia (Mukhopadhyay *et al.*, 1995). NO, however, which inhibits VEGF expression, has also been reported to activate p21[Ras] *via* S-nitrosylation of a critical cysteine residue which stimulates nucleotide exchange on this G-protein (Lander *et al.*, 1995), so the role of p21[Ras] in the O_2 or NO signal transduction pathways remains unclear.

A difficulty with interpreting studies of second messenger pathways by employing inhibitors is the inevitable lack of specificity of such inhibitors. Examination of the pattern of gene induction by hypoxia has provided additional information about mediators involved early in signaling. The following events are induced in endothelial cells by hypoxia and independently by the cytokines TNF-α/IL-1β: (1) reduction in *c-NOS* mRNA; (2) increase in *endothelin-1* mRNA; (3) increase in *PDGF-B* mRNA; (4) increase in platelet activating factor (PAF) release; (5) increase in c-*jun* mRNA; (6) increase in PGI_2 (and other prostaglandins); (7) increase in a number of leukocyte adhesion molecules; (8) increases in IL-8. In view of this striking congruence of these genes or activities regulated by IL-1/TNF and similarly by hypoxia, we hypothesized that hypoxia may act *via* an autocrine loop involving elaboration of IL-1/TNF, or *via* signaling pathway common to that used by TNF and IL-1. That is, the common signaling pathway utilized by these cytokines may be independently activated by hypoxia. Both of these cytokines transduce signals through the sphingomyelin-ceramide (SM/Cer) cycle (Hannun, 1994; Kolesnick and Golde, 1994). Activation of this pathway leads to the generation of Cer from membrane SM through a neutral phospholipase, and eventual activation of a Cer-activated serine/threonine kinase and a PP2A-like phosphatase. Utilizing [^{32}P]-radiolabeling of cellular ceramide with a specific lipid kinase, diacylglycerol kinase, and separation of the products with a thin layer chromatography assay, we have found that both hypoxia and TNF activate the phospholipase and increase intracellular ceramide (Rothendler *et al.*, 1997; Faller, 1997). We further hypothesized that, through activation of the ceramide pathway, hypoxia would activate transcription factor NFκB in endothelial cells. By activation of the SM-ceramide pathway with hypoxia, or activation beyond the TNFα receptor using cell-permeable ceramide analogues (C2-cer and C8-cer), we have indeed demonstrated activation of NFκB, a known intermediate in the SM-ceramide pathway. Additionally, we have shown that ceramide, as well as TNF, induces *endothelin-1* mRNA in these cells (Rothendler *et al.*, 1998; Faller, 1997).

Transcription Factors Regulated by Hypoxia

The genes controlled by hypoxia in endothelial or vascular smooth muscle cells can be regulated at either of two levels: transcription (e.g., *PDGF-B*, *endothelin-1*, *cNOS*) or mRNA stability (e.g., *thrombospondin-1*), or both (e.g., *VEGF*). Several transcription factors have been found to be regulated by hypoxia, including hypoxia-inducible transcription factor-1 (HIF-1) (Wang and Semenza, 1993; Wang *et al.*, 1995a; Maxwell *et al.*, 1993), AP-1 (Bandyopadhyay *et al.*, 1995; Bandyopadhyay and Faller, 1997; Yao *et al.*, 1994), the tumor suppressor p53 (Graeber *et al.*, 1994), NF-κB (Koong *et al.*, 1994a; Yao and O'Dwyer, 1995), and the heat shock trancription factor HSF (Benjamin *et al.*, 1990). Less is known about the mechanisms underlying hypoxic regulation of transcript stability. However, some indirect evidence exists to suggest that the signaling pathways leading to control at the level of transcription (and perhaps stability) may have some components in common at the level of the sensor and at the level of the signal transducing mediators.

Hypoxia-Inducible Transcription Factor-1 (HIF-1)

The transcriptional induction of erythropoietin by hypoxia is the best characterized at the molecular level. Expression of the *erythropoietin* gene is activated by low oxygen tension in renal cells and hepatoma cell lines (Goldberg *et al.*, 1987; Lacombe *et al.*, 1988). *Erythropoietin* gene transcription is induced under low oxygen conditions through the hypoxia-inducible transcription factor 1 (HIF-1), which is present in hypoxic but not normoxic nuclear extracts (Wang and Semenza, 1993; Wang *et al.*, 1995a; Maxwell *et al.*, 1993). HIF-1 itself is a heterodimeric complex of basic-helix-loop-helix proteins, consisting of a unique α subunit and HIF-1β, previously identified as ARNT, a subunit of the aryl hydrocarbon receptor nuclear complex. The HIF-1 complex binds to a promoter element 3' to the coding sequence of the *erythropoietin* gene (Semenza *et al.*, 1991). The induction of HIF-1 itself by hypoxia requires *de novo* protein synthesis. The expression of both subunits is upregulated by hypoxia and the proteins degrade rapidly upon reoxygenation. Treatment with a serine/threonine or a tyrosine protein kinase inhibitor blocks hypoxic induction of HIF-1 or HIF-1 DNA-binding activity (Wang *et al.*, 1995b), although the role of protein phosphorylation in induction of *erythropoietin* transcription is currently uncertain. HIF-1 was subsequently found to bind to hypoxia-responsive enhancer regions of a range of genes encoding glycolytic enzymes, including aldolase A, phosphoglycerate kinase

1, enolase 1, lactate dehydrogenase A, pyruvate kinase M and phosphofructokinase L (Semenza *et al.*, 1994).

Although HIF-1 is likely to participate in the transcriptional component of VEGF regulation by oxygen tension, whether the HIF-1 system participates in the regulation of the hypoxia-regulated endothelial cell vasoactive genes *PDGF-B*, *cNOS*, *thrombospondin-1* and *endothelin-1* is not clear. Although HIF-1 is widely expressed in mammalian cells, including endothelium and vascular smooth muscle (Wang and Semenza, 1993; Schmedtje *et al.*, 1996), no HIF-1 binding site has been identified in these genes. *Erythropoietin* gene regulation by hypoxia does share a number of similarities with these endothelial genes however. Tyrosine kinase inhibitors and certain serine-threonine kinase inhibitors prevent the induction of *endothelin-1* by hypoxia (DVF, unpublished). Carbon monoxide and nitric oxide mimic oxygen in their regulation of *endothelin-1*, *PDGF-B*, *eNOS*, *thrombospondin-1*, and *erythropoietin*. Finally, treatment with cobalt which can substitute for iron and constitutively mimic O_2 binding simulates hypoxia in inducing the expression of *endothelin-1*, *PDGF-B*, *thrombospondin-1*, *VEGF* and *erythropoietin* (in hepatoma cells) genes, and suppressing *cNOS genes*, suggesting a fundamental role for a heme protein in both systems (Phelan and Faller, 1996; Kourembanas *et al.*, 1993, 1990, 1991; Goldberg *et al.*, 1988; EyssenHernandez *et al.*, 1996; Phelan *et al.*, 1998).

Transcription Factor AP-1

A number of the endothelial cell vasoactive genes which are transcriptionally activated by hypoxia, including *PDGF-B*, *endothelin-1*, and *VEGF* are known to be inducible by phorbol esters through an AP-1 binding site (Lee *et al.*, 1991; Finkenzeller *et al.*, 1995; Pech *et al.*, 1989; Majack *et al.*, 1987). Modulations in the binding activities of the transcription factors AP-1 and NF-κB have been reported in response to anoxia in tumor cell lines (Meyer *et al.*, 1994, 1993; Toledano and Leonard, 1991). The transcription factor AP-1 is involved in regulating a number of genes, including c-*jun*, the product of which (c-Jun) can constitute part or all of this transcription factor. Anoxia, for example, has been reported to transcriptionally activate a set of genes, including c-*jun*, in certain cell types in culture (Ausserer *et al.*, 1994; Yao *et al.*, 1994; Webster *et al.*, 1994). The ability of changes in redox potential to regulate AP-1 activity (Abate *et al.*, 1990; Bannister *et al.*, 1991; Frame *et al.*, 1991; Meyer *et al.*, 1993) make it appealing as a possible mechanism of hypoxic gene regulation. The intracellular concentrations of reactive oxygen intermediates (ROIs) are finely regulated. Intracellular levels of ROIs become unphysiologically low in the absence of oxygen, for instance during ischemia, hypoxia or anoxia. Hypoxia results in downregulation of enzymes and antioxidative metabolites involved in controlling ROI levels. A "hypoxic" state can also be mimicked by exposure of cells to antioxidants or reducing agents. c-Jun and c-Fos contain conserved redox-sensitive cysteine residues in their DNA-binding domains (Abate *et al.*, 1990). In nuclear extracts from cells, these cysteine residues are found partially oxidized, which causes a loss in DNA-binding activity. Pre-existing c-Jun occurs in HeLa cells in an oxidized form, and antioxidant treatment of cells could directly or indirectly result in reduction of cysteine residues leading to increased DNA binding and transactivation by c-Jun. This post-translational mechanism is the primary cause for activation of AP-1 in response to antioxidants (Meyer *et al.*, 1993).

Activation of AP-1 in response to H_2O_2 and UV light has been shown to be independent of PKC (Buscher *et al.*, 1988; Nose *et al.*, 1991), suggesting that the ROI-dependent pathway does not require PKC. Rather, activation of PKC is only one of several mechanisms inducing oxidative stress in cells (reviewed in Cerutti, 1985). Likewise, the antioxidant response is apparently independent of PKC since PDTC does not influence PKC activity and PKC redistribution by PMA in intact cells.

AP-1 activation may thus be a common mediator of hypoxic induction of vasoactive genes in endothelial cells. We have recently demonstrated that other AP-1-inducible genes are activated by hypoxia in endothelial cells, including *collagenase IV* and c-*jun*, and sought to correlate the activation of genes by hypoxia with the activation of transcription factor AP-1 (Bandyopadhyay *et al.*, 1995; Bandyopadhyay and Faller, 1997). Rupec, *et al.* have reported similar induction of AP-1 genes and activity in HeLa cells in low oxygen tension (Rupec and Baeuerle, 1995). Depending upon the type of cell studied, hypoxic exposure resulted in the induction of AP-1 transcription factor DNA-binding activity with wide variations in levels of binding (Bandyopadhyay *et al.*, 1995; Bandyopadhyay and Faller, 1997). Because hypoxia has been reported to lower the intracellular redox potential, the effect of redox state changes on AP-1 transcription factor activity and on the activation of AP-1-inducible genes was also studied. PDTC, a potent reducing agent, activated the AP-1 transcription factor in HeLa cells, and also resulted in increased accumulation of c-*jun* mRNA in these cells. In contrast to PDTC-mediated activation of the AP-1 transcription factor and the subsequent induction of the AP-1-regulated c-*jun* gene, hypoxic activation of AP-1 transcription factor binding to its cognate DNA sequence did not activate the c-*jun* gene in HeLa cells, thus documenting distinct differences in signals generated by the reducing intracellular microenvironments created by hypoxia and PDTC. These results demonstrate the induction of AP-1 transcription factor activity by hypoxic environments, but suggest that additional factors or cell-specific signals are involved in the regulation of hypoxia-induced genes. Gene activation by hypoxia, however, did not always strictly correlate with the level of activation of transcription factor AP-1 in each cell type. For example, despite substantial increases in c-*jun* or *prepro-endothelin-1* mRNA levels in endothelial cells in response to hypoxia, no proportionate change in the DNA-binding activity of AP-1 was observed in nuclear extracts from these cells after hypoxia. This result may reflect the limitations of the EMSA. Standard gel-shift analysis essentially measures the DNA-binding activity of an isolated nuclear component or complex, which may not represent the functional behavior of that complex in conjunction with multiple other factors in cell. Another possible explanation for the observed lack of strict correlation between *in vitro* AP-1 activity and AP-1-inducible gene induction in some cell types could be that some other factor(s) are also necessary (in addition to AP-1) to activate the AP-1 inducible genes during hypoxia. Further evidence for this possibility comes from the EMSA and transfection analyses of AP-1-inducible genes in different cell lines. HeLa cells, for example, display strong activation of AP-1 transcription factor during hypoxia. Yet, two different *collagenase* promoter-CAT gene constructs, after transient transfection into normoxic or hypoxic HeLa cells, demonstrated an absolute requirement of an intact AP-1 binding site for the basal level expression of CAT gene, but no further activation of CAT gene expression in response to hypoxia was observed. The results of these transfection analyses thus also suggest a requirement for factors in addition to AP-1 for the activation of genes in hypoxic conditions. Furthermore, despite the fact that hypoxic environments resulted in significant increases in AP-1 DNA-binding activity in HeLa cells, no parallel increases in c-*jun* message levels was observed. The activation of AP-1 factor alone was not enough to lead to concomitant activation of c-*jun* gene in these circumstances. Hypoxia and TPA may induce AP-1 complexes with distinctly different hetero- or homo-dimeric components. It appears then that hypoxia may induce an additional, as yet unknown, signal in those hypoxia-responsive cell types (e.g., endothelial and hepatoma cells) capable of inducing specific sets of genes in response to changes in oxygen tension, which is independent of AP-1 activation but required for induction of AP-1 inducible genes. This additional signal is lacking in cell types which do not induce these specific genes in response to hypoxia (HeLa cells). We hypothesize that both signals are required for a complete oxygen-sensing pathway and gene response. The mechanism of c-*jun* gene and AP-1 activation by hypoxia in certain human cell types may be mediated by the hypoxia-induced activation of a protein kinase capable of phosphorylating the ATF-2 transcription factor (Laderoute *et al.*, 1996).

Transcription factor NFκB

NFκB complexes are sensitive to diverse environmental stimuli (Sen and Packer, 1996; Thanos and Maniatis, 1995), including exposure to phorbol esters or changes in intracellular redox potential, and rapid activation of NFκB has been demonstrated in Jurkat leukemia cells, NIH-3T3 cells, and other cells exposed to hypoxia (Koong *et al.*, 1994a, 1994b; Yao and O'Dwyer, 1995). These stimuli lead to the degradation of the IκBα chaperone protein which retains the NFκB p50–p65 cytoplasmic heterodimer in the cytoplasm, allowing nuclear translocation and binding of this complex to cognate elements in the promoters of specific genes leading to transcriptional activation. The induction of NFκB by hypoxia is therefore not dependent on new protein synthesis, but is dependent of phosphorylation of tyrosine residues within the IκBα protein, which target it for degradation. Exposure of certain human tumor cell lines for up to 6 hr of hypoxia increased NFκB DNA binding activity, consisting mainly of p65(Rel A) and p50 subunits. Very modest increases in NFκB-dependent promoter activity could also be observed in these cells (Laderoute *et al.*, 1996). We have demonstrated induction of NFκB DNA-binding activity in bovine and human endothelial cells by exposure to hypoxia, likely mediated through activation of the SM-ceramide pathway (Rothendler *et al.*, 1998; Faller, 1997).

The role of NFκB in mediating induction of hypoxia-responsive genes in endothelial cells remains unclear, however. This is due, at least in part, to the potential complexity and redundancy of the elements comprising the NFκB complex, as well as the reliance on the use of anti-oxidants to define NFκB-dependent pathways. The interpretation of such experiments is especially difficult when oxygen tension is the stimulus to be studied. In addition, the effects on NFκB of transient re-oxygenation during experimental manipulations may confound interpretation. In one tumor cell line, it was reported that while AP-1 activity was induced by hypoxia, NFκB DNA-binding activity, while not induced by a hypoxia environment alone, was induced upon re-oxygenation (Rupec and Baeuerle, 1995).

Conflicting reports exist regarding the ability of NO to regulate NFκB activity. A study using peripheral blood mononuclear cells reported that treatment with the NO donors sodium nitroprusside (SNP) or S-nitroso-N-acetylpenicillamine (SNAP) resulted in activation of the DNA-binding activity of the NFκB family proteins (Lander *et al.*, 1993). More recent studies in endothelial cells, however, demonstrated that exposure to NO donors could inhibit the activation of NFκB DNA-binding activity by tumor necrosis factor-α, by both stabilizing IκBα and increasing the transcription of the IκBα gene (Peng *et al.*, 1995a, 1995b). In addition, NO has been shown to directly inhibit the DNA-binding activity of the NFκB p50 subunit. NO appears to exert this inhibitory effect by modification of the conserved, redox-sensitive C62 residue in the molecule, perhaps through S-nitrosylation of this cysteine group (Matthews *et al.*, 1996). Because the human and murine iNOS genes are transcriptionally regulated by NFκB (Nunokawa *et al.*, 1996), and because the NFκB site in the iNOS gene promoter appears critical for inducible gene expression, the possibility of a feedback mechanism has been proposed whereby activation of NFκB leads to transcription of IκBα and iNOS, which then in turn serve to limit further transcriptional activation by NFκB. Similarly, NO produced by iNOS could serve to inhibit further iNOS transcription. The induction of iNOS gene expression by TNF-a and LPS can be inhibited by pretreatment of cultured cells with either SNP or a NO gas solution (Colasanti *et al.*, 1995).

A PHYSIOLOGICAL MODEL OF OXYGEN-REGULATED GENE EXPRESSION

Acute Responses to Hypoxia

Hypoxic pulmonary vasoconstriction is an important mechanism for matching ventilation and perfusion (Fishman, 1976). We have suggested in the past that the ability of hypoxia to induce the transcription and production of vasoconstrictors like PDGF-B and endothelin-1 may represent one molecular mechanism whereby low oxygen tension mediates vasoconstriction regionally (Kourembanas *et al.*, 1993, 1990, 1991; Faller, 1994). It is also possible that the decrease in cNOS transcripts and enzyme by hypoxia reported herein may also contribute to the vasoconstrictor response to hypoxia, by inhibiting production of the counteracting vasodilator NO. In addition, we have demonstrated previously that nitric oxide itself can prevent the induction of vasoconstrictor gene transcription by hypoxia, and that even the relatively low amounts of nitric oxide produced by endothelial cells in culture exert a tonic inhibitory force on the transcription of PDGF-B and endothelin-1 (Kourembanas *et al.*, 1993). While it is tempting to postulate a homeostatic role for these reciprocal interactions in the regional balancing of ventilation and perfusion, it is also possible that these same mechanisms may contribute to the pathology observed in the setting of global hypoxia (Faller, 1994).

There has already been some validation of this model in a clinical setting, that of acute hypoxic pulmonary artery hypertension following lung transplantation (Mentzer *et al.*, 1995). On the basis of the molecular feedback mechanisms proposed above (Phelan and Faller, 1996; Faller, 1992; Kourembanas *et al.*, 1993; Mentzer *et al.*, 1995; Faller, 1992; Faller *et al.*, 1996), we predicted that patients with this acute pulmonary shunt in response to global hypoxia would respond to a systemic nitric oxide donor (nitroprusside), despite the ongoing hypotensive shock resulting from the progressive hypoxemia. The clinical response to nitric oxide in this setting was indeed rapid resolution of the pulmonary hypertension, hypoxemia, and systemic hypotension. We hypothesize that nitric oxide broke the cycle of hypoxia-vasoconstriction-hypoxia in three ways: (a) nitric oxide served as a direct vasodilator of pulmonary vascular smooth muscle; (b) nitric oxide fed back on the endothelial cell through the oxygen sensor to turn off production of the vasoconstrictors endothelin-1 and PDGF-B; and, (c) perhaps the resulting increase in pulmonary pO_2 allowed more transcription and activity of cNOS and autocrine production of nitric oxide by pulmonary vascular smooth muscle.

Similarly, the physiological balance and postulated feedback control between vasoconstricting and vasodilating stimuli can become disrupted and pathologically reinforcing under another clinical situation, that of septic shock. In vascular smooth muscle cells, iNOS activity is expressed and NO is produced in response to exposure to bacterial lipopolysaccharide (LPS) (Fleming *et al.*, 1990; Lorsbach *et al.*, 1993; Sirsjo *et al.*, 1994; Radomski *et al.*, 1990) and interferon-γ (IFNγ) (Lorsbach *et al.*, 1993; Sirsjo *et al.*, 1994; Lorsbach *et al.*, 1993; Lowenstein *et al.*, 1993), as well as to proinflammatory cytokines such as tumor necrosis factor-α (TNFα) (Owens and Grisham, 1993; Koide *et al.*, 1993; Cunha *et al.*, 1994) and interleukin-1β (IL-1β) (Busse and Mulsch, 1990; Schini *et al.*, 1991; Kanno *et al.*, 1993; Owens and Grisham, 1993; Cunha *et al.*, 1994), all central mediators of the septic shock syndrome. It has been widely hypothesized that induction of NOS activity in vascular smooth muscle in response to these mediators causes the profound vasodilation and refractory hypotension seen in septic shock (Rees *et al.*, 1990; Beasley *et al.*, 1991; Warren *et al.*, 1992; Junquero *et al.*, 1992; Sirsjo *et al.*, 1994; Cunha *et al.*, 1994). The time course of iNOS induction by LPS in cultured smooth muscle cells appears too slow to fully account for the rapid vasodilation which results from LPS release in the intact organism (Faller *et al.*, 1996). It is possible that the presence of other mediators *in vivo* effect a more rapid induction of the smooth muscle enzyme by LPS. Alternatively, it is possible that rapid release of other mediators could produce the early phase of septic shock, with autocrine production of NO in smooth muscle cells being responsible for the prolonged and refractory hypotension which is observed. Production of NO by smooth muscle cells in response to LPS may have adverse paracrine, as well as autocrine, effects in this setting. For example, NO decreases the constitutive expression of the vasoconstrictors endothelin-1 and PDGF-B in vascular endothelial cells, and furthermore blocks the induction of these genes by stimuli such as hypoxia (Kourembanas *et al.*, 1993). Thus, the induction of iNOS and resulting elaboration of NO in response to LPS may not only cause widespread vasodilation directly, but also reciprocally inhibit the production of locally acting vasoconstrictor proteins.

Chronic Responses to Hypoxia

Chronic hypoxia elicits a unique pulmonary vascular pathology. The discovery of PDGF-B and endothelin-1 induction by hypoxia may mediate much of the structural remodeling that characterizes the chronic pulmonary disease of chronic hypoxia. In addition to being a potent vasoconstrictor of the pulmonary circulation acutely, endothelin-1 also stimulates the mitogenesis of fibroblast (Takuwa *et al.*, 1989) and smooth muscle cells (Komuro *et al.*, 1988; Bobik *et al.*, 1990) *in vitro*, and initiates pulmonary artery adventitial fibroblast replication and chemotaxis (Peacock *et al.*, 1992). PDGF-B is a vasoconstrictor (Berk *et al.*, 1986), a potent fibroblast chemoattractant (Seppa *et al.*, 1982) and vascular smooth muscle cell mitogen, and has also been implicated in vascular remodeling (Dzau and Gibbons, 1991). Media conditioned by hypoxic endothelial cells has been shown recently to exhibit chemoattractant and mitogenic activity towards pulmonary fibroblasts, and this activity was due to secreted endothelin-1 and PDGF (Dawes *et al.*, 1994). In contrast, nitric oxide, the production of which is inhibited by hypoxia, inhibits fibroblast and smooth muscle cell mitogenesis (Garg and Hassid, 1989; Nakaki *et al.*, 1990; Nunokawa and Tanaka, 1992; Scott-Burden *et al.*, 1992; Garg and Hassid, 1990), as well as endothelial cell migration, mitogenesis and proliferation (Sarkar *et al.*, 1995; Lau and Ma, 1996). The hypoxia-induced activation of vasoconstrictors, and hypoxia-induced inhibition of the capacity to synthesize opposing and balancing vasodilators like NO, may result in the pulmonary artery hypertension observed during hypoxic exposure. Increases in oxygen tension dilate the pulmonary circulation *via* nitric oxide (Tiktinsky and Morin 1993). Conversely, lack of nitric oxide production in the setting of hypoxia is likely to mediate the resulting persistent pulmonary hypertension (Fineman *et al.*, 1994). Over time, the mitogenic effects of PDGF-B, VEGF, thrombospondin-1 and endothelin-1 on vascular smooth muscle and connective tissue, facilitated by the concurrent upregulation of their receptors on vascular tissue, and unopposed by the antiproliferative effects of nitric oxide (see Garg and Hassid, 1989; Kariya *et al.*, 1989), might result in the structural remodeling and smooth muscle hypertrophy observed in chronic pulmonary artery hypertension (see Smith, 1994), whether due to hypoxia in the fetus and newborn or to the regional hypoxia caused secondarily by such diverse diseases as primary pulmonary hypertension (Giaid and Saleh, 1995), systemic sclerosis (scleroderma), or sickle cell anemia (Faller, 1994).

Peripheral tissues, as well as the lungs, may also be chronically and adversely affected by the effects of hypoxia and nitric oxide on the endothelial cell, and the feedback signaling pathways we have proposed (Figures 8-1A & B) might further exacerbate the pathogenesis of certain diseases. For example, a role for regional hypoxia in systemic sclerosis (scleroderma) has been postulated. Although the etiology of systemic sclerosis remains obscure, at least three pathological events are likely to be involved in its evolution: (1) Alterations of the blood vessel wall/endothelial cell damage; (2) Immunological changes/inflammatory reactions in the dermis; and (3) Disturbances in the production of extracellular proteins/ accumulation of collagen in the dermal layer (Kahari, 1993; LeRoy, 1992). In scleroderma patients whose condition was worsening, both endothelial and fibroblast activation preceded fibrosis. (Claman *et al.*, 1991) It is likely that endothelial cell-derived mediators create a chemotactic gradient that attracts fibroblasts towards the blood vessel, initiates their proliferation, and activates their cellular mechanisms including collagen synthetic capacity. Extravasation of plasma proteins to the interstitial area, including products released from platelets and other plasma cytokines, may further stimulate fibroblast activation. Endothelial cell expression of surface adhesion molecules could activate the trafficking of all types of circulating cells to the interstitium, where they could also activate fibroblasts. The deposition of PDGF in scleroderma tissues (Gay *et al.*, 1989; Moreland *et al.*, 1990) have been suggested to be related to vascular injury and endothelial responses to hypoxia (Blann *et al.*, 1993; Campbell and LeRoy, 1975). The result of the interaction between these growth factors, the perivascular extracellular matrix, and the resident fibroblasts of the reticular dermis maybe increased proliferation of these latter cells (Trojanowski *et al.*, 1988; Feghali *et al.*, 1992). The upregulation of collagen synthesis results in the deposition of new matrix in the lower dermis. The reduction of capillary blood flow following endothelial damage and deposition of excessive perivascular matrix may result in the low tissue O_2 tension which has been observed in patients with scleroderma (Silverstein *et al.*, 1988; Smiley, 1992; Piasecki *et al.*, 1995; Voelkel and Tuder, 1995). This low O_2 tension, in turn, is a strong inducer of production of growth factors for fibroblasts (PDGF-B, ET-1, IL-6), as well as production of collagenases and matrix proteins (e.g., thrombospondin-1) by endothelial cells, begetting more fibrosis and more tissue hypoxia.

Similarly, in sickle cell disease, the regulation of genes encoding vasoactive substances by nitric oxide and O_2 may at times adversely affect the course of the disease (Faller, 1994).

     Douglas V. Faller

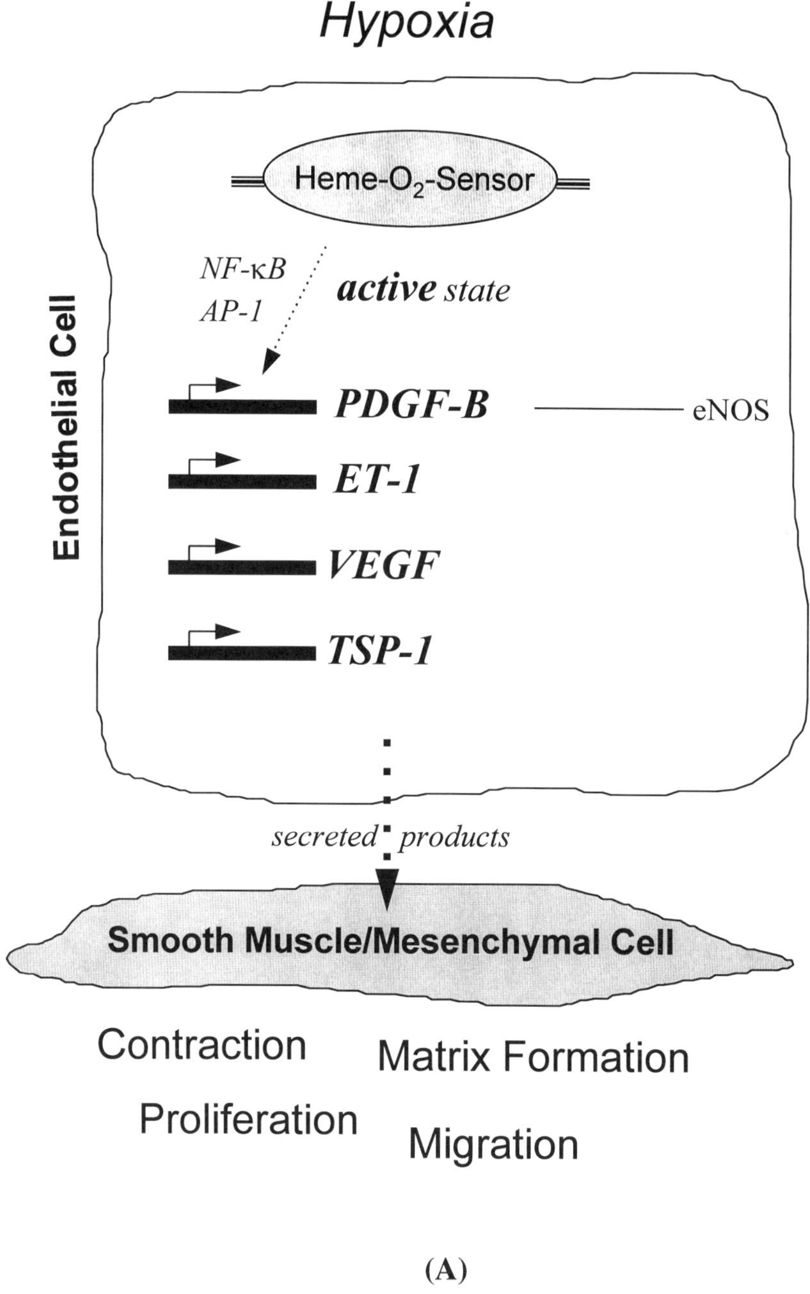

(A)

Figure 8-1(A). Schematic model of vasoactive gene regulation by O$_2$ and nitric oxide (NO). **(A)** Under hypoxic conditions, the heme-containing oxygen sensor is activated (unbound, tense conformation), leading to activation of genes whose products can act upon the underlying vascular smooth muscle and stromal cells to cause constriction, proliferation, migration and matrix deposition, and suppression of genes whose products counteract these effects.

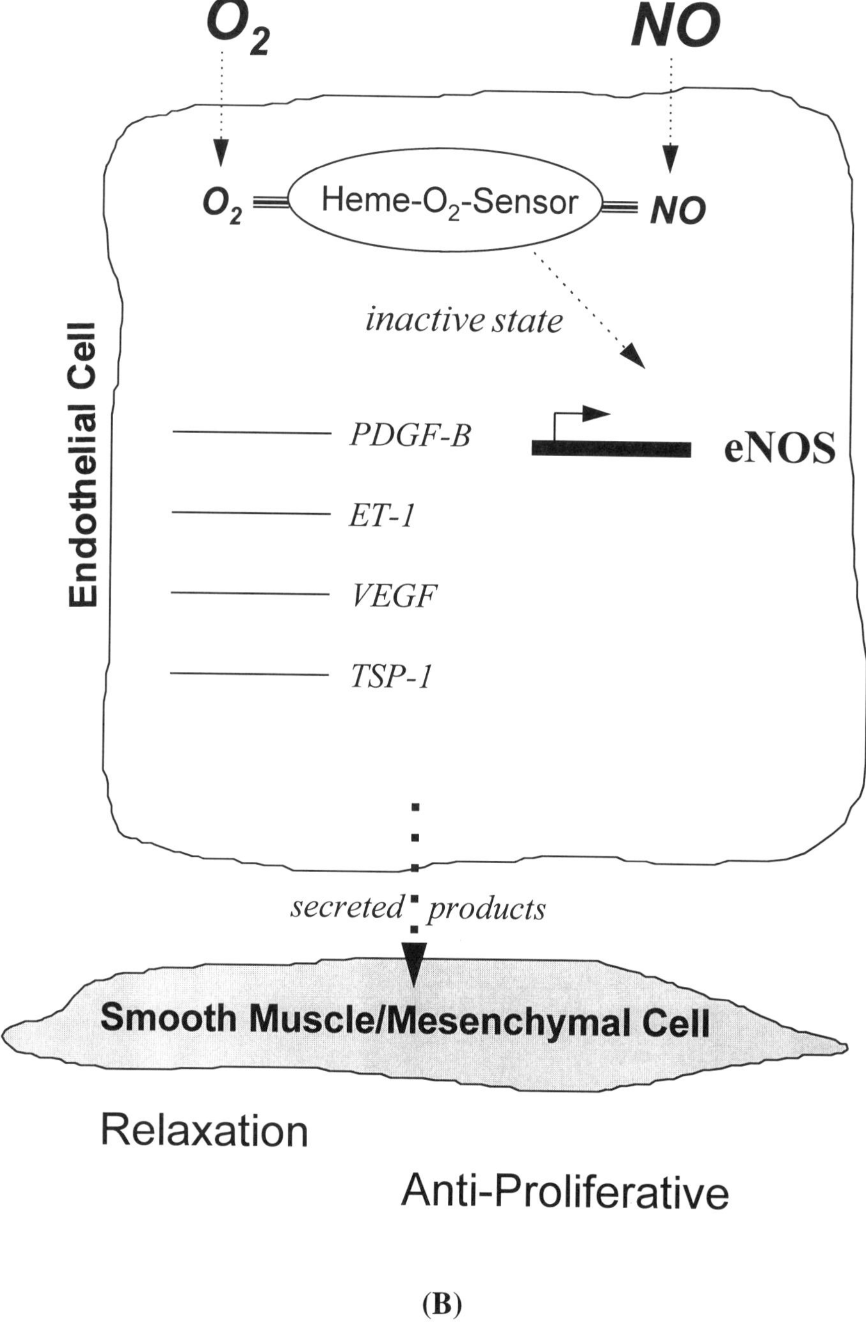

Figure 8-1(B). Conversely, in the presence of ligands for the heme-containing oxygen sensor, like O_2 or NO, the sensor attains an inactive state (bound, relaxed conformation), resulting in a decrease in the production of vasoconstrictor or mitogenic factors.

     Douglas V. Faller

The vasoconstriction induced by the vascular instability inherent in this disease, in combination with the tissue hypoxia resulting from prolonged transit times of erythrocytes through capillaries and capillary occlusion, may contribute in a major way to local sickling events. In a vicious cycle, sickling and sludging would result in further hypoxia, more local vasoconstriction (secondary to the hypoxia-induced elaboration of vasodilators and inhibition of vasoconstrictors) and stasis, resulting in more sickling. In addition, in response to the hypoxic signal, endothelial cells elaborate not only potent vasoconstrictors, but also growth factors which could result in the smooth muscle hyperplasia and fibrosis found in sickle cell chronic lung disease and cor pulmonale (Faller, 1994). Finally, adhesion of sickled erythrocytes to endothelium is mediated in large part by endothelial thrombospondin-1, binding to CD36 or sulfated proteoglycans on the erythrocyte (Sugihara *et al.*, 1992; Brittain *et al.*, 1993; Hillery *et al.*, 1996). Hypoxia not only facilitates erythrocyte sickling, but induces endothelial cell binding to a number of blood cell types (Ginis *et al.*, 1993a, 1995, 1993b), in addition to promoting vaso-constriction (Phelan and Faller, 1996; Phelan *et al.*, 1995). The induction of TSP-1 production during hypoxic episodes (Phelan *et al.*, 1997) in sickle cell disease might further facilitate sickle cell-endothelial cell adhesions, thus reinforcing vaso-occlusive events and contributing to the vicious cycle of sickling, sludging, vasoconstriction, and further local hypoxia.

REFERENCES

Abate, C., Patel, L., Rauscher, F.J. and Curran, T. (1990) Redox regulation of fos and jun DNA-binding activity *in vitro. Science*, **249**, 1157–1161.

Acker, H., Dufau, E., Huber, J. and Sylvester, D. (1989) Indications to an NADPH oxidase as a possible pO_2 sensor in the rat carotid body. *FEBS Lett.*, **265**, 75.

Acker, H. (1994) Mechanisms and meaning of cellular oxygen sensing in the organism. *Respir. Physiol.*, **95**, 1.

Adnot, S., Rajjestin, B., Eddahibi, S., Braquet, P. and Chabrier, P.-E. (1991) Loss of endothelium-dependent relaxant activity in the pulmonary circulation of rats exposed to chronic hypoxia. *J. Clin. Invest.*, **87**, 155–162.

Archer, S.L., Freude, K.A. and Shultz, P.J. (1995) Effect of graded hypoxia on the induction and function of inducible nitric oxide synthase in rat mesangial cells. *Circ. Res.*, **77**, 21–28.

Assender, J., Southgate, K., Hallett, M. and Newby, A. (1992) Inhibition of proliferation, but not of Ca^{2+} mobilization, by cAMP and GMP in rabbit aortic smooth muscle cells. *Biochem. J.*, **288**, 527–532.

Assreuy, J., Cunha, F.Q., Liew, F.Y. and Moncada, S. (1993) Feedback inhibition of nitric oxide synthase activity by nitric oxide. *Br. J. Pharmacol.*, **108**, 833–837.

Ausserer, W.A., Bourratfloeck, B., Green, C.J., Laderoute, K.R. and Sutherland, R.M. (1994) Regulation of c-jun expression during hypoxic and low- glucose stress. *Mol. Cell Biol.*, **14**, 5032–5042.

Baenziger, N.L., Brodie, G.N. and Majerus, P.W. (1971) A thrombin-sensitive protein of human platelet membranes. *Proc. Natl. Acad. Sci. USA*, **68**, 240–243.

Baenziger, N.L., Brodie, G.N. and Majerus, P.W. (1972) Isolation and properties of a thrombin-sensitive protein in human platelets. *J. Biol. Chem.*, **247**, 2723–2731.

Bandyopadhyay, R.S., Phelan, M. and Faller, D.V. (1995) Hypoxia induces AP-1-regulated genes and AP-1 transcription factor binding in human endothelial and other cell types. *Bba-Gene. Struct. Express.*, **1264**, 72–78.

Bandyopadhyay, R.S. and Faller, D.V. (1997) Regulation of c-*jun* expression in endothelial cells. *Endothelium*, **in press,**

Bannister, A.J., Cook, A. and Kouzarides, T. (1991) *In vitro* DNA binding activity of Fos/Jun and BZLF1 but not C/EBP is affected by redox changes. *Oncogene*, **6**, 1243–1250.

Beasley, D., Schwartz, J.H. and Brenner, B.M. (1991) Interleukin-1 induces prolonged L-arginine-dependent cyclic guanosine mono-phosphate and nitrite production in rat vascular smooth muscle cells. *J. Clin. Invest.*, **87**, 602–608.

Benitz, W., Kelley, R., Anderson, C., Lorant, D. and Bernfield, M. (1990) Endothelial heparan sulfate proteoglycan.I. Inhibitory effects on smooth muscle cell proliferation. *Am. J. Respir. Cell Mol. Biol.*, **2**, 13–24.

Benjamin, I.J., Kroger, B. and Williams, R.S. (1990) Activation of the heat shock transcription factor by hypoxia in mammalian cells. *Proc. Natl. Acad. Sci. USA*, **87**, 6263–6267.

Berk, B.C., Alexander, R.W., Brock, T.A., Gimbrone, M.A. and Webb, R.C. (1986) Vasoconstriction: A new activity for platelet-derived growth factor. *Science*, **32**, 87–89.

Bicknell, R. and Harris, A.L. (1991) Novel growth regulatory factors and tumour angiogenesis. *Eur. J. Cancer*, **27**, 785–789.

Blann, A.D., Illingworth, K. and Jayson, M. (1993) Mechanisms of endothelial cell damage in systemic sclerosis and Raynaud's phenomenon. *J. Rheumatol.*, **20**, 1325–1330.

Bobik, A., Grooms, A., Millar, J.A., Mitchell, A. and Grinpukel, S. (1990) Growth activity of endothelin on vascular smooth muscle. *Am. J. Physiol.*, **258**, C408–C415.

Brittain, H.A., Echman, J.R., Swerlick, R.A., Howard, R.J., and Wick, T.M. (1993) Thrombospondin from activated platelets promotes sickle erythrocyte adherence to human microvascular endothelium under physiologic flow: A potential role for platelet activation in sickle cell vaso-occlusion. *Blood*, **81**, 2137.

Buscher, M., Rahmsdorf, H.J., Litfin, M., Karin, M. and Herrlich, P. (1988) Activation of the c-fos gene by UV and phorbol ester: different signal transduction pathways converge to the same enhancer element. *Oncogene*, **3**, 301–311.

Busse, R. and Mulsch, A. (1990) Induction of nitric oxide synthase by cytokines in vascular smooth muscle cells. *FEBS Lett.*, **275**, 87–90.

Campbell, P.M. and LeRoy, E.C. (1975) Pathogenesis of systemic sclerosis: A vascular hypothesis. *Semin. Arthritis Rheum.*, **4**, 351–368.

Carville, C., Raffestin, B., Eddahibi, S., Blouquit, Y. and Adnot, S. (1993) Loss of endothelium-dependent relaxation in proximal pulmonary arteries from rats exposed to chronic hypoxia – effects of *in vivo* and *in vitro* supplementation with L-arginine. *J. Cardiovasc. Pharmacol.*, **22**, 889–896.

Cerutti, P.A. (1985) Prooxidant states and tumor promotion. *Science*, **227**, 375–381.

Claman, H.N., Giorno, R.C. and Seibold, J.R. (1991) Endothelial and fibroblastic activation in scleroderma. *Arthritis Rheum.*, **34**, 1495–1501.

Colasanti, M., Persichini, T., Menegazzi, M., Mariotto, S., Giordano, E., Caldarera, C.M. *et al.* (1995) Induction of nitric oxide synthase mRNA expression. Suppression by exogenous nitric oxide. *J. Biol. Chem.*, **270**, 26731–26733.

Crawley, D.E., Zhao, L., Giembycz, M.A., Liu, S., Barnes, P.J., Winter, R.J.P. *et al.* (1992) Chronic hypoxia impairs soluble guanylyl cyclase mediated pulmonary arterial relaxation in the rat. *Am. J. Physiol.*, **263**, L325–L332.

Cunha, F.Q., Assreuy, J., Moss, D.W., Rees, D., Leal, L.M.C., Moncada, S. *et al.* (1994) Differential induction of nitric oxide synthase in various organs of the mouse during endotoxaemia — Role of TNF-alpha and IL-1-beta. *Immunology*, **81**, 211–215.

Da Re, S., Bertagnoli, S., Fourment, J., Reyrat, J. and Kahn, D. (1994) Intramolecular signal transduction within the FixJ transcriptional activator: *in vitro* evidence for the inhibitory effect of the phosphorylatable regulatory domain. *Nucleic Acids Res.*, **22**, 1555–1561.

Dawes, K.E., Peacock, A.J., Gray, A.J., Bishop, J.E. and Laurent, G.J. (1994) Characterization of fibroblast mitogens and chemoattractants produced by endothelial cells exposed to hypoxia. *Amer. J. Respir. Cell Molec. Biol.*, **10**, 552–559.

De Mey, J.G. and Vanhoutte, P.M. (1983) Anoxia and endothelium-dependent reactivity of the canine femoral artery. *J. Physiol.*, **335**, 65–74.

Dinh-Xuan, A.T., Higenbottam, T.W., Clelland, C.A., Pepke-Zaba, J., Cremona, G., Butt, A.Y. *et al.* (1991) Impairment of endothelium-dependent pulmonary-artery relaxation in chronic obstructive lung disease. *N. Engl. J. Med.*, **324**, 1539–1547.

Dixit, V.M., Haverstick, D.M., O'Rourke, K.M., Hennessy, S.W., Grant, G.A., Santoro, S.A. *et al.* (1985) A monoclonal antibody against human thrombospondin inhibits platelet aggregation. *Proc. Natl. Acad. Sci. USA*, **82**, 3472–3476.

Dobrina, A. and Rossi, F. (1983) Metabolic properties of freshly isolated bovine endothelial cells. *Biochim. Biophys. Acta*, **762**, 295–301.

Dzau, V.J. and Gibbons, G.H. (1991) Endothelium and growth factors in vascular remodelling of hypertension. *ypertension*, **18**, 115–121.

Elton, T.S., Oparil, S., Taylor, G.R., Hicks, P.H., Yang, R.H., Jin, H.K. *et al.* (1992) Normobaric hypoxia stimulates endothelin-1 gene expression in the rat. *Am. J. Physiol.*, **263**, R1260–R1264.

EyssenHernandez, R., Ladoux, A. and Frelin, C. (1996) Differential regulation of cardiac heme oxygenase-1 and vascular endothelial growth factor mRNA expressions by hemin, heavy metals, heat shock and anoxia. *FEBS Lett.*, **382**, 229–233.

Faller, D.V. (1992). Normal endothelial/smooth muscle cell interactions, growth factors, and the regulation of cellular growth in the fetus and newborn. In *Pulmonary Circulatory Disorders in the Newborn, Infant, and Child*, edited by Anonymous, pp. 9–21. Gronigen, Netherlands.

Faller, D.V. (1994). Vascular Modulation in Sickle Cell Anemia. In *Sickle Cell Disease: Basic Principles and Clinical Practice*, edited by S.H. Embury, R.P. Hebbel, N. Mohandas and M.H. Steinberg, pp. 235–246.New York: Raven Press, Ltd.

Faller, D.V., Barnett, H., Weisbrod, R. and Cohen, R.A. (1996) Regulation of nitric oxide synthase induction in cultured vascular smooth muscle cells by lipopolysaccharide and interferon. *Endothelium*, **4**, 99–114.

Faller, D.V. (1997). Hypoxia, Nitric Oxide and Vasoactive Gene Transcription. In *Nitric Oxide, Cytochrome P450 and Sexual Steroid Hormones*, edited by J. Parkinson, pp. 75–116. New York: Springer-Verlag.

Fandrey, J., Frede, S. and Jelkmann, W. (1994) Oxygen sensing by H_2O_2-generating heme proteins. *Annals New York Acad. Sci.*, 341–343.

Feghali, C.A., Bost, K.L., Boulware, D.W. and Levy, L.S. (1992) Mechanism of pathogenesis in scleroderma: I. Overproduction of interleukin 6 by fibroblasts cultured from affected skin sites of patients with scleroderma. *J. Rheumatol.*, **19**, 1207–1211.

Ferrara, N., Houck, K., Jakeman, L. and Leung, D.W. (1992) Molecular and biological properties of the vascular endothelial growth factor family of proteins. *Endocrine. Rev.*, **13**, 18–32.

Ferri, C., Bellini, C., Deangelis, C., Desiati, L., Perrone, A., Properzi, G. *et al.* (1995) Circulating endothelin-1 concentrations in patients with chronic hypoxia. *J. Clin. Pathol.*, **48**, 519–524.

Ferris, J.A., Falanga, V. and Leitzel, K.E. (1990) Elevated plasma platelet-derived growth factor β chain levels in scleroderma patients (abstr). *Arthritis Rheum.*, **33**, S64.

Fineman, J.R., Wong, J., Morin, I, Wild, L.M. and Soifer, S.J. (1994) Chronic nitric oxide inhibition in utero produces persistent pulmonary hypertension in newborn lambs. *J. Clin. Invest.*, **93**, 2675–2683.

Finkenzeller, G., Marme, D., Weich, H.A. and Hug, H. (1992) Platelet-derived growth factor-induced transcription of the vascular endothelial growth factor gene is mediated by protein kinase-C. *Cancer Res.*, **52**, 4821–4823.

Finkenzeller, G., Technau, A. and Marme, D. (1995) Hypoxia-induced transcription of the vascular endothelial growth factor gene is independent of functional AP-1 transcription factor. *Biochem. Biophys. Res. Commun.*, **208**, 432–439.

Fiscus, R.R. (1988) Molecular mechanisms of endothelium-mediated vasodilation. *Semin. Thromb. Hemostasis.*, **14**, 12–22.

Fishman, A.P. (1976) Hypoxia on the pulmonary circulation. *Circ. Res.*, **38**, 221–231.

Fleming, I., Gray, G.A., Julou-Schaeffer, G., Parratt, J.R. and Stoclet, J. (1990) Incubation with endotoxin activates the L-arginine pathway in vascular tissue. *Biochem. Biophys. Res. Commun.*, **171**, 562–568.

Frame, M.C., Wilkie, N.M., Darling, A.J., Chudleigh, A., Pintzas, A., Lang, J.C. *et al.* (1991) Regulation of AP-1/DNA complex formation *in vitro*. *Oncogene*, **6**, 205–209.

Frazier, W.A. (1991) Thrombospondins. *Curr. Opin. Cell Biol.*, **3**, 792–799.

Furchgott, R.F. and Zawadzki, J.V. (1980) The obligatory role of endothelial cells in the relaxation of arterial smooth muscle by acetylcholine. *Nature*, **288**, 373–376.

Garg, U.C. and Hassid, A. (1989) Nitric-oxide-generating vasodilators and 8-bromo-cyclic guanosine monophosphate inhibit mitogenesis and proliferation of culture rat smooth muscle cells. *J. Clin. Invest.*, **83**, 1774–1777.

Garg, U.C. and Hassid, A. (1990) Nitric oxide-generating vasodilators inhibit mitogenesis and proliferation of BALB/c 3T3 fibroblasts by a cyclic GMP-independent mechanism. *Biochem. Biophys. Res. Commun.*, **171**, 474–479.

Gay, S., Jones, J., Huang, G. and Gay, R.E. (1989) Immunohistologic demonstration of platelet-derived growth factor (PDGF) and sis-oncogene expression in scleroderma. *J. Invest. Dermatol.*, **92**, 301–303.

Gerritsen, M.E. and Bloor, C.M. (1993) Endothelial cell gene expression in response to injury. *FASEB J.*, **7**, 523–532.

Giaid, A., Yanagisawa, M., Langleben, D., Michel, R.P., Levy, R., Shennib, H. *et al.* (1993) Expression of endothelin-1 in the lungs of patients with pulmonary hypertension. *N. Engl. J. Med.*, **328**, 1732–1739.

Giaid, A. and Saleh, D. (1995) Reduced expression of endothelial nitric oxide synthase in the lungs of patients with pulmonary hypertension. *N. Engl. J. Med.*, **333**, 214–221.

Ginis, I., Mentzer, S.J. and Faller, D.V. (1993a) Hypoxia induces lymphocyte adhesion to human mesenchymal cells via an LFA-1-dependent mechanism. *Am. J. Physiol.*, **264**, C617–C624.

Ginis, I., Mentzer, S.J. and Faller, D.V. (1993b) Oxygen tension regulates neutrophil adhesion to human endothelial cells via an LFA-1-dependent mechanism. *J. Cell Physiol.*, **157**, 569–578.

Ginis, I., Mentzer, S.J., Li, X.P. and Faller, D.V. (1995) Characterization of a hypoxia-responsive adhesion molecule for leukocytes on human endothelial cells. *J. Immunol.*, **155**, 802–810.

Gleadle, J.M., Ebert, B.L., Firth, J.D. and Ratcliffe, P.J. (1995a) Regulation of angiogenic growth factor expression by hypoxia, transition metals, and chelating agents. *Amer. J. Physiol. Cell Physiol.*, **37**, C1362–C1368.

Gleadle, J.M., Ebert, B.L. and Ratcliffe, P.J. (1995b) Diphenylene iodonium inhibits the induction of erythropoietin and other mammalian genes by hypoxia – Implications for the mechanism of oxygen sensing. *Eur. J. Biochem.*, **234**, 92–99.

Gnarra, J.R., Zhou, S.B., Merrill, M.J., Wagner, J.R., Krumm, A., Papavassiliou, E. *et al.* (1996) Post-transcriptional regulation of vascular endothelial growth factor mRNA by the product of the VHL tumor suppressor gene. *Proc. Natl. Acad. Sci. USA*, **93**, 10589–10594.

Goldberg, M.A., Glass, G.A., Cunningham, J.M. and Bunn, H.F. (1987) The regulated expression of erythropoietin by two hepatoma cell lines. *Proc. Natl. Acad. Sci. USA*, **84**, 7972.

Goldberg, M.A., Dunning, S.P. and Bunn, H.F. (1988) Regulation of the erythropoietin gene: evidence that the oxygen sensor is a heme protein. *Science*, **242**, 1412–1415.

Goldberg, M.A. and Schneider, T.J. (1994) Similarities between the oxygen-sensing mechanisms regulating the expression of vascular endothelial growth factor and erythropoietin. *J. Biol. Chem.*, **269**, 4355–4359.

Good, D.J., Polverini, P.J., Rastinejad, F., LeBeau, M.M., Lemmons, R.S., Frazier, W.A. *et al.* (1990) A tumor suppressor dependent inhibitor of angiogenesis is immunologically and functionally indistinguishable from a fragment of thrombospondin. *Proc. Natl. Acad. Sci. USA*, **87**, 6624–6628.

Graeber, T.G., Peterson, J.F., Tsai, M., Monica, K., Fornace, A.J. and Giaccia, A.J. (1994) Hypoxia induces accumulation of p53 protein, but activation of a G_1-phase checkpoint by low-oxygen conditions is independent of p53 status. *Mol. Cell Biol.*, **14**, 6264–6277.

Graser, J. and Vanhoutte, P.M. (1991) Hypoxic contraction of canine coronary arteries: role of endothelium and cGMP. *Am. J. Physiol.*, **261**, H1769–H1777.

Griscavage, J.M., Rogers, N.E., Sherman, M.P. and Ignarro, L.J. (1993) Inducible nitric oxide synthase from a rat alveolar macrophage cell line is inhibited by nitric oxide. *J. Immunol.*, **151**, 6329–6337.

Grugel, S., Finkenzeller, G., Weindel, K., Barleon, B. and Marme, D. (1995) Both v-Ha-ras and v-raf stimulate expression of the vascular endothelial growth factor in NIH 3T3 cells. *J. Biol. Chem.*, **270**, 25915–25919.

Halliwell, B. and Gutteridge, J.M. (1990) Role of free radicals and catalytic metal ions in human disease: an overview. *Meth. Enzymol.*, **186**, 1–85.

Hannan, R., Kourembanas, S., Flanders, K., Rogelj, S., Roberts, A., Faller, D.V. *et al.* (1988) Endothelial cells synthesize basic fibroblast growth factor and transforming growth factor beta. *Growth Factors*, **1**, 7–17.

Hannun, Y.A. (1994) The sphingomyelin cycle and the second messenger function of ceramide. *J. Biol. Chem.*, **269**, 3125–3128.

Harpel, P.E., Silerstein, R.L., Pannel, R., Gurewich, V. and Nachman, R.L. (1999) Thrombospondin forms complexes with single chain and two-chain forms of urokinase. *J. Biol. Chem.*, **265**, 11289–11294.

Hieda, H. and Gomez-Sanchez, C. (1990) Hypoxia increases endothelin release in bovine endothelial cells in culture, but epinephrine, norepinephrine, serotonin, histamine and angiotensin II do not. *Life Sci.*, **247**, 247–251.

Hillery, C.A., Du, M.C., Montgomery, R.R. and Scott, J.P. (1996) Increased adhesion of erythrocytes to components of the extracellular matrix: Isolation and characterization of a red blood cell lipid that binds thrombospondin and laminin. *Blood*, **87**, 4879–4886.

Holden, W.E. and McCall, E. (1984) Hypoxia-induced contractions of porcine pulmonary artery strips depends on intact endothelium. *Exp. Lung Res.*, **7**, 101–112.

Hosokawa, T., Muraishi, A., Rothman, V.L., Papale, M. and Tuszynski, G.P. (1993) The effect of thrombospondin on invasion of fibrin gels by human A5-19 lung carcinoma. *Oncol. Res.*, **5**, 183–189.

Humphries, D., Lee, S., Fanburg, B. and Silbert, J. (1986) Effects of hypoxia and hyperoxia on proteoglycan production by bovine pulmonary artery endothelial cells. *J. Cell Physiol.*, **126**, 249–253.

Hwang, S.M., Wilson, P.D., Laskin, J.D. and Denhardt, D.T. (1994) Age and development-related changes in osteopontin and nitric oxide synthase mRNA levels in human kidney proximal tubule epithelial cells: contrasting responses to hypoxia and reoxygenation. *J. Cell Physiol.*, **160**, 61–68.

Ignarro, L.J., Burke, T.M., Wood, K.S., Wolin, M.S. and Kadowitz, P.J. (1984) Association between cyclic GMP accumulation and acetylcholine-elicited relaxation of bovine intrapulmonary artery. *J. Pharmacol. Exp. Ther.*, **228**, 682–690.

Ignarro, L.J., Byrns, R.E., Buga, G.M. and Woods, K.S. (1987) Endothelium-derived relaxing factor from pulmonary artery and vein possessed pharmacologic and chemical properties identical to those of nitric oxide radical. *Circ. Res.*, **61**, 866–879.

Incardona, F., Lewalle, J.M., Morandi, V., Lambert, S., Legrand, Y., Foidart, J.M. *et al.* (1995) Thrombospondin modulates human breast adenocarcinoma cell adhesion to human vascular endothelial cells. *Cancer Res.*, **55**, 166–173.

Johns, R.A., Linden, J. and Peach, M.J. (1989) Endothelium-dependent relaxation and cyclic GMP accumulation in rabbit pulmonary artery are selectively impaired by moderate hypoxia. *Circ. Res.*, **65**, 1508–1515.

Joseph, P., Thompson, B. and Hales, C. (1990) Effect of hypoxia on the secretion of platelet derived growth factor by bovine pulmonary artery endothelial cells. *Am. Rev. Resp. Dis.*, **141**, A346.

Junquero, D.C., Scottburden, T., Schini, V.B. and Vanhoutte, P.M. (1992) Inhibition of cytokine-induced nitric oxide production by transforming growth factor-beta1 in human smooth muscle cells. *J. Physiol. (London)*, **454**, 451–465.

Kahari, V.M. (1993) Activation of dermal connective tissue in scleroderma. *Ann. Med.*, **25**, 511–518.

Kanno, K., Hirata, Y., Imai, T. and Marumo, F. (1993) Induction of nitric oxide synthase gene by interleukin in vascular smooth muscle cells. *Hypertension*, **22**, 34–39.

Kariya, K., Kawahara, Y., Araki, S., Fukuzaki, H. and Takai, Y. (1989) Antiproliferative action of cyclic GMP elevating vasodilators in cultured rabbit aortic smooth muscle cells. *Atherosclerosis*, **80**, 143–147.

Katayose, D., Ohe, M., Yamauchi, K., Ogata, M., Shirato, K., Fujita, H. *et al.* (1993) Increased expression of PDGF A-chain and B-chain genes in rat lungs with hypoxic pulmonary hypertension. Am. J. Physiol., **264**, L100–L106.

Kiechle, F.L. and Maliski, T. (1993) Nitric oxide: biochemistry, pathophysiology, and detection. *Am. J. Clin. Pathol.*, **100**, 567–575.

Klatt, P., Schmidt, K. and Mayer, B. (1992) Brain nitric oxide synthase is a haemoprotein. *Biochem. J.*, **288**, 15–17.

Koide, M., Kawahara, Y., Tsuda, T. and Yokoyama, M. (1993) Cytokine-Induced Expression of an Inducible Type of Nitric Oxide Synthase Gene in Cultured Vascular Smooth Muscle Cells. *FEBS Lett.*, **318**, 213–217.

Kolesnick, R. and Golde, D.W. (1994) The sphingomyelin pathway in tumor necrosis factor and interleukin-1 signaling. *Cell*, **77**, 325–328.

Komuro, I., Kurihara, H., Sugiyama, T., Takaku, F. and Yazaki, Y. (1988) Endothelin stimulates c-fos and c-myc expression and proliferation of vascular smooth muscle cells. *FEBS Lett.*, **238**, 249–252.

Koong, A.C., Chen, E.Y. and Giaccia, A.J. (1994a) Hypoxia causes the activation of Nuclear Factor-kappa-B through the phosphorylation of I-kappa-B alpha on tyrosine residues. *Cancer Res.*, **54**, 1425–1430.

Koong, A.C., Chen, E.Y., Mivechi, N.F., Denko, N.C., Stannbrook, P. and Giaccia, A.J. (1994b) Hypoxic activation of nuclear factor-kappa B is mediated by a Ras and Raf signaling pathway and does not involve MAP Kinase (ERK1 or ERK2). *Cancer Res.*, **54**, 5273–5279.

Kourembanas, S., Hannan, R. and Faller, D.V. (1990) Oxygen tension regulates the expression of platelet-derived growth factor-B chain gene in human endothelial cells. *J. Clin. Invest.*, **86**, 670–674.

Kourembanas, S., Marsden, P.A., Mcquillan, L.P. and Faller, D.V. (1991) Hypoxia induces endothelin gene expression and secretion in cultured human endothelium. *J. Clin. Invest.*, **88**, 1054–1057.

Kourembanas, S., Mcquillan, L.P., Leung, G.K. and Faller, D.V. (1993) Nitric oxide regulates the expression of vasoconstrictors and growth factors by vascular endothelium under both normoxic and hypoxic conditions. *J. Clin. Invest.*, **92**, 99–104.

Kourembanas, S. and Faller, D.V. (1989) Platelet-derived growth factor production by human umbilical vein endothelial cells is regulated by basic fibroblast growth factor. *J. Biol. Chem.*, **264**, 4456–4459.

Kourembanas, S. and Faller, D.V. (1991) Oxygen tension regulates the expression of the platelet-derived growth factor-B chain and endothelin genes in human endothelial cells. *Sem. Perinat.*, **5**, 346–350.

Kupper, T.S. (1995) Adhesion molecules in scleroderma: collagen binding integrins. *Int. Rev. Immunol.*, **12**, 217–225.

Kuwabara, K., Ogawa, S., Matsumoto, M., Koga, S., Clauss, M., Pinsky, D.J. *et al.* (1995) Hypoxia-mediated induction of acidic/basic fibroblast growth factor and platelet-derived growth factor in mononuclear phagocytes stimulates growth of hypoxic endotheliad cells. *Proc. Natl. Acad. Sci. USA*, **92**, 4606–4610.

Lacombe, C., Da Silva, J., Bruneval, P., Fournier, J., Wendling, F., Casadevall, N. *et al.* (1988) Peritubular cells are the site of erythropoietin synthesis in the murine hypoxic kidney. *J. Clin. Invest.*, **81**, 620–623.

Laderoute, K.R., Grant, T.D., Murphy, B.J. and Sutherland, R.M. (1992a) Enhanced epidermal growth factor receptor synthesis in human squamous carcinoma cells exposed to low levels of oxygen. *Int. J. Cancer*, **52**, 428–432.

Laderoute, K.R., Grant, T.D., Murphy, B.J. and Sutherland, R.M. (1992b) Enhancement of transforming growth factor-α synthesis in multicellular tumor spheroids of A431 squamous carcinoma cells. *Br. J. Cancer*, **65**, 157–162.

Laderoute, K.R., Calaoagan, J.M., Mendonca, H.L., Ausserer, W.A., Chen, E.Y., Giaccia, A.J. *et al.* (1996) Early responses of SiHa human squamous carcinoma cells to hypoxic signals: Evidence of parallel activation of NF-κB and AP-1 transcriptional complexes. *Int. J. Oncol.*, **8**, 875–882.

Lander, H.M., Sehajpal, P.K., Levine, D.M. and Novogrodsky, A. (1993) Activation of human peripheral blood mononuclear cells by nitric oxide-generating compounds. *J. Immunol.*, **150**, 1509–1516.

Lander, H.M., Ogiste, J.S., Pearce, S.F.A., Levi, R. and Novogrodsky, A. (1995) Nitric oxide-stimulated guanine nucleotide exchange on p21(ras). *J. Biol. Chem.*, **270**, 7017–7020.

Lau, Y.T. and Ma, W.C. (1996) Nitric oxide inhibits migration of cultured endothelial cells. *Biochem. Biophys. Res. Commun.*, **221**, 670–674.

Le Cras, T.D., Xue, C., Rengasamy, A. and Johns, R.A. (1996) Chronic hypoxia upregulates endothelial and inducible NO synthase gene and protein expression in rat lung. *Amer. J. Physiol. Lung Cell M. Ph.*, **14**, L164–L170.

Lee, M., Dhadly, M.S., Temizer, D.H., Clifford, J.A., Yoshizumi, M. and Quertermous, T. (1991) Regulation of endothelin-1 gene expression by fos and jun. *J. Biol. Chem.*, **266**, 19034–19039.

LeRoy, E.C. (1992) A brief overview of pathogenesis of scleroderma (systemic sclerosis). *Ann. Rheum. Dis.*, **51**, 286–288.

Leung, D.W., Cachianes, G., Kuang, W.-J., Goeddel, D.V. and Ferrara, N. (1989) Vascular endothelial growth factor is a secreted angiogenic mitogen. *Science*, **246**, 1306–1308.

Levy, A.P., Levy, N.S., Wegner, S. and Goldberg, M.A. (1995) Transcriptional regulation of the rat vascular endothelial growth factor gene by hypoxia. *J. Biol. Chem.*, **270**, 13333–13340.

Li, H., Elton, T.S., Chen, Y.F. and Oparil, S. (1994) Increased endothelin receptor gene expression in hypoxic rat lung. *Am. J. Physiol.*, **266**, L553–L560.

Li, H.B., Chen, S.J., Chen, Y.F., Meng, Q.C., Durand, J., Oparil, S. *et al.* (1994) Enhanced endothelin-1 and endothelin receptor gene expression in chronic hypoxia. *J. Appl. Physiol.*, **77**, 1451–1459.

Liu, Y.X., Cox, S.R., Morita, T. and Kourembanas, S. (1995) Hypoxia regulates vascular endothelial growth factor gene expression in endothelial cells – Identification of a 5' enhancer. *Circ. Res.*, **77**, 638–643.

Lockhart, A. and Saiag, B. (1981) Altitude and the human pulmonary circulation. *Clin. Sci.*, **60**, 599–605.

Lorsbach, R.B., Murphy, W.J., Lowenstein, C.J., Snyder, S.H. and Russell, S.W. (1993) Expression of the Nitric Oxide Synthase Gene in Mouse Macrophages Activated for Tumor Cell Killing – Molecular Basis for the Synergy Between Interferon-gamma and Lipopolysaccharide. *J. Biol. Chem.*, **268**, 1908–1913.

Lowenstein, C.J., Alley, E.W., Raval, P., Snowman, A.M., Snyder, S.H., Russell, S.W. *et al.* (1993) Macrophage Nitric Oxide Synthase Gene – Two Upstream Regions Mediate Induction by Interferon-gamma and Lipopolysaccharide. *Proc. Natl. Acad. Sci. USA*, **90**, 9730–9734.

Lynch, D., Ansell, P. and Levene, R. (1988) Effect of anoxia on gene expression in human endothelial cells. *J. Cell Biol.*, **107**, 581a.

MacCumber, M.W., Ross, C.A., Glaser, B.M. and Snyder, S.H. (1989) Endothelin: Visualization of mRNAs by situ hybridization provides evidence for local action. *Proc. Natl. Acad. Sci. USA*, **86**, 7285–7289.

Majack, R.A., Milbrant, J. and Dixit, V.M. (1987) Induction of thrombospondin messenger RNA levels occurs as a immediate primary response to platelet-derived growth factor. *J. Biol. Chem.*, **262**, 8821–8825.

Majack, R.A., Goodman, L.V. and Dixit, V.M. (1988) Cell surface thrombospondin in functionally essential for vascular smooth muscle cell proliferation. *J. Cell Biol.*, **106**, 415–422.

Majack, R.A., Majesky, M.W. and Goodman, L.V. (1990) Role of PDGF-A expression in the control of vascular smooth muscle cell growth by transforming growth factor-beta. *J. Cell Biol.*, **111**, 239–248.

Marshall, C., Cooper, D. and Marshall, B. (1988) Reduced availability of energy initiates pulmonary vasoconstriction. *Proc. Soc. Exp. Biol. Med.*, **187**, 282–286.

Martinet, Y., Rom, W.R. and Grotendorst, G.R. (1987) Exaggerated spontaneous release of platelet-derived growth factor by alveolar macrophages from patients with idiopathic pulmonary fibrosis. *N. Engl. J. Med.*, **317**, 202–209.

Matthews, J.R., Botting, C.H., Panico, M., Morris, H.R. and Hay, R.T. (1996) Inhibition of NF-kappa B DNA binding by nitric oxide. *Nucleic Acids Res.*, **24**, 2236–2242.

Maxwell, P.H., Pugh, C.W. and Ratcliffe, P.J. (1993) Inducible operation of the erythropoietin 3' enhancer in multiple cell lines: evidence for a widespread oxygen-sensing mechanism. *Proc. Natl. Acad. Sci. USA*, **90**, 2423–2427.

Mentzer, S.J., Reilly, J.J., DeCamp, M., Sugerbaker, D.J. and Faller, D.V. (1995) Potential mechanisms of vasomotor dysregulation after lung transplantation for primary pulmonary hypertension. *J. Heart. Lung. Transplant.*, **14**, 387–393.

Mertens, S., Noll, T., Spahr, R., Krutzfeldt, A. and Piper, H. (1990) Energetic response of coronary endothelial cells to hypoxia. *Am. J. Physiol.*, **258**, H689–H694.

Meyer, M., Schreck, R. and Baeuerle, P.A. (1993) H2O2 and Antioxidants Have Opposite Effects on Activation of NF-kappa-B and AP-1 in Intact Cells – AP-1 as Secondary Antioxidant-Responsive Factor. *EMBO J.*, **12**, 2005–2015.

Meyer, M., Pahl, H.L. and Baeuerle, P.A. (1994) Regulation of the transcription factors NF-kappa B and AP-1 by redox changes. *Chem. Biol. Interact.*, **91**, 91–100.

Meyrick, B. and Reid, L. (1980) Hypoxia-induced structural changes in the media and adventitia of the rat hilar pulmonary artery and their regression. *Am. J. Pathol.*, **100**, 151–178.

Michiels, C., De Leener, F., Arnould, T., Dieu, M. and Remacle, J. (1994) Hypoxia stimulates human endothelial cells to release smooth muscle cell mitogens: Role of prostaglandins and bFGF. *Exp. Cell Res.*, **213**, 43–54.

Minchenko, A., Salceda, S., Bauer, T. and Caro, J. (1994) Hypoxia regulatory elements of the human vascular endothelial growth factor gene. *Cell Mol. Biol. Res.*, **40**, 35–39.

Minkes, R.K., Bellan, J.A., Saroyan, R.M., Kerstein, M.D., Coy, D.H., Murphy, W.A. *et al.* (1990) Analysis of cardiovascular and pulmonary responses to endothelin-1 and endothelin-3 in the anaesthetized cat. *J. Pharmacol. Exp. Ther.*, **253**, 1118–1125.

Monson, E., Veinstein, M., Ditta, G. and Helinski, D. (1992) The fixL protein of Rhizobium metiloti can be separated into a heme-binding oxygen-sensing domain and a functional C-terminal kinase domain. *Proc. Natl. Acad. Sci. USA*, **89**, 4280–4284.

Monson, E.K., Ditta, G.S. and Helinski, D.R. (1995) The oxygen sensor protein, FixL, of Rhizobium meliloti — Role of histidine residues in heme binding, phosphorylation, and signal transduction. *J. Biol. Chem.*, **270**, 5243–5250.

Moreland, L.W., Huang, G. and Gay, R. (1990) Immunohistologic demonstration of platelet-derived growth factor (PDGF) and fibroblast growth factor (FGF) in scleroderma (SCL) skin. *Arthritis Rheum.*, **33**, S36.

Moscatelli, D. and Rifkin, D.B. (1994) Membrane and matrix localization of proteases: a common theme in tumor cell invasion and angiogenesis. *Biochem. Biophys. Acta.*, **948**, 67–85.

Mosher, D.F. (1990) Physiology of thrombospondin. *Annu. Rev. Med.*, **41**, 85–97.

Mukhopadhyay, D., Tsiokas, L., Zhou, X.-M., Foster, D., Brugge, J.S. and Sukhatme, V.P. (1995) Hypoxic induction of human vascular endothelial growth factor expression through c-Src activation. *Nature*, **375**, 577–581.

Mumby, S.M., Abbot-Brown, G.L. and Bornstein, P. (1984) Regulation of thrombospondin secretion by cells in culture. *J. Cell Physiol.*, **120**, 280–288.

Nakaki, T., Nakayama, M. and Kato, R. (1990) Inhibition by nitric oxide and nitric oxide-producing vasodilators of DNA synthesis in vascular smooth muscle cells. *Eur. J. Pharmacol.*, **189**, 347–353.

Namiki, A., Brogi, E., Kearney, M., Kim, E.A., Wu, T.G., Couffinhal, T. *et al.* (1995) Hypoxia induces vascular endothelial growth factor in cultured human endothelial cells. *J. Biol. Chem.*, **270**, 31189–31195.

Nathan, C. (1992) Nitric oxide as a secretory product of mammalian cells. *FASEB J.*, **6**, 3051–3064.

Nathan, C. and Xie, Q. (1994) Regulation of biosynthesis of nitric oxide. *J. Biol. Chem.*, **269**, 13725–13728.

Nelin, L.D., Rickaby, D.A., Linehan, J.H. and Dawson, C.A. (1994) The vascular site of action of hypoxia in the neonatal pig lung. *Pediatr. Res.*, **35**, 25–29.

Nose, K., Shibanuma, M., Kikuchi, K., Kageyama, H., Sakiyama, S. and Kuroki, T. (1991) Transcriptional activation of early-response genes by hydrogen peroxide in a mouse osteoblastic cell line. *Eur. J. Biochem.*, **201**, 99–106.

Nunokawa, Y., Oikawa, S. and Tanaka, S. (1996) Human inducible nitric oxide synthase gene is transcriptionally regulated by nuclear factor-kappa B dependent mechanism. *Biochem. Biophys. Res. Commun.*, **223**, 347–352.

Nunokawa, Y. and Tanaka, S. (1992) Interferon-gamma inhibits proliferation of rat vascular smooth muscle cells by nitric oxide generation. *Biochem. Biophys. Res. Commun.*, **188**, 409–415.

O'Shea, K.S. and Dixit, V.M. (1988) Unique distribution of the extracellular matrix component thrombospondin in the developing mouse embryo. *J. Cell Biol.*, **107**, 2737–2748.

Ohkawa, F., Ikeda, U., Kawasaki, K., Kusano, E., Igarashi, M. and Shimada, K. (1994) Inhibitory effect of interleukin 6 on vascular smooth muscle contraction. *Am. J. Physiol.*, **266**, H898–H902.

Oppenheimer, E.H. and Esterly, J.R. (1971) Pulmonary changes in sickle cell disease. *Am. Rev. Resp. Dis.*, **103**, 858–859.

Orton, E., Reeves, J. and Stenmark, K. (1988) Pulmonary vasodilation with structurally altered pulmonary vessels and pulmonary hypertension. *J. Appl. Physiol.*, **65**, 2459–2467.

Owens, M.W. and Grisham, M.B. (1993) Nitric Oxide Synthesis by Rat Pleural Mesothelial Cells – Induction by Cytokines and Lipopolysaccharide. *Am. J. Physiol.*, **265**, L110–L116.

Palmer, R.M.J., Ferrige, A.G. and Moncada, S. (1987) Nitric oxide release accounts for the biological activity of endothelium-derived relaxing factor. *Nature*, **327**, 524–526.

Patel, J.M., Zhang, J.L. and Block, E.R. (1996) Nitric oxide-induced inhibition of lung endothelial cell nitric oxide synthase via interaction with allosteric thiols: Role of thioredoxin in regulation of catalytic activity. *Amer. J. Respir. Cell Molec. Biol.*, **15**, 410–419.

Peach, M.J., Johns, R.A. and Rose, C.E. (1989). The potential role of interactions between endothelium and smooth muscle in pulmonary vascular physiology and pathophysiology. In *Pulmonary Vascular Physiology and Pathophysiology*, edited by E.K. Weir and J.T. Reeves, pp. 643–697. New York: Marcel Dekker, Inc.

Peacock, A.J., Dawes, K.E., Shock, A., Gray, A.J., Reeves, J.T. and Laurent, G.J. (1992) Endothelin-1 and Endothelin-3 Induce Chemotaxis and Replication of Pulmonary Artery Fibroblasts. *Amer. J. Respir. Cell Molec. Biol.*, **7**, 492–499.

Pech, M., Rao, C.D., Robbins, K.C. and Aaronson, S.A. (1989) Functional identification of regulatory elements within the promoter region of platelet-derived growth factor 2. *Mol. Cell Biol.*, **9**, 396–405.

Peng, H.B., Libby, P. and Liao, J.K. (1995a) Induction and stabilization of I kappa B alpha by nitric oxide mediates inhibition of NF-kappa B. *J. Biol. Chem.*, **270**, 14214–14219.

Peng, H.B., Rajavashisth, T.B., Libby, P. and Liao, J.K. (1995b) Nitric oxide inhibits macrophage-colony stimulating factor gene transcription in vascular endothelial cells. *J. Biol. Chem.*, **270**, 17050–17055.

Perkett, E., Badesch, D., Roessler, M., Stenmark, K. and Meyrick, B. (1992) Insulin-like growth factor-I and pulmonary hypertension induced by continuous air enbolization in sheep. *Am. Rev. Resp. Dis.*, **6**, 82–87.

Phelan, M., Perrine, S.P., Brauer, M. and Faller, D.V. (1995) Sickle erythrocytes, after sickling, regulate the expression of the endothelin-1 gene and protein in human endothelial cells in culture. *J. Clin. Invest.*, **96**, 1145–1151.

Phelan, M.W., Forman, L.W. and Faller, D.V. (1998) Regulated expression of thrombospondin-1 gene expression and release by hypoxia in cultured endothelial cells. *J. Lab. Clin. Med.*, in press.

Phelan, M.W. and Faller, D.V. (1996) Hypoxia decreases constitutive nitric oxide synthase transcript and protein in cultured endothelial cells. *J. Cell Physiol.*, **167**, 469–476.

Piasecki, C., Chin, J., Greenslade, L., McIntyre, N., Burroughs, A.K. and McCormick, P.A. (1995) Endoscopic detection of ischaemia with a new probe indicates low oxygenation of gastric epithelium in portal hypertensive gastropathy. *Gut*, **36**, 654–656.

Plate, K.H., Breier, G., Millauer, B., Ullrich, A. and Risau, W. (1993) Up-regulation of vascular endothelial growth factor and its cognate receptors in a rat glioma model of tumor angiogenesis. *Cancer Res.*, **53**, 5822–5827.

Pohl, U. and Busse, R. (1989) Hypoxia stimulates release of endothelium-derived relaxant factor. *Am. J. Physiol.*, **256**, H1595–H1600.

Rabinovitch, M., Konstam, M., Gamble, W., Papanicolaou, N., Aronovitz, M., Treves, S. *et al.* (1983) Changes in pulmonary blood flow affect vascular response to chronic hypoxia in rats. *Circ. Res.*, **52**, 432–441.

Radomski, M.W., Palmer, R.M.J. and Moncada, S. (1990) Glucocorticoids inhibit the expression of an inducible, but not the constitutive, nitric oxide synthase in vascular endothelial cells. *Proc. Natl. Acad. Sci. USA*, **87**, 10043–10047.

Rakugi, H., Tabuchi, Y., Nakamaru, M., Nagano, M., Higashmori, K., Mikami, H. *et al.* (1990) Evidence for endothelin-1 release from resistance vessels of rats in response to hypoxia. *Biochem. Biophys. Res. Commun.*, **169**, 973–977.

Ravichandran, L.V., Johns, R.A. and Rengasamy, A. (1995) Direct and reversible inhibition of endothelial nitric oxide synthase by nitric oxide. *Amer. J. Physiol. Heart Circ. Phy.*, **37**, H2216–H2223.

Rees, D.D., Cellek, S., Palmer, R.M.J. and Moncada, S. (1990) Dexamethasone prevents the induction by endotoxin of a nitric oxide synthase and the associated effects of vascular tone: an insight into endotoxin shock. *Biochem. Biophys. Res. Commun.*, **173**, 541–547.

Rothendler, J., Bandyopadhyay, R. and Faller, D.V. (1998) Hypoxia activates the ceramide signaling pathway in endothelial cells. *Submitted,*

Rubanyi, G.M. and Vanhoutte, P.M. (1985) Hypoxia releases a vasoconstrictor substance from the canine vascular endothelium. *J. Physiol-London.*, **364**, 4-5-56.

Rubler, S. and Fleischer, R.A. (1967) Sickle cell states and cardiomyopathy: Sudden death due to pulmonary thrombosis and infarction. *Am. J. Cardiol.*, **19**, 867–873.

Rupec, R.A. and Baeuerle, P.A. (1995) The genomic response of tumor cells to hypoxia and reoxygenation – Differential activation of transcription factors AP-1 and NF-kappa B. *Eur. J. Biochem.*, **234**, 632–640.

Ryseck, R.-P., Hirai, S.I., Yaniv, M. and Bravo, R. (1988) Transcriptional activation of c-jun-1 during the G0/G1 transition in mouse fibroblasts. *Nature*, **334**, 535–537.

Sachinidis, A., Locher, R., Vetter, W., Tatje, D. and Hoppe, J. (1990) Different effects of platelet-derived growth factor isoforms on rat vascular smooth muscle cells.. *J. Biol. Chem.*, **265**, 10238–10243.

Sarkar, R., Webb, R.C. and Stanley, J.C. (1995) Nitric oxide inhibition of endothelial cell mitogenesis and proliferation. *Surgery*, **118** 274–279.

Schini, V.B., Junquero, D.C., Scott-Burden, T. and Vanhoutte, P.M. (1991) Interleukin-1 beta induces the production of an L-arginine-derived relaxing factor from cultured smooth muscle cells from rat aorta. *Biochem. Biophys. Res. Commun.*, **176M**, 114–121.

Schmedtje, J.F., Liu, W.L., Ji, Y.S., Thompson, T.M. and Runge, M.S. (1996) Evidence of hypoxia-inducible factor-1 in vascular endothelial and smooth muscle cells. *Biochem. Biophys. Res. Commun.*, **220**, 687–691.

Scott-Burden, T., Schini, V.B., Elizondo, E., Junquero, D.C. and Vanhoutte, P.M. (1992) Platelet-derived growth factor suppresses and fibroblast growth factor enhances cytokine-induced production of nitric oxide by cultured smooth muscle cells – Effects on cell proliferation. *Circ. Res.*, **71**, 1088–1100.

Seko, Y., Tobe, K., Takahashi, N., Kaburagi, Y., Kadowaki, T. and Yazaki, Y. (1996) Hypoxia and hypoxia/reoxygenation activate Src family tyrosine kinases and p21(ras) in cultured rat cardiac myocytes. *Biochem. Biophys. Res. Commun.*, **226**, 530–535.

Semenza, G.L., Nejfelt, M.K., Chi, S.M. and Antonarakis, S.E. (1991) Hypoxia-inducible nuclear factors bind to an enhancer element located 3' to the human erythropoietin gene. *Proc. Natl. Acad. Sci. USA*, **88**, 5680–5684.

Semenza, G.L., Roth, P.H., Fang, H.M. and Wang, G.L. (1994) Transcriptional regulation of genes encoding glycolytic enzymes by hypoxia-inducible factor 1. *J. Biol. Chem.*, **269**, 23757–23763.

Sen, C.K. and Packer, L. (1996) Antioxidant and redox regulation of gene transcription. *FASEB J.*, **10**, 709–720.

Seppa, H., Grotendorst, G., Seppa, S., Sciffman, E. and Martin, G.R. (1982) Platelet-derived growth factor is chemotactic for fibroblasts. *J. Cell Biol.*, **92**, 584–588.

Shaul, P.W., Wells, L.B. and Horning, K.M. (1993) Acute and prolonged hypoxia attenuate endothelial nitric oxide production in rat pulmonary arteries by different mechanisms. *J. Cardiovasc. Pharmacol.*, **22**, 819–827.

Shaul, P.W., North, A.J., Brannon, T.S., Ujiie, K., Wells, L.B., Nisen, P.A. *et al.* (1995) Prolonged *in vivo* hypoxia enhances nitric oxide synthase type I and type III gene expression in adult rat lung. *Amer. J. Respir. Cell Molec. Biol.*, **13**, 167–174.

Shima, D.T., Deutsch, U. and D'Amore, P.A. (1995a) Hypoxic induction of vascular endothelial growth in human epithelial cells is mediated by increases in mRNA stability. *Fed. Eur. Biochem. Soc.*, **370**, 203–208.

Shima, D.T., Deutsch, U. and Damore, P.A. (1995b) Hypoxic induction of vascular endothelial growth factor (VEGF) in human epithelial cells is mediated by increases in mRNA stability. *FEBS Lett.*, **370**, 203–208.

Shirakami, G., Nakao, K., Saito, Y., Magaribuchi, T., Jougasaki, M., Mukoyama, M. *et al.* (1991) Acute pulmonary alveolar hypoxia increases lung and plasma endothelin-1 levels in conscious rats. *Life Sci.*, **48**, 969–976.

Shweiki, D., Itin, A., Soffer, D. and Keshet, E. (1992) Vascular endothelial growth factor induced by hypoxia may mediate hypoxia-initiated angiogenesis. *Nature*, **359**, 843–845.

Shweiki, D., Neeman, M., Itin, A. and Keshet, E. (1995) Induction of vascular endothelial growth factor expression by hypoxia and by glucose deficiency in multicell spheroids: Implications for tumor angiogenesis. *Proc. Natl. Acad. Sci. USA*, **92**, 768–772.

Silverstein, J.L., Steen, V.D., Medsger, T.A., Jr. and Falanga, V. (1988) Cutaneous hypoxia in patients with systemic sclerosis (scleroderma). *Arch. Dermatol.*, **124** 1379–1382.

Sirsjo, A., Soderkvist, P., Sundqvist, T., Carlsson, M., Ost, M. and Gidlof, A. (1994) Different Induction Mechanisms of mRNA for Inducible Nitric Oxide Synthase in Rat Smooth Muscle Cells in Culture and in Aortic Strips. *FEBS Lett.*, **338**, 191–196.

Smiley, J.D. (1992) The many faces of scleroderma. *Am. J. Med. Sci.*, **304**, 319–333.

Smith, P. (1994) Ultrastructure of the lung in chronic hypoxia. *Thorax*, **49**, S27–S32.

Stavri, G.T., Hong, Y., Zachary, I.C., Breier, G., Baskerville, P.A., Ylaherttuala, S. *et al.* (1995a) Hypoxia and platelet-derived growth factor-BB synergistically upregulate the expression of vascular endothelial growth factor in vascular smooth muscle cells. *FEBS Lett.*, **358**, 311–315.

Stavri, G.T., Zachary, I.C., Baskerville, P.A., Martin, J.F. and Erusalimsky, J.D. (1995b) Basic fibroblast growth factor upregulates the expression of vascular endothelial growth factor in vascular smooth muscle cells: Synergistic interaction with hypoxia. *Circulation*, **92**, 11–14.

Stein, I., Neeman, M., Shweiki, D., Itin, A. and Keshet, E. (1995) Stabilization of vascular endothelial growth factor mRNA by hypoxia and hypoglycemia and coregulation with other ischemia-induced genes. *Mol. Cell Biol.*, **15**, 5363–5368.

Stenmark, K., Fasules, J., Voelkel, N., Henson, J., Tucker, A., Wilson, H. *et al.* (1987) Severe pulmonary hypertension and arterial adventitial changes in newborn calves at 4300 m. *J. Appl. Physiol.*, **62**, 821–830.

Stevens, T. and Rodman, D.M. (1995) The effect of hypoxia on endothelial cell function. *Endothelium*, **3**, 1–12.

Stewart, D.J., Levy, R.D., Cernacek, P. and Langleben, D. (1991) Increased plasma endothelin-1 in pulmonary hypertension: maker or mediator of disease?. *Ann. Intern. Med.*, **114**, 464–469.

Storch, T.G. and Talley, G.D. (1988) Oxygen concentration regulates the proliferative response of human fibroblasts to serum and growth factors. *Exp. Cell Res.*, **175**, 317–325.

Sugihara, K., Sugihara, T., Mohandas, N. and Hebbel, R.P. (1992) Thrombospondin Mediates Adherence of CD36+ Sickle Reticulocytes to Endothelial Cells. *Blood*, **80**, 2634–2642.

Sylvester, J. and McGowan, C. (1978) The effects of agents that bind to cytochrome P-450 on hypoxic pulmonary vasoconstriction. *Circ. Res.*, **43**, 429–437.

Takagi, H., King, G.L., Ferrara, N. and Aiello, L.P. (1996) Hypoxia regulates vascular endothelial growth factor receptor KDR/Flk gene expression through adenosine A(2) receptors in retinal capillary endothelial cells. *Invest. Ophthalmol. Visual Sci.*, **37**, 1311–1321.

Takuwa, N., Takuwa, Y., Yanagisawa, M., Yamashita, K. and Masaki, T. (1989) A novel vasoactive peptide endothelin stimulates mitogenesis through inositol lipid turnover in Swiss 3T3 fibroblasts. *J. Biol. Chem.*, **264**, 7856–7861.

Thanos, D. and Maniatis, T. (1995) NF-κB: a lesson in family values. *Cell*, **80**, 529–532.

Tiktinsky, M.H. and Morin, F.C. (1993) Increasing oxygen tension dilates fetal pulmonary circulation via endothelium-derived relaxing factor. *Am. J. Physiol.*, **265**, H376–H380.

Toledano, M.B. and Leonard, W.J. (1991) Modulation of transcription factor NF-kappa B binding activity by oxidation-reduction *in vitro*. *Proc. Natl. Acad. Sci. USA*, **88**, 4328–4332.

Trojanowski, M., Wu, L. and LeRoy, E.C. (1988) Elevated expression of c-myc proto-oncogene in scleroderma fibroblasts. *Oncogene*, **3**, 477–481.

Tuder, R.M., Flook, B.E. and Voelkel, N.F. (1995) Increased gene expression for VEGF and the VEGF receptors KDR/Flk and Flt in lungs exposed to acute or to chronic hypoxia-Modulation of gene expression by nitric oxide. *J. Clin. Invest.*, **95**, 1798–1807.

Vairo, G., Innes, K.M. and Adams, J.M. (1996) Bcl-2 has a cell cycle inhibitory function separable from its enhancement of cell survival. *Oncogene*, **13**, 1511–1519.

Vane, J.R., Anggard, E.E. and Botting, R.M. (1990) Mechanisms of disease: regulatory functions of the vascular endothelium. *N. Engl. J. Med.*, **323**, 27–36.

Vanhoutte, P.M., Auch-Schwelk, W., Boulanger, C., Janssen, P.A., Katusik, Z.S., Komori, K. *et al.* (1989) Does endothelin-1 mediate endothelium-dependent contractions during anoxia?. *J. Cardiovasc. Pharmacol.*, **13**, S124–S128.

Vender, R.L., Clemmons, D.R., Kwock, L. and Friedman, M. (1987) Reduced oxygen tension induces pulmonary endothelium to release a pulmonary smooth muscle cell mitogen(s). *Am. Rev. Resp. Dis.*, **135**, 622–627.

Voelkel, N.F. and Tuder, R.M. (1995) Cellular and molecular mechanisms in the pathogenesis of severe pulmonary hypertension. *Eur. Resp. J.*, **8**, 2129–2138.

Von Euler, U. and Lilestrand, G. (1946) Observations on the pulmonary arterial blood pressure in the cat. *Acta Phys. Scandinav.*, **12**, 301–320.

Waltenberger, J., Mayr, U., Pentz, S. and Hombach, V. (1996) Functional upregulation of the vascular endothelial growth factor receptor KDR by hypoxia. *Circulation*, **94**, 1647–1654.

Wang, G. and Semenza, G. (1993) General involvement of hypoxia-inducible factor 1 in transcriptional response to hypoxia. *Proc. Natl. Acad. Sci. USA*, **90**, 4304–4308.

Wang, G.L., Jiang, B.H., Rue, E.A. and Semenza, G.L. (1995a) Hypoxia-inducible factor 1 is a basic-helix-loop-helix-PAS heterodimer regulated by cellular O-2 tension. *Proc. Natl. Acad. Sci. USA*, **92**, 5510–5514.

Wang, G.L., Jiang, B.H. and Semenza, G.L. (1995b) Effect of protein kinase and phosphatase inhibitors on expression of hypoxia-inducible factor 1. *Biochem. Biophys. Res. Commun.*, **216**, 669–675.

Wang, T.N., Qian, X.H., Granick, M.S., Solomon, M.P., Rothman, Y.L. and Tuszynski, G.P. (1995) The effect of thrombospondin on oral squamous carcinoma cell invasion of collagen. *Am. J. Surg.*, **170**, 502–505.

Wang, T.N., Qian, X.H., Granick, M.S., Solomon, M.P., Rothman, V.L., Berger, D.H. *et al.* (1996) Thrombospondin-1 (TSP-1) promotes the invasive properties of human breast cancer. *J. Surg. Res.*, **63**, 39–43.

Warren, J.B., Coughlan, M.L. and Williams, T.J. (1992) Endotoxin-Induced Vasodilatation in Anaesthetized Rat Skin Involves Nitric Oxide and Prostaglandin Synthesis. *Br. J. Pharmacol.*, **106**, 953–957.

Webster, K.A., Discher, D.J. and Bishopric, N.H. (1994) Regulation of fos and jun immediate-early genes by redox or metabolic stress in cardiac myocytes. *Circ. Res.*, **74**, 679–686.

Wegner, R.H., Marti, H.H., Schuerer-Maly, C.C., Kvietikova, I., Bauer, C., Gassman, M. *et al.* (1996) Hypoxic induction of gene expression in chronic granulomatous disease-derived B-cell lines: Oxygen sensing is independent of the cytochrome-b558-containing nicotinamide adenine dinucleotide phosphate oxidase. *Blood*, **87**, 756–761.

White, F.C., Carroll, S.M. and Kamps, M.P. (1995) VEGF mRNA is reversibly stabilized by hypoxia and persistently stabilized in VEGF-overexpressing human tumor cell lines. *Growth Factors*, **12**, 289–301.

Wing, D.A., Talley, G.D. and Storch, T.G. (1988) Oxygen concentration regulates EGF-induced proliferation and EGF-receptor down regulation. *Biochem. Biophys. Res. Commun.*, **153**, 952–958.

Xue, C., Rengasamy, A., Lecras, T.D., Koberna, P.A., Dailey, G.C. and Johns, R.A. (1994) Distribution of NOS in normoxic vs hypoxic rat lung: Upregulation of NOS by chronic hypoxia. *Amer. J. Physiol. Lung Cell M. Ph.*, **11**, L667–L678.

Yabkowitz, R., Mansfield, P.J., Dixit, V.M. and Suchard, S.J. (1993) Motility of human carcinoma cells in response to thrombospondin: Relationship to metastatic potential and thrombospondin domains. *Cancer Res.*, **53**, 378–387.

Yamane, K., Miyauchi, T., Suzuki, N., Yuhara, T., Akama, T., Suzuki, H. *et al.* (1992) Significance of plasma endothelin-1 levels in patients with systemic sclerosis. *J. Rheumatol.*, **19**, 1566–1571.

Yamane, K. (1994) Endothelin and collagen vascular disease: a review with special reference to Raynaud's phenomenon and systemic sclerosis. *Internal Med.*, **33**, 579–582.

Yan, S.F., Tritto, I., Pinsky, D., Liao, H., Huang, J., Fuller, G. *et al.* (1995) Induction of interleukin 6 (IL-6) by hypoxia in vascular cells -Central role of the binding site for nuclear factor-IL-6. *J. Biol. Chem.*, **270**, 11463–11471.

Yanagisawa, M., Kurihara, H., Kimura, S., Tomobe, Y., Kobayashi, M., Mitsui, Y. *et al.* (1988) A novel potent vasoconstrictor peptide produced by vascular endothelial cells. *Nature*, **332**, 411–415.

Yang, B.C. and Mehta, J.L. (1994) Prior episode of anoxia attenuates vasorelaxation in response to subsequent episode of anoxia. *Am. J. Physiol.*, **266**, H974–H979.

Yao, K.S., Xanthoudakis, S., Curran, T. and O'Dwyer, P.J. (1994) Activation of AP-1 and of a nuclear redox factor, Ref-1, in the response of HT29 colon cancer cells to hypoxia. *Mol. Cell Biol.*, **14**, 5997–6003.

Yao, K.S. and O'Dwyer, P.J. (1995) Involvement of NF-kappa B in the induction of NAD(P)H:quinone oxidoreductase (DT-diaphorase) by hypoxia, oltipraz and mitomycin C. *Biochem. Pharmacol.*, **49**, 275–282.

Yoshimoto, S., Ishizaki, Y., Sasaki, T. and Murota, S. (1991) Effect of carbon dioxide and oxygen on endothelin production by cultured porcine cerebral endothelial cells. *Stroke*, **22**, 378–383.

Zachariae, H., Heickendorff, L., Bjerring, P., Halkier-Sorensen, L. and Sondergaard, K. (1994) Plasma endothelin and the aminoterminal propeptide of type III procollagen (PIIINP) in systemic sclerosis. *Acta Dermato-Vener.*, **74**, 368–370.

Zamora, M., Dempsey, E., Walchak, S. and Stelzner, T. (1993) BQ 123, an ET_A receptor antagonist, inhibits endothelin-1-mediated proliferation of human pulmonary artery smooth muscle cells. *Am. J. Respir. Cell Mol. Biol.*, **9**, 429–433.

Zamora, M.R., Stelzner, T.J., Webb, S., Panos, R.J., Ruff, L.J. and Dempsey, E.C. (1996) Overexpression of endothelin-1 and enhanced growth of pulmonary artery smooth muscle cells from fawn-hooded rats. *Amer. J. Physiol. Lung Cell M. Ph.*, **14**, L101–L109.

Zubiaga, A.M., Munoz, E. and Huber, B.T. (1991) Superinduction of IL-2 gene transcription in the presence of cyclohexamide. *J. Immunol.*, **146**, 3857–3863.

9 Cyclooxygenase: An Important Transduction System for the Multifaceted Roles of Nitric Oxide

Daniela Salvemini

Discovery Pharmacology, G.D. Searle Co., 800 N. Lindbergh Boulevard,
St. Louis, Missouri 63167, USA
Tel: (314) 694-5705; Fax: (314) 694-8949; E-mail: DDSALV@ccmail.monsanto.com

Nitric oxide derived from L-arginine (L-Arg) by the enzyme nitric oxide synthase (NOS) is involved in the regulation of several important physiological and pathophysiological functions. Most of the effects exerted by NO are shared by products of the cyclooxygenase pathway (COX). COX is the enzyme that converts arachidonic acid to prostaglandins (PG), prostacyclin (PGI_2) and thromboxane A_2.

Two major forms of NOS and COX have been identified to date. Under normal circumstances, the constitutive isoforms of these enzymes are found in virtually all organs. Their presence accounts for the regulation of several important physiological effects (e.g. antiplatelet activity, vasodilation, cytoprotection). On the other hand, in an inflammatory setting, these enzymes are induced in a variety of cells resulting in the production of large amounts of proinflammatory and cytotoxic NO and PG. The release of NO and PG has been associated with the pathological roles of these mediators in several disease states.

The mechanisms by which NO exerts some of its beneficial or detrimental effects include activation of guanylate cyclase, formation of peroxynitrite, apoptosis, and regulation of cyclooxygenase (COX). Thus an important link between the NOS and COX pathways is that NO activates the COX enzymes and this results in an augmented production of prostaglandins.

The role of NO in the regulation of constitutive and inducible cyclooxygenase (COX-1 and COX-2 respectively) will be reviewed and evidence presented to suggest that these enzymes are targets for the physiopathological roles of NO. Once activated, they represent important transduction mechanisms for the multifaceted roles of nitric oxide in physiopathological events.

BIOSYNTHESIS OF NITRIC OXIDE

The mechanisms involved in the biosynthesis, release and inhibition of NO will be discussed briefly since this topic has been addressed in great details in previous chapters of this book. Nitric oxide is synthesized in mammalian systems via a five-electron oxidation of the terminal guanidino nitrogen of the amino acid, L-arginine, by a family of enzymes known collectively as nitric oxide synthases (NOS; for recent reviews see Griffith and Stuehr, 1995; Forstermann *et al.*, 1995; Knowles and Moncada, 1994; Nathan, 1992). Three distinct human NOS isoforms have been identified and the cDNAs for each of these enzymes have been isolated: the endothelial constitutive NOS (ecNOS or NOS III; 135

kDa) (Janssens *et al.*, 1992; Marsden *et al.*, 1992), the neuronal constitutive NOS (ncNOS, bNOS or NOS I; 150–160 kDA) (Schmidt and Murad, 1991; Nakane *et al.*, 1993) and the inducible NOS isoform (iNOS, mac-NOS or NOS II; 130 kDa) (Geller *et al.*, 1993; Sherman *et al.*, 1993; Charles *et al.*, 1993). All of the NOS isoforms are iron-containing heme proteins which utilize L-arginine (D-arginine is not a substrate), molecular oxygen and NADPH as cosubstrates and require an array of cofactors including flavin adenine dinucleotide (FAD), flavin mononucleotide (FMN), (6R)-tetrahydro-L-biopterin (H_4B) and calcium-activated calmodulin for activity. The three isoforms are distinct from each other based on their primary amino acid sequence (only 50–60% identity), tissue and cellular distribution and mode of regulation. With respect to their regulation, the NOS enzymes fall into two distinct categories: (i) constitutively-expressed isoforms (ecNOS and ncNOS) which are regulated primarily by Ca^{2+} and calmodulin and, (ii) a cytokine (or endotoxin)-inducible isoform (iNOS) which is regulated primarily at the level of *de novo* protein synthesis.

The distinct properties of each of the NOS isoforms have important implications since, it is the magnitude, the duration, and the cellular sites of NO production which determine the overall physiological or pathophysiological effect of NO. For example, release of NO from cNOS occurs in small amounts and for a short period of time. NO released under these circumstances plays a crucial role in the cardiovascular system where it controls organ blood flow distribution, inhibits the aggregation and adhesion of platelets to the vascular wall, inhibits leukocyte adhesion and smooth muscle cell proliferation (Moncada and Higgs, 1995). In contrast, the inducible NOS isoform is not continuously present but is expressed in a wide variety of cells in response to inflammatory stimuli such as cytokines and lipopolysaccharide (LPS). The net result is a delayed (typically 4–6 hours) but very prolonged synthesis of high levels of NO. NO released from iNOS is thus involved in several pathological events (Clancy and Abramson, 1995; Kroncke *et al.*, 1995; Moncada and Higgs, 1995). This sustained synthesis of NO derived from the inducible NOS isoform has been implicated in acute and chronic inflammation and host defense.

The nitric oxide pathway shares a number of similarties with cyclooxygenase, another critical enzyme involved in the regulation of physiological and pathological events.

BIOSYNTHESIS OF THE PROSTAGLANDINS

The prostaglandins are formed by the action of the prostaglandin synthase in a two step conversion of arachidonic acid. First, the enzyme converts arachidonic acid to a cyclic endoperoxide (PGG_2) by the action of a cyclooxygenase (COX) activity, which is then followed by a peroxidase that cleaves the peroxide to yield the endoperoxide (PGH_2) (Needleman *et al.*, 1986). These unstable intermediate products of arachidonic acid metabolism by COX are then rapidly converted to the prostaglandins (e.g. PGE_2, PGF_2, TxA_2, PGI_2) by specific isomerase enzymes (Needleman *et al.*, 1986). COX was first purified from the sheep seminal vesicle (Hemler *et al.*, 1976; Miyamoto *et al.*, 1976; Van der Ouderaa *et al.*, 1977; DeWitt and Smith, 1988; Merlie *et al.*, 1988) as a homodimer of approximate molecular mass of 140kD and subsequently cloned from the same tissue (DeWitt and Smith, 1988; Merlie *et al.*, 1988). With the availablity of the cDNA encoding the protein and specific antibodies, numerous studies were performed to evaluate the distribution, expression, and regulation of COX both *in vitro* and *in vivo*.

For a number of years, it was thought that COX was a single enzyme that produced prostaglandins in most tissues and cell types. However, a number of studies have illustrated that COX activity is increased in certain inflammatory states and is induced in cells by pro-inflammatory cytokines and growth factors *in vitro* (DeWitt, 1991; Bailey *et al.*, 1985; Sano *et al.*, 1992; Jonas-Whitley and Needleman, 1984; Habenicht *et al.*, 1985; Lin *et al.*, 1989; Maier *et al.*, 1990). For instance, incubation of dermal or synovial fibroblasts with interleukin-1b (IL-1b) revealed an induction of *de novo* synthesized COX enzyme and a significant increase in the PG production by these cells *in vitro*; this induction was blocked by inhibitors of mRNA and protein synthesis. The potent anti-inflammatory drug, dexamethasone, selectively blocked the synthesis of the "inducible" COX enzyme without altering basal COX activity in the cells (Raz *et al.*, 1988, 1989). Similar experiments were carried out in both human monocytes and murine macrophages *in vitro* demonstrating that COX activity could be "induced" by LPS and selectively blocked by glucocorticoids (Fu *et al.*, 1990). Furthermore, administration of LPS to mice resulted in a significant increase in prostaglandin production *in vivo* by peritoneal macrophages that could be blocked by dexamethasone (Masferrer *et al.*, 1992a,b). This suggested the existence of a second, "inducible" form of the enzyme at the inflammatory site. Futhermore, this induction of COX activity was the result of new protein synthesis and could be selectively blocked by anti-inflammatory agents, like the glucocorticoids. While dexamethasone blocked "*de novo*" COX synthesis, it had no effect on basal, constitutive COX expression or prostaglandin production in inflammatory cells, either *in vitro* or *in vivo*. In, 1971, Vane demonstrated that the anti-inflammatory mechanism of action of aspirin is to inhibit prostaglandin production *in vivo* by inhibiting the prostaglandin synthase or COX enzyme (Vane, 1971). Since then, in addition to aspirin, the non-steroidal anti-inflammatory drugs (NSAIDs) have been shown to inhibit the COX enzyme, providing the therapeutically beneficial anti-inflammatory and analgesic activity observed with these drugs *in vivo*. A second gene expressing COX was identified in the chick and mouse, and subsequently in rat and human cell lines (Xie *et al.*, 1991; Kujubu *et al.*, 1991; Hla and Neilson, 1992; Sirois and Richards, 1992; Kennedy *et al.*, 1993). The transcript for this "COX-2" is induced both *in vitro* and *in vivo* by cytokines, growth factors, and LPS, and is selectively inhibited by dexamethasone (Kujubu *et al.*, 1993; O'Bannion *et al.*, 1992; Kujubu and Herschman, 1992; Lee *et al.*, 1992; Phillips *et al.*, 1993; Masferrer *et al.*, 1994a; Isakson *et al.*, 1995). The inducible "COX-2" is approximately 60% identical at the amino acid level from the constitutive "COX-1", with the most prominent difference in the sequences being a stretch approximately 20 novel amino acids inserted in the carboxy terminus of the COX-2. COX-2 is rapidly induced in response to numerous pro-inflammatory mediators (e.g. endotoxin, growth factors, IL-1b, TNFa). The induction of COX-2 transcripts is rapid, transient, and selectively inhibited by glucocorticoids. The closely related transcript encoding COX-1 is not induced by inflammatory mediators nor is its expression altered by glucocorticoids (Kujubu *et al.*, 1993). There is now substantial evidence to support the concept that production of PG from the constitutive enzyme accounts for the beneficial effects attributed to the PG (renal/gastroprotection) whereas PG production from the induced enzyme accounts for the proinflammatory effects (Isakson *et al.*, 1995). In this respect, we have presented evidence demonstrating that inhibition of COX-2 activity by selective COX-2 inhibitors is antiinflammatory (Masferrer *et al.*, 1994b), analgesic (Seibert *et al.*, 1994) and non-ulcerogenic (a COX-1 mediated effects seen for instance with non-selective COX inhibitors; Masferrer *et al.*, 1994b).

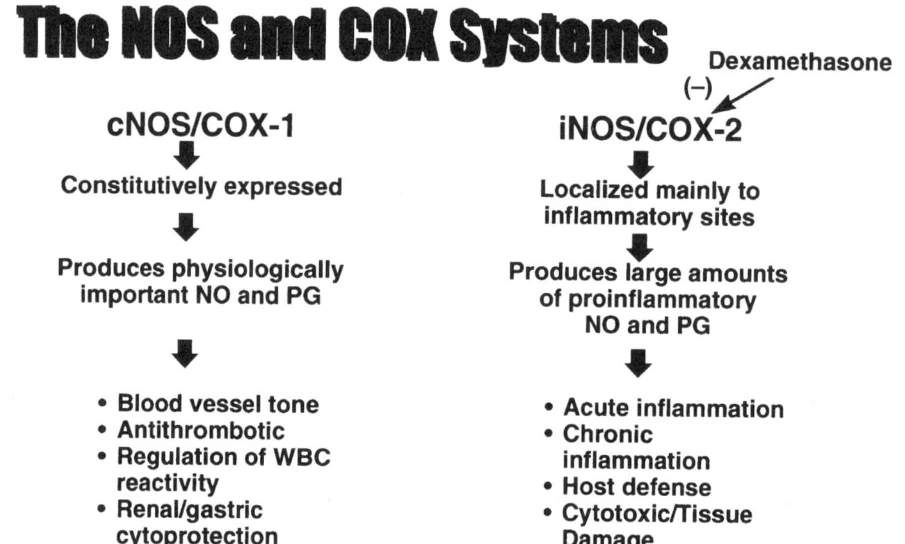

Figure 9-1. Similarities between the NOS and COX pathways. NO and PG derived from the constitutive isoform of NOS and COX (cNOS and COX-1) plays a critical function in a variety of physiological processes whereas the sustained release of large amounts of NO and PG from the inducible NOS and COX isoform (iNOS and COX-2) in a variety of disease states. The ability of steroids including dexamethasone to inhibit the induction of both iNOS and COX-2 may explain their potent anti-inflammatory effects in several diseases.

In summary it is therefore clear that the NOS and COX systems are often present together, share a number of similarities and play fundamental roles in similar physiopathological conditions (Figure 9-1). The ability of the steroids to inhibit induction of both iNOS and COX-2 may explain their potent anti-inflammatory effects (Figure 9-1). An additional feature that links the NOS and COX pathways is that NO can markedly enhance the production of PG (Salvemini *et al.*, 1993). This property of NO is attributable to its ability to activate the COX enzymes (Salvemini *et al.*, 1993).

ACTIVATION OF THE COX ENZYMES BY NITRIC OXIDE: DISCOVERY

The discovery that NO regulates COX activity was originally made using cellular systems and purified enzymes. Microsomal sheep vesicles are a rich source of COX-1 and can thus be used to explore whether the exogenous application of NO can augment further COX-1 activity (Salvemini *et al.*, 1996a). NO gas directly increases COX-1 activity of microsomal sheep seminal vesicles as well as murine recombinant COX-1; this leads to a remarkable 7 fold increase in PGE_2 formation (Salvemini *et al.*, 1993, 1995a). COX-2 is also activated by NO. COX-2 but not iNOS is induced in human fetal fibroblast by IL-1β. Therefore IL-1β stimulated fibroblasts can be used as a cellular model to investigate the effects of exogenous NO on COX-2 activity. Exposure of IL-1β stimulated fibroblasts to either NO gas or two NO-donors sodium nitroprusside (SNP) and glyceryl trinitrate (GTN) increased

COX-2 activity by at least 4 fold; this resulted in increased production of PG. This phenomenon is independent of the known effects of NO on the soluble guanylate cyclase. Methylene blue, an inhibitor of the soluble guanylate cyclase inhibited the increase in cGMP by NO in the fibroblast but did not prevent its ability to stimulate COX activity and hence PG production (Salvemini *et al.*, 1993). The ability of NO to directly activate COX-2 was supported by the findings that NO increases the activity of purified recombinant COX-2 enzymes. Having observed that NO activates COX-1 and COX-2 enzymes, we then asked the question as to whether COX-2 activity was affected by endogenously produced NO. In this respect, the mouse macrophage cell line RAW-264.7 was stimulated with endotoxin so as to induce iNOS and COX-2 enzymes. This results in the production of large amounts of NO and PG. Inhibition of iNOS activity by non-selective NOS inhibitors such L-NMMA or NO_2Arg or more selective iNOS inhibitors such as L-NIL or aminoguanidine (AG), (Corbett *et al.*, 1992; Misko *et al.*, 1993; Moore *et al.*, 1994; Connor *et al.*, 1995) attenuated as expected the release of NO from these cells. The remarkable finding was that when NO release was inhibited, there was a simultaneous inhibition of PG release (Salvemini *et al.*, 1993). The NOS inhibitors did not behave as non steroidal antiinflammatory drugs (NSAIDs) for they did not inhibit COX activity (Salvemini *et al.*, 1993) nor did they affect the induction of COX-2 (unpublished observations). These results suggested that endogenously released NO from the macrophages exerted a stimulatory action on COX-2 activity enhancing the production of PG. Thus, inhibition of NOS activity reduced the output of PG from COX-2.

ACTIVATION OF COX ENZYME BY NITRIC OXIDE: MOLECULAR MECHANISMS

Over the last few years experimental evidence has been reported that may help us to understand how NO activates the COX enzymes: these possibilities are summarized in Figure 9-2.

NO interacts with several heme proteins which can either be inhibited or activated as a consequence of reacting with NO. For instance, NO mediates some of its toxic effects by interacting with iron-sulfur centers in key enzymes of the respiratory cycle and DNA synthesis (Kroncke *et al.*, 1995 for review) and the oxygen-carrier function of hemoglobin/ myoglobin is inhibited by NO following nitrosation of their heme (Murad *et al.*, 1978). One of the best examples for a NO-mediated activation of an iron-containing enzyme is the soluble guanylate cyclase and indeed for many years has been considered to be the principle target for the physiological and pathological roles of NO. Thus, the ability of NO to inhibit platelet aggregation and to relax vascular smooth muscle is the result of NO binding to the heme-Fe^{2+} prosthetic group of the soluble GC leading to its stimulation. Cyclooxygenase is a heme protein having a protoporphirin IX group that is required for both its cyclooxygenase and hydroperoxidase activities (Van der Ouderaa *et al.*, 1979; Roth *et al.*, 1981; Lambier *et al.*, 1985; Karthein *et al.*, 1987). One of the principle mechanism for direct activation of the COX enzyme is reaction of fatty acid hydroperoxides with the heme prosthetic group to generate a protein radical (probably Tyr-385). This protein radical then serves as the catalytic oxidant of arachidonic acid (Karthein *et al.*, 1988; Tsai *et al.*, 1995). Evidence has emerged which suggests that the mechanism by which NO activates COX is unlikely to be the consequence of its binding to the ferric heme in the enzyme

                    Daniela Salvemini

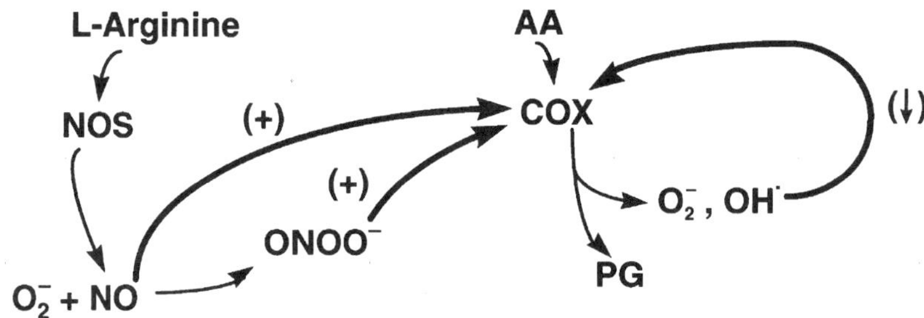

Figure 9-2. Proposed mechanisms implicated in the activation of COX by nitric oxide (see text for description).

(Tsai *et al.*, 1994). Rather it has been shown that formation of nitrosothiols is important for increased COX activity (Hajjar *et al.*, 1995). Thus, Hajjar and coworkers demonstrated that NO nitrosylates cysteine residues in the catalytic domain of COX enzymes leading to the formation of nitrosothiols; these can produce changes in the structure of the enzyme which results in increased catalytic efficiency (Hajjar *et al.*, 1995).

O_2^- is generated during COX activation and has been postulated to be involved in the auto-inactivation of the COX enzyme (Egan *et al.*, 1976). NO interacts with O_2^- and limits the amounts of the radical necessary for autoinactivation (Gryglewski *et al.*, 1987). One possibility could therefore be that NO augments COX activity by acting as an antioxidant (removal of O_2^-) preventing the auto-inactivation of COX.

When the amounts of NO and O_2^- are increased such as during an inflammatory response, NO and O_2^- interact to form the cytotoxic molecule peroxynitrite (ONOO-, Beckman *et al.*, 1990). ONOO- is an inorganic hydroperoxide and as such is a potential substrate for the peroxidase activity of COX and an activator of COX activity. In this respect, using purified COX-1 and COX-2 enzymes as well as sheep seminal vesicles, Landino and coworkers reported that ONOO- increased COX-1 and COX-2 activity (Landino *et al.*, 1996).

Peroxynitrite decomposes to OH· (Beckman *et al.*, 1990). The participation of the hydroxyl radical in COX activation has recently been suggested by demonstrating that OH· relaxes isolated rabbit aorta via an indomethacin-sensitive pathway indicative for the participation of COX-metabolites in this effect (Bharadwaj and Prasad, 1997).

The molecular mechanims that are involved in the NO mediated COX activation require to be defined further.

NO-MEDIATED COX-1 ACTIVATION

The activation of COX-1 by NO released from the constitutive form of NOS has important consequences in normal physiological conditions. This could indeed represent an important mechanism through which NO and PGs exert their beneficial cytoprotective effects in the cardiovascular system and in the central nervous systems.

The production of small amounts of NO and PGs from the constitutive enzymes regulate various physiological processes including the inhibition of platelet aggregation and white blood cell adhesion, regulation of blood vessel tone, cytoprotection in the kidney and intestinal mucosa. The main difference between the effects of NO and PGs in mediating these effects lies at the level of their respective intracellular transduction mechanisms. Thus, NO activates the soluble guanylate cyclase leading to increased cGMP and PG activate the adenylate cyclase leading to increase in cAMP. This raises the possibility that the increased platelet aggregation, vasoconstriction and elevation of systemic blood pressure induced by inhibition of endogenous NO production by non-selective NOS inhibitors could be due not only to removal of endogenous NO but also to a concurrent reduction of antiplatelet and vasodilator COX products. Dual inhibition of NO and PG by non-selective NOS inhibitors may well explain the deleterious effects observed with these drugs in organs such as the kidneys and the gastrointestinal tract where both NO and PG are cytoprotective.

The discovery that NO directly activates COX-1 enzyme has given a new dimension to the usefulness of exogenous NO therapy and has uncovered additional mechanisms by which NO-donors might exert their beneficial effects in the clinic. It is known that NO release from clinically useful NO-donors such as GTN or SIN-1 accounts for their vasodilatory and antithrombotic actions and that these properties are pertinent to the therapeutic effects of the NO-donors in conditions such as myocardial ischemia, thrombosis and atherosclerosis that are associated with a failing endogenous NO pathway (Anggard, 1991). For instance, since NO-donors do not require an intact endothelium in order to be effective (Luscher *et al.*, 1990) they can restore the desired vascular dilation and suppress platelet aggregation despite advanced atherosclerosis (Parker, 1987; Stamler and Loscalzo, 1991). The cardiovascular effects exerted by endogenously produced NO are often mediated in conjunction with prostacyclin (PGI_2) a potent vasodilator and platelet inhibitory COX metabolite released from the endothelium. NO released from the NO-donors activates COX-1 *in vitro* in endothelial cells as well as *in vivo* promoting the release of PGI_2 (Davidge *et al.*, 1995; Salvemini *et al.*, 1996b). Activation of COX is a cGMP-independent process. Our work indicates that the vasodilator and antithrombotic effects of NO (and NO-donors) occur via two steps; (a) direct NO-stimulated soluble guanylate cyclase activation and cGMP elevation and (b) NO-mediated activation of cyclooxygenase in the endothelial cells leading to PGI_2 release and cAMP elevation (Salvemini *et al.*, 1996b). During the progression of atherosclerosis or hypertension, defects at the level of the soluble guanylate cyclase have been observed. This may theoretically hamper the effectiveness of NO donors in restoring their vasodilator and antithrombotic (cGMP dependent) properties. However, since NO does not have to rely on the soluble guanylate cyclase to activate COX, NO-donors will still be able to provide adequate satisfactory vasodilation and antithrombotic effects. Thus, dual formation of NO and PGI_2 and their subsequent interactions at the levels of the vasculature and circulating cells such as platelets may represent a powerful mechanism used by therapeutically administered nitrovasodilators to restore abnormal vascular tone and platelet reactivity in pathological states associated with an altered endothelium.

An interesting effect of NO on COX-1 is perhaps in the regulation of neuropeptide release. Norepinephrine (NE) mediates the release of luteinizing hormone-releasing hormone (LHRH) from LHRH terminals and it is known that LHRH release from these terminals requires increased release of PGE_2 (Rettori *et al.*, 1992). In an elegant study, Rettori and collegues demonstrated that the release LHRH following NE-stimulated

                                    Daniela Salvemini

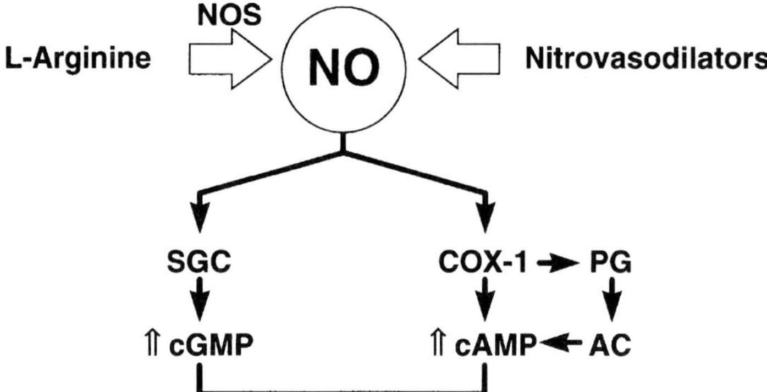

Figure 9-3. Nitric oxide mediated COX-1 activation and its implications (see text for further details).

hypothalamic slices was prevented by NOS inhibition and subsequent blockade of NE-mediated PGE_2 release (Rettori *et al.*, 1992). These results have opened the possibility that through COX activation, NO may mediate exocytosis of secretory granules not only for LHRH but also for other neuropeptides that are released by PGE_2. NO-driven COX-1 activation also seems to have a role in reproduction in that NO also modulates uterine motility (Franchi *et al.*, 1994) by activating COX-1 and releasing PGE_2. The potential interaction of NO and COX-1 and its implications in reproduction remains an exciting area for future investigation.

Physiological implications of NO-driven COX-1 activation are summarized in Figure 9-3.

NO-MEDIATED COX-2 ACTIVATION AND INFLAMMATION

iNOS and COX-2 are induced in a number of inflammatory models including rabbit hydronephrotic kidney (Salvemini *et al.*, 1994), endotoxin-induced septic shock (Salvemini *et al.*, 1995b) and carrageenan-induced pouch and paw inflammation (Salvemini *et al.*, 1995c,, 1996c; Vane *et al.*, 1994). The prolonged release of large amounts of NO and PG subsequently lead to deleterious effects and inhibition of either iNOS or COX-2 activity is protective in several diseases. *In vivo* studies revealed that the regulation of COX-2 by NO is a powerful mechanism that is used by NO to amplify the course of the inflammatory response. To give but one of several examples, in an acute model of inflammation, namely carrageenan-induced paw edema in rats, inhibition of edema with selective iNOS inhibitors such as aminoguanidine or L-NIL was associated with a dose-dependent inhibition of NO release and pro-inflammatory PG (Salvemini *et al.*, 1996c). The anti-inflammatory potency of the iNOS inhibitors correlate with their ability to block both NO and PG (see all papers mentioned above). That the proinflammatory roles of NO have a PG component has also been demonstrated by showing that the injection of SNP elicits edema in the footpad of rats and the formation of edema is blocked by NSAID (Sautebin *et al.*, 1995). Therefore, the effects of endogenously released NO on COX-2 are mimicked by exogenous NO. Inhibition of NO in lungs taken from endotoxin-treated rats resulted in an inhibition of PGI_2 release (Sautebin and DiRosa, 1994). In acute pancreatis, the increased production of PG was inhibited by NOS inhibitors indicative of the participation of NO in this process (Closa *et al.*, 1994).

Furthermore, injection of carrageenan into the preformed air pouch of a rat induces an inflammatory response characterised by iNOS and COX-2 induction, white blood cell infiltration, edema and protein leakage into the pouch (Salvemini *et al.*, 1995c). Selective inhibition of iNOS by L-NIL and AG inhibited not only NO but also PG production (Salvemini *et al.*, 1995c). All other parameters of inflammation were attenuated and histological examination of the pouch lining taken from animals treated with the iNOS inhibitor revealed a lack of inflammation (Salvemini *et al.*, 1995c). In this scenario, we were able to demonstrate that in the presence of L-NIL, COX-2 was activated at least 7 fold by exogenous injection of NO-donors such as sodium nitroprusside or nitroglycerin; this activation resulted in a profound increase in PGE_2 release (Salvemini *et al.*, 1995c). Similar results have been observed in the hydronephrotic rabbit model of inflammation (Salvemini *et al.*, 1994). Non-selective NOS inhibitors such as L-NMMA are also anti-inflammatory in these models and block both NO and PG. Nevertheless, when compared to a more selective iNOS inhibitor their antiinflammatory profile is weaker. This may stem from the fact that non-selective NOS inhibitors also block release of NO and PG from the constitutive enzymes which are known to play key cytoprotective roles. This would mask to some degree protective effects that would arise from their inhibitory effects on iNOS. This adds support to the notion that it is imperative to preserve cNOS and COX-1 activity in an inflammatory setting and this can be achieved by the use of glucocorticoids. The potent anti-inflammatory glucocorticoid dexamethasone is a good example of an agent able to inhibit the induction of both iNOS (Moncada and Higgs, 1995) and COX-2 (Masferrer and Seibert, 1994). Unfortunately, serious side effects of steroids, independent of their ability to block iNOS and COX-2 expression, limit their clinical usage. Selective inhibition of iNOS or COX-2 is also antiinflammatory in a number of models. Besides its proinflammatory role, a feature of NO that is not shared by PG is its potent cytotoxic effect and this could explain why in arthritis a NSAIDs such as indomethacin, by blocking PG but not NO production, alleviates the symptoms associated with the inflammatory insult but does not modify the course of the disease (Flynn, 1994). Since selective iNOS inhibitors have the potential of reducing both NO and PG, it is then exciting to conceive that the utilization of these selective iNOS inhibitors may have the advantage of not only alleviating inflammatory symptoms through dual inhibition of NO and NO-driven COX-2 activation, but would also eliminate further cytotoxic damage of NO (Figure 9-4).

NO-MEDIATED COX-2 IN OTHER SYSTEMS

NO has also been reported to modulate COX-2 activity in other systems. Rat islet cells or vascular smooth muscle cells stimulated with IL-1β release both NO and PG; inhibition of NO production with NOS inhibitors blocked the production of PGE_2 from immunostimulated cells (Corbett *et al.*, 1993; Inoue *et al.*, 1993). Recent studies also indicate that the release of PG following induction of COX-2 in astroglial cells stimulated

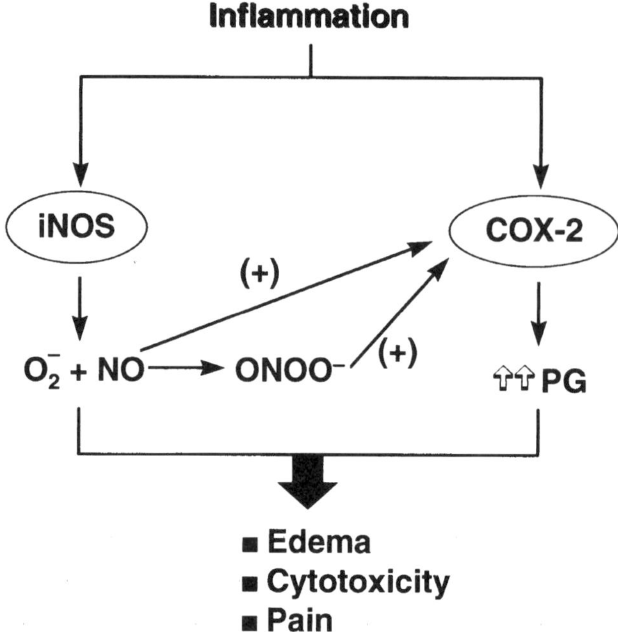

Figure 9-4. NO-driven COX-2 activation in inflammation. Regulation of COX-2 by NO itself or by a product formed from its interaction with other radicals is a powerful mechanism which amplifies the course of the inflammatory response. In this scenario it is easy to see how, by inhibiting not only NO but also PG, NOS inhibitors exert their antiinflammatory effects.

with NMDA is inhibited by iNOS inhibition (Mollace *et al.*, 1995). Stimulation of RAW 264.7 macrophages with conditioned medium of activated encephalitogenic lymphoid cells is associated with induction of both iNOS and COX-2 (Misko *et al.*, 1995). Inhibition of iNOS activity by the selective iNOS inhibitor aminoguanidine inhibited not only NO but also the increased production of PGE_2 (Misko *et al.*, 1995). Aminoguanidine had no effects on the induction of COX-2 enzyme (Misko *et al.*, 1995). Cultured T67 human astrocytoma cells induced with the HIV coating gp120 glycoprotein, induced iNOS and COX-2 which in turn released large amounts of NO and PGE_2 in the cell culture media (Mollace *et al.*, 1994). The NOS inhibitor, nitroarginine, inhibited both NO and PGE_2 (Mollace *et al.*, 1994). The interactions between the iNOS and COX-2 system may play an important modulatory role in pathophysiological mechanisms such as cerebral ischemia, stroke, multiple sclerosis and AIDS (Mollace *et al.*, 1994,, 1995; Misko *et al.*, 1995).

INHIBITION OF PG BY NITRIC OXIDE

The above *in vitro* as well as *in vivo* data indicate that NO activates COX enzymes. A few investigators have however reported an inhibitory effect of NO on both the expression and the activity COX in cells including mouse macrophage cell line J774.2, rat kupffer cells and rat peritoneal macrophages (Swierkosz *et al.*, 1995; Stadler *et al.*, 1993; Habib *et al.*, 1994). In endotoxin-stimulated J774.2 macrophages sodium nitroprusside inhibits expression of COX-2 as well as its activity and the inhibition of endogenous NO by NOS inhibitor increased PGI_2 release and COX-2 expression (Swierkosz *et al.*, 1995). Nevertheless, when these cells were treated first with a NOS inhibitor and then stimulated with a NO-donor, an increased release of PGI_2 was observed, indicative of the ability of NO to increase COX activity (Swierkosz *et al.*, 1995). Most of the data collected has been performed in *in vitro* cells or tissue. Overall, the experimental data obtained so far on an inhibitory effect of NO on the COX pathway is controversial and insufficient. Thus the *in vivo* significance of a possible negative effect of NO on COX needs to be assessed in future studies.

MODULATION OF NITRIC OXIDE SYNTHASE BY PG

Products of the COX pathway may inhibit or increase the release of NO. Large amounts of PGI_2 inhibit LPS-stimulated iNOS induction in J774.2 macrophages or murine peritoneal macrophages without affecting the enzyme once expressed (Marotta *et al.*, 1992; Milano *et al.*, 1995; Raddassi *et al.*, 1993; Bulut *et al.*, 1993).

A role for PG as enhancers of NO production as been reported. In rat Kupffer cells stimulated with endotoxin, inhibition of endogenously produced PG by indomethacin resulted in a concommittant inhibition of NO release (Gaillard *et al.*, 1992). On the other hand, indomethacin did not inhibit NO release in mouse RAW 264.7 cells (Salvemini *et al.*, 1993) or in J774.2 macrophages (Swierkosz *et al.*, 1995).

The mechanism of action of the PG on the NOS pathway has been attributed in most instances to activation of adenylate cyclase system with subsequent increase in cyclic AMP (cAMP) levels. Increases in intracellular cAMP has been suggested as mediating either the inhibitory role of PG on NOS pathway (Raddassi *et al.*, 1993; Bulut *et al.*, 1993) or the stimulatory effect of PG on the NOS pathway. For instance, in rat vascular smooth muscle, agent such as forskolin or ditutyryl-cAMP increase iNOS mRNA as well as the formation of NO (Ima *et al.*, 1994; Hirokawa *et al.*, 1994). Also, in rat glomerular mesangial cells it has been shown that cAMP upregulates iNOS expression (Muhl *et al.*, 1994). In rat aorta smooth muscle cells an elevation of intracellular cAMP positively regulates NO production at the level of the iNOS mRNA expression (Koide *et al.*, 1993). In the same cells, the iNOS gene has been shown to be up-regulated by cAMP (Imai *et al.*, 1994). The molecular mechanisms by which cAMP uregulates iNOS has been related to an increased mRNA stability and/or transcriptional activation (Koide *et al.*, 1993; Muhl *et al.*, 1994; Imai *et al.*, 1994).

Again, the experimental data obtained so far on the inhibitory/stimulatory effects of PG on the NOS pathways is controversial and insufficient. The *in vivo* significance of possible effects of PG on the NOS system both in physiology and pathology needs to be evaluated. We and others have found that *in vivo* inhibition of PG release by selective inhibition of COX-2 in inflammation has only minimal effects on the NO production indicating that products of the COX pathway do not play a major role in the regulation of NOS activity.

CONCLUSIONS

This review was to highlight the importance of the discovery that cyclooxygenase enzymes are activated by NO. The molecular mechanisms involved in this activation require further investigation. Further understanding on how these two critical systems interact will undoubtelly provide us with a better understanding of the complex roles described so far for NO. The discovery that NO may regulate physiological and pathological effects via the activation of the COX enzymes raised the novel possibility that the COX enzymes are important endogenous transduction mechanisms for the physiological and pathological effects of NO.

REFERENCES

Anggard, E.E. (1991) Endogenous nitrates-implications for treatment and prevention. *Eur. Heart, J.*, **12**, 5–8.

Bailey, J.M., Muza, B., Hla, T. and Salata, K. (1985) Restoration of prostacyclin synthase in vascular smooth muscle cells after aspirin treatment: regulation by epidermal growth factor. *J. Lipid Res.*, **26**, 54–61.

Beckman, J.S., Beckman, T.W., Chen, T.W., Marshall, P.A. and Freeman, B.A. (1990) Apparent hydroxyl radical production by peroxynitrite: implications for endothelial injury from nitric oxide and superoxide. *Proc. Natl. Acad. Sci. USA*, **87**, 1620-1624.

Bharadwaj, L.A. and Prasad, K. (1997) Mechanism of hydroxyl radical-induced modulation of vascular tone. *Free. Radical. Biol. Med.*, **22**, 381–390.

Bulut, V., Severn, A. and Liew, F.Y. (1993) Nitric oxide production by murine macrophages is inhibited by prolonged elevation of cyclic AMP. *Biochem. Biophy. Res. Commun.*, **195**, 1134–1138.

Clancy, R.M. and Abramson, S.B. (1995) Nitric oxide: a novel mediator of inflammation. *Soc. Exp. Biol. Med.*, **23**, 93–101.

Closa, D., Hotter, G., Prats, N., Bulbena, O., Rosello-Catafau, J., Fernandez-Cruz, L. and Gelpi, E. (1994) Prostanoid generation in early stages of acute pancreatis: a role for nitric oxide. *Inflammation*, **18**, 469–480.

Charles, I.G., Palmer, R.M.J., Hickery, M.S., Bayliss, M.T., Chubb, A.P., Hall, V.S., Moss, D.W. and Moncada, S. (1993) Cloning, characterization, and expression of a cDNA encoding an inducible nitric oxide synthase from the human chondrocyte. *Proc. Natl. Acad. Sci. USA*, **90**, 11419–11423.

Connor, J., Manning, P.T., Settle, S.L., Moore, W.M., Jerome, G.M., Webber, R.K., Tjoeng, F.S and Currie, M. G. (1995) Suppression of adjuvant-induced arthritis by selective inhibition of inducible nitric oxide synthase. *Eur. J. Pharmacol.*, **273**, 15–24.

Corbett, J.A., Tilton, R.G., Chang, K., Hasan, K.S., Ido, Y., Wang, J.L., Sweetland, M.A., Lancaster, J.R.Jr., Williamson, J.R. and McDaniel, M. L. (1992) Aminoguanidine, a novel inhibitor of nitric oxide formation, prevents diabetic vascular dysfunction. *Diabetes*, **41**, 552–556.

Corbett, J.A., Kwon, G., Turk, J. and McDaniel, M.L. (1993) IL1ß induces the coexpression of both nitric oxide synthase and cyclooxygenase by islets of langerhans: activation of cyclooxygenase by nitric oxide. *Biochem.*, **32**, 13767–13770.

Davidge, S.T., Baker, P.N., Mclaughlin, M.K. and Roberts, J.M. (1995) Nitric oxide produced by endothelial cells increases production of eicosanoids through activation of prostaglandin H synthase. *Circ. Res.*, **77**, 274–283.

DeWitt D.L. and. Smith, W.L. (1988) Primary structure of prostaglandin G/H synthase from sheep vesicular gland determined from the complementary DNA sequence, *Proc. Natl. Acad. Sci. USA*, **85**,1412–1416.

DeWitt, D.L. Prostaglandin endoperoxide synthase: regulation of enzyme expression. (1991) *Biochim. Biophys. Acta*, **1083**, 121–134.

Egan, R.W., Paxton, J. and Kuehl, F.A.Jr. (1976) Mechanism for irreversible self-deactivation of prostaglandin synthetase. *J. Biol. Chem.*, **251**, 7329–7335.

Flynn, B. L. (1994) Rheumatoid arthritis and osteoarthritis: current and future therapies. *Am. Pharm.*, **NS34**, 31–42.

Forstermann, U., Closs, E.I., Pollock, J.S., Nakane, M., Schwarz, P., Gath, I. and Kleinart, H. (1995) Nitric oxide synthase isozymes. Characterization, purification, molecular cloning, and functions. *Hypertension*, **23**, 1121–1131.

Franchi, A.M, Chaud, M., Rettori, V., Suburu, A., McCann, S.M. and Gimeno, M. (1994) Role of nitric oxide in eicosanoid synthesis and uterine motility in estrogen-treated rat uteri. *Proc. Natl. Acad. Sci. USA*, **91**, 539–543.

Fu, J.-Y., Masferrer, J.L., Seibert, K., Raz, A. and Needleman, P. (1990) The Induction and suppression of prostaglandin H_2 synthase (cyclooxygenase) in human monocytes. *J. Biol. Chem.*, **265**, 16737–16740.

Gaillard, T., Mulsch, A., Klein, H. and Decker, K. (1992) Regulation by prostaglandin E_2 of cytokine-elicited nitric oxide synthesis in rat liver macrophages. *Biol. Chem.*, **373**, 897–902.

Geller, D.A., Lowenstein, C.J., Shapiro, R.A., Nussler, A.K. and DiSilvo, M. (1993) Molecular cloning and expression of inducible nitric oxide synthase from human hepatocytes. *Proc. Natl. Acad. Sci. USA*, **90**, 3491–3495.

Griffith, O. W. and Stuehr, D. J. (1995) Nitric Oxide Synthases: Properties and Catalytic Mechanism. In *Annu. Rev. Physiol.*, Hoffman, J. F. and De Weer, P., Eds. (Annual Reviews Inc., Palo Alto, CA), Vol., **57**, pp. 707–736.

Gryglweski, R.J., Palmer, R.M.J. and Moncada, S. (1987) Superoxide anion is involved in the breakdown of endothelium-derived vascular relaxing factor. *Nature*, **320**, 454–456.

Habib, A., Bernard, C., Tedgui, A. and Maclouf, J. (1994) Evidence of cross-talks between inducible nitric oxide synthase and cyclooxygenase II in rat peritoneal macrophages. *Abstracts of the 9th International Conference on Prostaglandins and Related Compounds*, Florence (Italy), June 6–10: 57.

Habenicht, A.J.R., Goerig, M., Grulich, J., Ruthe, D., Gronwald, R., Loter, U., Schettler, G., Kommerell, B. and Ross, R. (1985) Human platelet-derived growth factor stimulates prostaglandin synthesis by activation and by rapid *de novo* synthesis of cyclooxygenase. *J. Clin. Invest.*, **75**, 1381–1387.

Hajjar, D.P., Lander, H.M., Pearce, F.S., Upmacis, R.K. and Pomerantz, K.B. (1995) Nitric oxide enhances prostaglandin-H synthase activity by a heme-independent mechanism: evidence implicating nitrosothiols. *J. Am. Chem. Soc.*, **117**, 3340–3346.

Hemler, M., Lands, W.E.M. and Smith, W.L. (1976) Purification of the cyclooxygenase that forms prostaglandins. *J. Biol. Chem.*, **251**, 5575–5579 .

Hirokawa, K., O'Shaughnessy, K., Moore, K., Ramrakha, P. and Wilkins, M.R. (1994) Induction of nitric oxide synthase in cultured vascular smooth muscle cells: the role of cyclic AMP. *Br. J. Pharmacol.*, **112**, 396–402.

Imai, T., Hirata, Y., Kanno, K. and Marumo, F. (1994) Induction of nitric oxide synthase by cyclic AMP in rat vascular smooth muscle cells. *J. Clin. Invest.*, **93**, 543–549.

Inoue, T., Fukuo, K., Morimoto, S., Koh, E. and Ogihara, T. (1993) Nitric oxide mediates interleukin-1-induced prostaglandin E_2 production by vascular smooth muscle cells. *Biochem. Biophys. Res. Commun.*, **194**, 420–424.

Isakson, P., Seibert, K., Masferrer, J.L., Salvemini, D., Lee, L. and Needleman, P. (1995) In *Proceedings of the Ninth International Conference on Prostaglandins and Related Compounds*, ed. B. Samuelsson, R. Paoletti, and P. Ramwell. New York, Raven Press.

Janssens, S.P., Shimouchi, A., Quertermous, T., Bloch, D.B. and Bloch, K.D. (1992) Cloning and expression of a cDNA encoding human endothelium-derived relaxing factor/nitric oxide synthase. *J. Biol. Chem.*, **267**, 14519–14522 .

Jonas-Whitley, P. and Needleman, P. (1984) Mechanism of enhanced fibroblast arachidonic acid metabolism by mononuclear cell factor. *J. Clin. Invest.*, **74**, 2249–2253.

Karthein, R., Nastaincyzk, W. and Ruf, H.H. (1987) EPR study of ferric native prostaglandin H synthase and its ferrous NO derivative. *Eur. J. Biochem.*, **166**, 173–180.

Knowles, R.G. and Moncada, S. (1994) Nitric oxide synthases in mammals. *Biochem. J.*, **298**, 249–258.

Koide, M., Kawahara, Y., Nakayama, I., Tsuda, T. and Yokoyama, M. (1993) Cyclic AMP-elevating agents induce an inducible type of nitric oxide synthase in cultured vascular smooth muscle cells. *J. Biol. Chem.*, **268**, 24959–24966.

Kroncke, K.-D., Fehsel, K. and Kolb-Bachofen, V. (1995) Inducible nitric oxide synthase and its product nitric oxide, a small molecule with complex biological activities. *Biol. Chem.*, **376**, 327–343.

Kujubu, D.A., Reddy, S.T., Fletcher, B.S. and Herschman, H.R. (1993) Expression of the protein product of the prostaglandin synthase-2/TIS10 gene in mitogen-stimulated Swiss 3T3 Cells. *J. Biol. Chem.*, **268**, 5425–5430.

Kujubu, D.A. and Herschman, H.R. (1992) Dexamethasone inhibits mitogen induction of the TIS10 prostaglandin synthase/cyclooxygenase gene. *J. Biol. Chem.*, **267**, 7991–7994.

Lambier, A.M., Markey, C.M., Dunford, H.B. and Marnett, L.J. (1985) Spectral properties of the higher oxidation states of prostaglandin H synthase. *J. Biol. Chem.*, **260**, 14894–14896.

Landino, L.M., Crews, B.C., Timmons, M.D., Morrow, J.D. and Marnett, L.J. (1996) Peroxynitrite, the coupling product of nitric oxide and superoxide, activates prostaglandin biosynthesis. *Proc. Natl. Acad. Sci.*, **93**, 15069–15074.

Lee, S.H., Soyoola, E., Chanmugam, P., Hart, S., Sun, W., Zhong, H., Liou, S., Simmons, D. and Hwang, D. (1992) Selective expression of mitogen-inducible cyclooxygenase in macrophages stimulated with lipopolysaccharide. *J. Biol. Chem.*, **267**, 25934–25938.

Lin, A.H., Bienkowski M.J. and Gorman, R.R. (1989) Regulation of prostaglandin H synthase mRNA levels and prostaglandin biosynthesis by platelet-derived growth factor. *J. Biol. Chem.*, **264**, 17379–17383.

Luscher, T.F., V. Richard. and Z. Yang. (1990) Interaction between endothelium-derived nitric oxide and SIN-1 in human and porcine blood vessels. *J. Cardiovasc. Pharmacol.*, **14**, 76–80.

Maier, J.A.M., Hla, T. and Maciag, T. (1990) Cyclooxygenase is an immediate-early gene induced by interleukin-1 in human endothelial cells. *J. Biol. Chem.*, **265**, 10805–10808.

Marotta, P., Sautebin, L. and Di Rosa, M. (1992) Modulation of the induction of nitric oxide synthase by eicosanoids in the murine macrophage cell line J774. *Br. J. Pharmacol.*, **107**, 640-641.

Marsden, P.A., Schappert, K.T., Chen, H.S., Flowers, M., Sundell, C.L., Wilcox, J.N., Lamas, S. and Michel, T. (1992) Molecular cloning and characterization of human endothelial nitric oxide synthase. *FEBS Lett.*, **207**, 287–293.

Masferrer, J.L., Zweifel, B.S., Seibert, K. and Needleman, P. (1992a) Selective regulation of cellular cyclooxygenase by dexamethasone and endotoxin in mice. *J. Clin. Invest.*, **86**, 1375–1379.

Masferrer, J.L., Seibert, K., Zweifel, B. and Needleman, P. (1992b) Endogenous glucocorticoids regulate an inducible cyclooxygenase enzyme. *Proc. Natl. Acad. Sci. USA*, **89**, 3917–392.

Masferrer, J., Reddy, S., Zweifel, B., Seibert, K., Needleman, P., Gilbert, R. and Herschman, H. J. (1994a) *In vivo* regulation of cyclooxygenase-2 but not cyclooxygenase-1 by glucocorticoids in peritoneal macrophages. *Pharm. Exp. Ther.*, **270**, 1340.

Masferrer, J.L., Zweifel, B.S., Manning, P.T., Hauser, S.D., Leahy, K.M., Smith, W.G., Isakson, P.C. and Seibert, K. (1994b) Selective inhibition of inducible cyclooxygenase 2 *in vivo* is antiinflammatory and nonulcerogenic, *Proc. Natl. Acad. Sci., USA*, **91**, 3228–3232.

Masferrer, J.L. and Seibert, K. (1994) Regulation of prostaglandin synthesis by glucocorticoids. *Receptor*, **94**, 25–30.

Merlie, J.P., Fagan, D., Mudd, J. and Needleman, P. (1988) Isolation and characterization of the complementary DNA for sheep seminal vesicle prostaglandin endoperoxide synthase (cyclooxygenase), *J. Biol. Chem.*, **263**, 3550–3553.

Milano, S., Arcoleo, F., Dieli, M., D'Agostino, R., D'Agostino, P., De Nucci, G. and Cillari, E. (1995) Prostaglandin E$_2$ regulates inducible nitric oxide synthase in the murine macrophage cell line J774. *Prostaglandins*, **49**, 105–115.

Misko, T.P., Moore, W. M., Kasten, T.P., Nickols, G. A ., Corbett, J. A., Tilton, R.G., McDaniel, M.L., Williamson, J.R. and Currie, M.G. (1993) Selective inhibition of the inducible nitric oxide synthase by aminoguanidine. *Eur. J. Pharmacol.*, **233**, 119–125.

Misko, T.P., Trotter, J.L. and Cross, A.H. (1995) Mediation of inflammation by encephalitogenic cells: interferon γ induction of nitric oxide and cyclooxygenase 2. *J. Neuroimmunol.*, **61**, 195–204.

Miyamoto, T., Ogino, N., Yamammoto, S. and Hayaishi, O. (1976) Purification of Prostaglandin Endoperoxide Synthetase from Bovine Vesicular Gland Microsomes, *J. Biol. Chem*, **251**, 2629–2636.

Mollace, V., Colasanti, V., Rodino, P., Lauro, G.M., Rotiroti, D. and Nistico, G. (1995) NMDA-dependent prostaglandin E$_2$ release by human cultured astroglial cells is driven by nitric oxide. *Biochem. Biophys. Res. Commun*, **215**, 793–799.

Mollace, V., Colasanti, M., Rodino, P., Lauro, G.M. and Nistico, G. (1994) HIV coating gp 120 glycoprotein-dependent prostaglandin E$_2$ release by human cultured astrocytoma cells is regulated by nitric oxide formation. *Biochem. Biophys. Res. Commun.*, **203**, 87–92.

Moncada, S. and Higgs, E.A. (1995) Molecular mechanisms and therapeutic strategies related to nitric oxide. *Faseb. J.*, **9**, 1319–1330.

Moore, W.M., Webber, R.K., Jerome, G.M., Tjoeng, F.S., Misko, T.P. and Currie, M.G. (1994) L-N6-(1-Iminoethyl)lysine: A selective inhibitor of inducible nitric oxide synthase. *J. Med. Chem.*, **37**, 3886–3888.

Muhl, H., Kunz, D. and Pfeilschifter. Expression of nitric oxide synthase in rat glomerular mesangial cells mediated by cyclic AMP. *Br. J. Pharmacol.*, **112**, 1–8.

Murad, F., Mittal, C.K., Arnold, W.P., Katsuki, S. and Kimura, H. (1978) Guanylate cyclase: activation by azide, nitro compounds, nitric oxide and hydroxyl radical and inhibition by hemoglobin and myoglobin. *Adv. Cyclic. Nucleotide. Res.*, **9**, 145–158.

Nathan, C. (1992) Nitric oxide as a secretory product of mammalian cells. *FASEB. J.*, **6**, 3051–3064.

Nakane, M., Schmidt, H.H.H.W., Pollock, J.S., Forstermann, U. and Murad, F. (1993) Cloned human brain nitric oxide synthase is highly expressed in skeletal muscle. *FEBS. Lett.*, **316**, 175–180.

Needleman, P., Turk, J., Jakschik, B.A., Morrison, A.R. and Lefkowith, J.B. (1986) Arachidonic Acid Metabolism, *Annu. Rev. Biochem.*, **55**, 69–102.

O'Banion, M.K., Winn, V.D. and Young, D.A. (1992) cDNA cloning and functional activity of a glucocorticoid-regulated inflammatory cyclooxygenase. *Proc. Natl. Acad. Sci. USA*, **89**, 4888–4892.

Parker, J.O. Nitrate therapy in stable angina pectoris. (1987) *New. Eng. J. Med.*, **31**, 1635–1642.

Phillips, T.A., Kujubu, D.A., Mackay, R.J., Herschman, H.R., Russell, S.W. and Pace, J.L. (1993) The mouse macrophage activation-associated marker protein, p71/73, is an inducible prostaglandin endoperoxide synthase (cyclooxygenase). *J. Leukocyte. Biol.*, **53**, 411–419.

Raddassi, K., Petit, J.F. and Lemaire, G. (1993) LPS-induced activation of primed murine pertoneal macrophages is modulated by prostaglandins and cyclic nucleotides. *Cell. Immunol.*, **149**, 50–64.

Raz, A., Wyche, A., Siegel, N. and Needleman, P. (1988) Regulation of fibroblast cyclooxygenase synthesis by interleukin-1. *J. Biol. Chem.*, **263**, 3022–3028.

Raz, A., Wyche, A. and Needleman, P. (1989) Temporal and pharmacological division of fibroblast cyclooxygenase expression into transcriptional and translational phases. *Proc. Natl. Acad. Sci, USA*, **86**, 1657–1661.

Rettori, V., Gimeno, M., Lyson, K., and McCann, S.M. (1992) Nitric oxide mediates norepinephrine-induced prostaglandin E_2 release from the hypothalamus. *Proc. Natl. Acad. Sci. USA*, **89**, 11543–11546.

Roth, G.J., Machuga, E.T., Strittmatter, P. (1981) The heme-binding properties of prostaglandin synthetase from sheep vesicular gland. *J. Biol. Chem.*, **256**, 10018–10022.

Salvemini, D, Misko, T.P., Masferrer, J.L., Seibert, K., Currie, M.G. and Needleman, P. (1993) Nitric oxide activates cyclooxygenase enzymes. *Proc. Natl. Acad. Sci. USA*, **90**, 7240–7244.

Salvemini, D , Seibert, K., Masferrer, J.L., Misko, T.P., Currie, M.G. and Needleman, P. (1994) Endogenous nitric oxide enhances prostaglandin production in a model of renal inflammation. *J. Clin. Invest.*, **93**, 1940–1947.

Salvemini, D, Misko, T.P., Masferrer, J., Seibert, K., Currie, M.G. and Needleman, P. (1995a) Role of nitric oxide in the regulation of cyclo-oxygenase. In: S. Moncada, M. Feelish and R. Busse (eds) *Biology Of Nitric Oxide, Enzymology, Biochemistry and Immunology}*. **4**, 304–309. London: Portland Press.

Salvemini, D., Settle, S.L., Masferrer, J.L., Seibert, K., Currie, M.G. and Needleman, P. (1995b) Regulation of prostaglandin production by nitric oxide; an *in vivo* analysis. *Br. J. Pharmacol.*, **114**, 1171–1178.

Salvemini, D , Manning, P.T., Zweifel, B.S., Seibert, K., Connor, J., Currie, M.G., Needleman, P. and Masferrer, J.L. (1995c) Dual inhibition of nitric oxide and prostaglandin production contributes to the antiinflammatory properties of nitric oxide synthase inhibitors. *J. Clin. Invest.*, **96**, 301–308.

Salvemini, D. and Masferrer, J.L. (1996a) Interactions of nitric oxide with cyclooxygenase: *in vitro, ex vivo* and *in vivo* studies. In *Methods in Enzymology*, ed. L. Packer, **269**, 15–26. San Diego, CA: Academic Press, Inc.

Salvemini, D., Currie, M.G. and Mollace, V. (1996b) Nitric oxide-mediated cyclooxygenase activation. A key event in the antiplatelet effects of nitrovasodilators. *J. Clin. Invest.*, **97**, 2562–2568.

Salvemini, D, Wang, Z.Q., Wyatt, P.S., Bourdon, D.M., Marino, M.H., Manning P.T. and Currie, M.J. (1996c) Nitric oxide: a key mediator in the early and late phase of carrageenan-induced rat paw inflammation. *Br. J. Pharmacol.*, **118**, 829–838.

Sano, H., Hla, T., Maier, J.A.M., Crofford, L.J., Case, J.P, Maciag, T. and Wilder, R.L. (1992) *In vivo* cyclooxygenase expression in synovial tissues of patients with rheumatoid arthritis and osteoarthritis and rats with adjuvant and streptococcal cell wall arthritis, *J. Clin. Invest.*, **89**, 97–10.

Sautebin, L. and Di Rosa, M. (1994) Nitric oxide modulates prostacyclin biosynthesis in the lung of endotoxin-treated rats. *Eur. J. Pharmacol.*, **262**, 193–196.

Sautebin, L., Ialenti, A., Ianaro, A. and Di Rosa, M. (1995) Modulation by nitric oxide of prostaglandin biosynthesis in the rat. *Br. J. Pharmacol.*, **114**, 323–328.

Schmidt, H.H.H.W. and Murad, F. (1991) Purification and characterization of a human NO synthase. *Biochem. Biophys. Res. Commun.*, **181**, 1372–1377.

Seibert, K., Zhang, Y., Leahy, K., Hauser, S., Masferrer, J.L., Perkins, W., Lee, L. and Isakson, P. (1994) Pharmacological and biochemical demonstration of the role of cyclooxygenase 2 in inflammation and pain. *Proc. Natl. Acad. Sci. USA*, **91**, 12013–12017.

Sherman, P.A., Laubach, V.E., Reep, B.R. and Wood, E.R. (1993) Purification and cDNA sequence of an inducible nitric oxide synthase from a human tumor cell line. *Biochemistry*, **32**, 11600–11605.

Stamler, J.S. and Loscalzo, J. (1991) The antiplatelet effects of organic nitrates and related nitrous compounds *in vitro* and *in vivo* and their relevance to cardiovascular disorders. *J. Am. Coll. Cardiol.*, **18**, 1529–1536.

Stadler, J., Harbrecht, B.C., DiSilvio, M., Curran, R.D., Jordan, M.L., Simmons, R.L. and Billiar, T.R. (1993) Endogenous nitric oxide inhibits the synthesis of cyclooxygenase products and interleukin-6 by rat Kuppfer cells. *J. Leukocyte. Biol.*, **53**, 165–172.

Swierkosz, T.A., Mitchell, J.A., Warner, T.D., Botting, R.M. and Vane, R. (1995) Co-induction of nitric oxide synthase and cyclooxygenase: interactions between nitric oxide and prostanoids. *Br. J. Pharmacol.*, **114**, 1335–1342.

Tsai, A.L., Wei, C. and Kulmacz, R.J. (1994) Interaction between nitric oxide and prostaglandin H synthase. *Arch. Biochem. Biophys.*, **313**, 367–372.

Vane, J.R. (1971) Inhibition of prostaglandin synthesis as a mechanism of action for aspirin-like drugs. *Nature (New Biol.)*, **231**, 232–233.

Vane, J.R., Mitchell, J. A., Appleton, I., Tomlinson, A., Bishop-Bailey, D., Croxtall, J. and Willoughby, D.A. (1994) Inducible isoforms of cyclooxygenase and nitric-oxide synthase in inflammation. *Proc. Nat. Acad. Sci. USA*, **91**, 2046–2050.

Van der Ouderaa, F., Buyenhek, M., Nugteren, D. and VanDorp, D. (1977) Purification and characterisation of prostaglandin endoperoxide synthase from sheep vesicular glands, *Biochim. Biophys. Acta*, **367**, 315.

Van der Ouderaa, F.J., Buytenk, M., Slikkerveer, F.J. and Van Dorp D.A. (1979) On the haemprotein character of prostaglandin endoperoxide synthetase. *Biochem. Biophys. Res. Commun.*, **572**, 29–42.

Wu, K.K. (1995) Inducible cyclooxygenase and nitric oxide synthase. *Adv. Pharmacol.*, **33**, 179–207.

Xie, W., Chipman, J.G., Robertson, D.L., R.L. Erikson, R.L. and Simmons, D.L. (1991) Expression of a mitogen-responsive gene encoding prostaglandin synthase is regulated by mRNA splicing, *Proc. Natl. Acad. Sci. USA*, **88**, 2692–2696.

10 Regulation of Nitric Oxide Synthase Expression and Activity by Hemodynamic Forces

Bauer E. Sumpio, Babalola O. Oluwole, Xiujie Wang
and Mark Awolesi

*Department of Surgery (Vascular), Yale University School of Medicine, 333, Cedar Street,
New Haven, CT 06510, USA*

INTRODUCTION

During life, the blood vessel wall is subjected continually to a variety of external physical forces, which are the result of the circulatory system. These various forces have been characterized and studied *in vivo* and *in vitro*. Their influence on the vessel wall has also been studied by examining their specific effects on vascular endothelial and smooth muscle cell morphology and expression. From most of these studies, there is clear evidence that hemodynamic forces not only stimulate endothelial and smooth muscle cell expression phenotype, but that the function of these cells *in vivo* may be regulated by hemodynamic forces. The preponderance of these studies has been directed to determine how these hemodynamic forces influence and regulate endothelial cells.

The focus of this review will be on the effect of hemodynamic forces on endothelial cell nitric oxide synthase (NOS) activity and gene expression. The implications of these recent findings will be discussed from the standpoint of the contribution of NOS to vascular remodeling, early developmental angiogenesis and atherosclerosis

PHYSICAL FORCES THAT THE ENDOTHELIUM IS SUBJECTED TO *IN VIVO*

The endothelium is subject to various types of physical deformation, of which three have been well characterized: shear stress, cyclic strain, and pressure.

Shear stress is described as the frictional force applied tangentially across the surface of the endothelium resulting from the flow of blood. The pulsatile profile of blood flow, the fact that blood does not fully conform to Newtonian characteristics and the compliant nature of the blood vessel wall, result in a complex pattern of shear stress *in vivo* (Fung and Liu, 1995). The branching pattern of blood vessels and the presence of atherosclerotic plaques (Glagov *et al.*, 1988; Glagov *et al.*, 1992) in vessel walls add to the complexity of the characterization of shear stress patterns at the endothelial cell surface (Langille, 1993; Kassab and Fung, 1995; Xie *et al.*, 1995). However use of rigid tube models, accurate *in vivo* measurements of flow, and the application of the principles of Poiseuille fluid dynamics provide reasonable estimates of flow characteristics and hemodynamic forces (Liu and Fung, 1993). Through the use of tracers and sophisticated laser tracking systems, it is now known that the endothelium experiences a range of shear stress at various points in the blood vessel and at different phases of the cardiac cycle (Ojha, 1993, 1994). At points

171

                        Bauer E. Sumpio et al.

A   SHEAR PATTERN

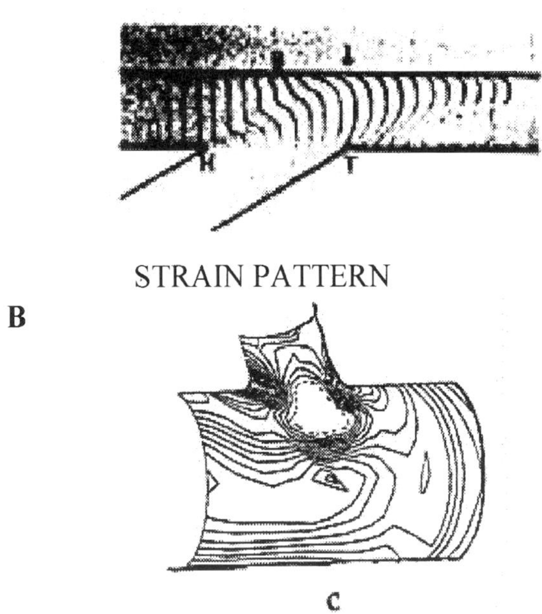

Figure 10-1. Mechanical forces in the circulation.
A. At points of vessel bifurcation and branches, eddy currents are generated by the flow of blood and this results in a complex pattern of shear stress at these points in the vessel.
B. Strain patterns at bifurcations and branches have been demonstrated by finite analysis. The greatest amount of wall strain appears to be at the toe and heel of a bifurcation.

of vessel bifurcation and branches, eddy currents are generated by the flow of blood and this results in a complex pattern of shear stress at these points in the vessel (Figure 10-1A).

Cyclic strain is a deformation that endothelial cells and other vessel wall elements experience resulting from the wall stress induced by the pulsatile blood flow across the luminal surface of the vessel. Studies on animals have shown that the aortic wall can be cyclically stretched as much as 30% during the pulsatile flow of blood (Deng *et al.*, 1994; Fung and Liu, 1995). Recent studies on vessel wall tone suggest that there is an underlying positive residual strain pattern in the vessel wall in fresh cadaveric specimens (Han and Fung, 1995). The amount of estimated residual strain varied depending on the site along the length of the aorta, but was as much as 10%. The magnitude of strain that the vessel wall actually experiences *in vivo* depends on other physiological factors such as the rate of blood flow or the distensibility of the vessel wall. In addition, arteries and veins will experience different degrees of strain not only because there is a difference in the rate of flow, but also because the elasticity differs in arteries and veins. This also holds true when one compares large arteries or veins to arterioles or venules respectively. Pathological factors that can alter the degree of strain a vessel experiences include the presence of intimal hyperplasia or severe atherosclerosis with resultant calcification of the media. Strain patterns at bifurcations and branches have been demonstrated by finite analysis (Salzar *et al.*, 1995; Thubrikar and Robicsek, 1995). The greatest amount of wall strain appears to be at the toe and heel of a bifurcation. This may have clinical implications, as these sites are the most common for the development of atherosclerotic plaques (Figure 10-1B).

A third component of mechanical deformation *in vivo*, is pulsatile pressure, which is a tangentially applied force on the cell surface as a pressure wave propagates down the vessel wall. This force directly impacts on the endothelial cell to that endothelial cells experience *in vivo*. Different sites within the same vessel experience different pulse pressure waves the further it is from the heart. There is also a difference in the pulse pressure between the left and right sides of the circulatory system. Some of the pathological factors described for differences in strain patterns in the vessel wall also hold true for pulsatile pressure. In addition, diseases such as hypertension result in even greater pulse pressures on the vascular endothelium.

Of the three major forms of mechanical deformation *in vivo*, the only one that significantly impacts only on the endothelium is shear stress. There may be some slow transmural flow around smooth muscle cells that can produce shear stress, however the order of magnitude of shear stress on the endothelium is such that the effect on smooth muscle cells may be negligible.

EFFECTS OF HEMODYNAMIC FORCES ON NOS *IN VIVO*

The role of hemodynamic forces *in vivo* in influencing the production of NOS has been an area of active inquiry. The problem in analyzing the data, however, is the inability to define the nature or character of the hemodynamic force *in vivo*. We may know that the forces involved include shear stress, cyclic strain and pulsatile pressure, however, we can not define the primary force at play. Some investigators have ascribed the effects seen with high flow states as being primarily shear stress related. At this time there can be no certainty as to which of the many hemodynamic forces is primarily responsible as the biomechanics of these forces *in vivo* are too complex to be merely ascribed to just one component.

The importance of hemodynamic-induced increases in eNOS is not yet known but it has been postulated that it may contribute to the effect of sustained dynamic exercise on cardiovascular disease. Recent studies have shown that dynamic exercise reduces the incidence of cardiovascular events (Posner *et al.*, 1990; Hellenius *et al.*, 1993) and hypertension in patients (Paffenbarger *et al.*, 1983). In addition, exercise also lowers resting blood pressure in hypertensive (Nakamura *et al.*, 1992) and normotensive (Poehlman *et al.*, 1992) adults and improves exercise tolerance in patients with ischemic heart disease (Ehsani *et al.*, 1981) and peripheral vascular disease (Hiatt *et al.*, 1990). Although exercise is currently prescribed as a therapy in the management and prevention of cardiovascular disease, the mechanisms underlying the benefits of exercise are not known. Several *in vivo* studies have shown that increases in hemodynamic forces can alter endothelium blood vessel tone and reactivity. Creation of arteriovenous anastomosis, which results in an increase in regional blood flow, enhances endothelium-dependent relaxations (Delp *et al.*, 1993). Human studies also show that sustained exercise leads to increased urinary nitrate excretion that indicates an increased NO production (Leaf *et al.*, 1990).

ANIMAL MODELS

There have been different models used to study the effect of hemodynamic forces *in vivo*, on NOS expression. Such models include that of a) spontaneously hypertensive rats (to see if there is an alteration in NOS expression between normotensive arterial EC and hypertensive arterial EC), b) congestive heart failure models (to determine if the low flow state of CHF results in a defect in NOS expression), c) the use of chronic exercise in dogs, or d) the creation of arteriovenous fistula to create a high flow state.

Increasing coronary blood flow by chronic cardiac pacing in dogs enhances endothelium-dependent dilation. Conversely, in experimental models of congestive heart failure, where blood flow is compromised, endothelium-dependent relaxations *in vitro* and flow-mediated dilation *in vivo* are depressed. These data seem to identify chronic alterations in blood flow as a possible regulatory stimulus on the expression of NO-mediated endothelium-dependent dilation. It is well recognized that a physiological stimulus that increases intracoronary blood flow is acute or chronic exercise. Acute exercise increases coronary blood flow and flow velocity, thereby causing dilation of epicardial coronary arteries. This is supported by experiments demonstrating that restriction of flow (by a critical stenosis) prevents exercise induced epicardial vasodilatation. The flow velocity mediated epicardial dilation is likely mediated by NO, because inhibitors of NOS blunt this response.

Using the model of chronic exercise in dogs, Sessa (1994) and coworkers conducted a study where dogs were conditioned by acute exercise (running at 9.5 km/in). These chronically instrumented, conscious dogs had significantly increased ventricular pressure, left ventricular dP/dt, mean arterial pressure, heart rate, and peak coronary blood flow and decreased late-diastolic coronary resistance. This level of exercise was repeated twice daily for 10 days. Exercise training for 10 days did not affect resting heart rate, coronary blood flow, or other cardiac parameters, suggesting that classic "cardiac conditioning" did not occur. The basal production of nitrite was not significantly different in large coronary vessels or microvessels from control or exercised dogs. Acetylcholine caused a dose-dependently increased release of nitrite from coronary arteries and microvessels in both groups of dogs. Moreover, acetylcholine-stimulated nitrite production was enhanced in vessels prepared from hearts of exercised dogs compared with hearts from control dogs. The generation of nitrite was completely inhibited by preincubation of the vessels with nitro-L-arginine demonstrating that the nitrite was derived via the metabolism of L-arginine by ECNOS. The agonist-stimulated production of nitrite in vessels from exercised dogs was postulated to be due to an induction of the calcium-dependent ECNOS gene. In a separate group of dogs (aortas not used), they measured cardiac output before and during exercise. Exercise increased cardiac output by 2 fold, demonstrating that the proximal aorta was clearly being exposed to high-flow velocity. Steady-state mRNA levels for ECNOS from aortic samples were significantly increased 2- to 3-fold than mRNA levels for vWF and GAPDH in exercised dogs.

The rationale for the use of arteriovenous fistula stems from the recognition of the fact that shunts increase the cardiac output by increasing venous return, thereby producing a hyperdynamic circulatory system. Miller et al first demonstrated that a chronic high flow states resulting from an arteriovenous fistula created between the femoral artery and vein resulted in enhanced endothelium-dependent vasorelaxation and NO release (Miller and Vanhoutte, 1988; Miller and Burnett, 1992). Nadaud *et al.* studied the effect of *in vivo* hemodynamic forces on the regulation of eNOS (Nadaud, Philippe *et al.*, 1996). Their hypothesis was that an increased flow state would modulate eNOS expression. Their experimental design involved using a model of an arteriovenous fistula — an aortocaval fistula (in rats) to simulate a high flow state. The rats were sacrificed after six weeks. A control group of rats underwent a sham operation with exposure and temporary cross clamping of the aorta and inferior vena cava. The fistula rats had a 2.2 fold increase in total RNA, while NOS protein levels were increased 3.5 fold in fistula rats. NOS activity as measured by the citrulline assay was 1.6 fold increased in fistula rats compared to control. In using the model of an arteriovenous fistula, the investigators recognized the possibility of inducing congestive heart failure.

In a study from Koller (Koller and Huang, 1994), arterioles dissected from the gracilis muscle of spontaneously hypertensive (SH) rats and control normotensive (NW) rats were subjected to varying flow rates using a physiologic salt solution as perfusate. The systolic blood pressure in the SH and NW rats were 200 ± 2.9 mmHg and 105 ± 3.1 mmHg respectively. In the presence of constant intravascular pressure (80 mmHg) and static flow conditions, there was no significant difference in the diameter of the vessels from either rat population. With stepwise increase in the flow rate, the diameters of the arterioles SH rats were less than that of NW rats. In the presence of acetylcholine and sodium nitroprusside, there were no significant differences in the dilator response of SH rats and NW rats. However in the presence of L-NNA, there was a significant reduction in flow-induced arteriolar dilatation in NW rats compared to control. L-NNA did not significantly alter the flow-induced arteriolar dilatation in SH rats. In the presence of L-NNA, a prostaglandin inhibitor, indomethacin, caused an additional 20% decrease in basal diameter in NW rats while practically eliminating the dilatation to increase in flow rate in SH rats. Thus the response to prostacyclin mediated dilatation was intact in both SH and NW rats but the NO mediated response to flow increase was curtailed in SH rats. This indicated that hypertension may be a response to defective NOS expression.

In a more recent study by Koller and Huang (1995), they examined the hypothesis that chronic exercise could alter flow-dependent vasodilatation and the mediation of this dilation by nitric oxide. They demonstrated an increased sensitivity to flow dependent dilation in gracilis arterioles from exercised rats compared to sedentary rats. When the vessel was denuded of endothelium, this increased sensitivity was lost, indicating that the phenomenon was endothelium dependent. In the presence of NOS inhibitor L-NNMA, there was a reduction in the flow dependent dilation in arterioles from both exercised and sedentary rats. In addition acetylcholine mediated dilation was also reduced in both groups of rats, in the presence of L-NNMA. The increased sensitivity to flow and acetylcholine mediated dilation was also found to be dependent on prostaglandin production. They concluded that the sensitivity of gracilis muscle arterioles of rats to wall shear stress was upregulated after short-term daily exercise, resulting in an augmented dilator response that is due to an increased release of both endothelium-derived nitric oxide and prostaglandins.

An interesting report on the expression of NOS in cirrhotic rats demonstrated an increase in eNOS and iNOS expression in the aorta from adult male Wistar rats in whom cirrhosis had been induced by administration of hepatoxin and phenobarbital (Rinaldi and Bohr, 1989; Delp *et al.*, 1993). The animals were exposed to the hepatoxin, carbon tetrachloride, twice a week. Upon the induction of cirrhosis with ascites, the animals were sacrificed and the thoracic and abdominal aortae, and mesenteric arteries dissected out. Vascular smooth muscle cells from the aorta were grown in culture, while arterial vessel homogenates were prepared from the thoracic aorta and eNOS assays carried out on these vessel homogenates.

Induction of iNOS was demonstrated in the smooth muscle cells by RT-PCR and Southern blotting. They also demonstrated eNOS mRNA induction by ribonuclease protection assays in aorta from cirrhotic rats compared to control. Western blotting of the vascular tissue homogenate revealed a 3 fold induction of eNOS protein in cirrhotic rats. The investigators postulate that iNOS and eNOS induction may account for the arterial vasodilatation in cirrhosis, however a more plausible explanation may be that eNOS induction is the result of a hyperdynamic circulation present in cirrhotics. A hallmark of cirrhosis is the development of shunts between the portal and arterial circulations. This may account for the high flow state which in turn induces the upregulation of eNOS and perhaps iNOS expression.

SIMULATION OF HEMODYNAMIC FORCES *IN VITRO*

While the above *in vivo* studies demonstrate the effect of hemodynamics on production of NO, it is difficult to analyze the contribution of the different hemodynamic forces. To better study the effects of the various mechanical forces on the vascular endothelium, various mechanical devices have been designed to simulate what happens *in vivo*.

For shear stress studies, two widely used models have been designed. The first (Figure 10-2A) is a simple assembly of parallel-plate flow chamber, in which laminar flow is generated by a pump over a confluent endothelial monolayer grown on a transparent coverslip (Eskin *et al.*, 1984; Frangos *et al.*, 1985). The shear stress which is imparted by the movement of a viscous fluid over the luminal endothelial surface, is a linear function of the volume flow rate through the chamber. Cells can be maintained under defined flow for several days in a sterile closed system with periodic medium replenishment. This system can be coupled with a phase-contrast microscope or microfluorimeter, thus permitting the visualization of shear-stress induced changes in cell morphology. The second system (Figure 10-2B) is a modified cone-plate viscometer, in which shear stress is produced in a layer of fluid contained between a stationary base plate and a rotating cone (Dewey *et al.*, 1981; Remuzzi *et al.*, 1984). Adjusting the cone angle, medium viscosity and cone rotation speed delivers a wide range of shear stress. Endothelial cells grown to confluence on small covers slips can be mounted on the baseplate at different angles to allow multiple samplings in the same experiment. Alternatively, a single large plate can be used instead to increase the biological sample size. The cone-plate viscometer can also be used to simulate shear stress forces present in areas of turbulent flow by tipping the cone with respect to its angle of rotation thereby creating eddy currents or, by adding a barrier within the flow pattern, create small defined areas of disturbed laminar flow as obtains at the sites of vessel bifurcation.

In contrast to shear stress, many models of cyclic strain have also been developed (Table 10-1) (Sumpio, 1993), but the two most widely used are based on applying strain to an elastic membrane. In the first method strain is applied londitudinally (Vandenburgh and Kaufman, 1979; Buck, 1980; Terracio *et al.*, 1988). For example Figure 10-3A illustrates a device that is constructed from aluminum and Plexiglass. A four-phase linear activator stepper motor was mounted in a horizontal position and attached to a moveable linear positioning translation stage. Three pairs of aluminum bars are positioned on top of the translation stage to which the culture wells (4 per pair) were attached by stainless steel screws. The rear bars stayed fixed in position while the front bars were attached to the

A.

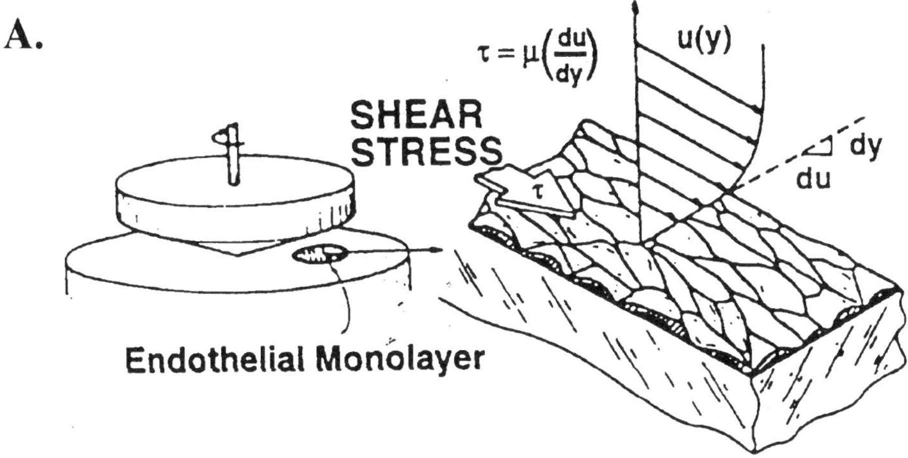

B.

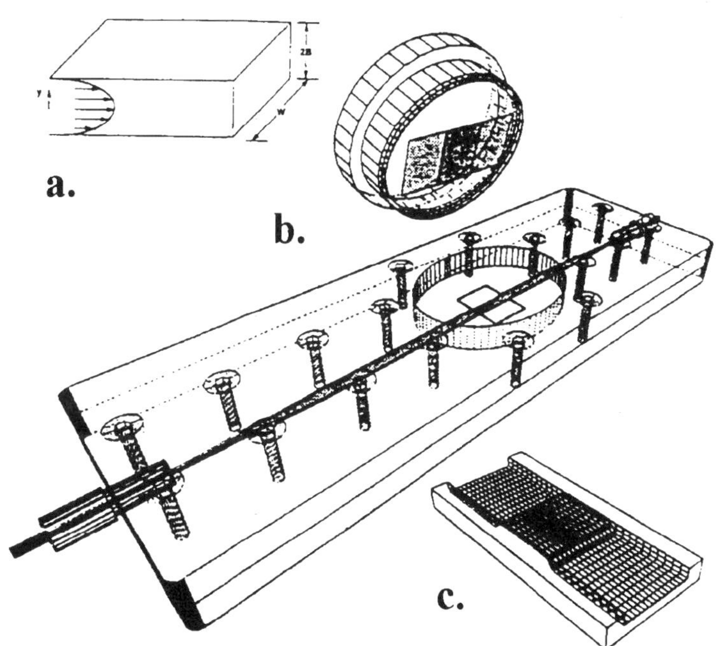

Figure 10-2. *In vitro* methods for studying shear stress.
A. A modified cone-plate viscometer, in which shear stress is produced in a layer of fluid contained between a stationary base plate and a rotating cone.
B. A parallel-plate flow chamber, in which laminar flow is generated by a pump over cells grown on a transparent coverslip. a) The shear stress which is imparted by the movement of a viscous fluid over the luminal endothelial surface, is a linear function of the volume flow rate through the chamber. b) The parallel plate can be made with eccentric stenosis. A plastic cover slip is placed on the stenosis shoulder. c) Profile of stenosis. Blood is pumped through the perfusion chamber at well defined shear rates.

                                    Bauer E. Sumpio et al.

Table 10-1. Models of Cyclic Strain.

Model	Year/ (Ref)	Substrate/ Frame	Static/ Cyclic/ Continuous	Mechanism of Deformation	Direction of Force	Advantages
A	1977	Bovine aortic elastin/Teflon	Cyclic	Membrane mounted in a frame such that one end fixed and the other movable. This end is moved by a rod coupled to a motor driven shaft	uniaxial	Range of mechanical combinations possible including imposing different initial tension
B	1978	Nylon mesh	cyclic	cells seeded on nylon mesh, the mesh is anchored to a polystyrene ring in Petri and mechanically deformed with piezoelectric ceramic element	uniaxial	Area of mesh remains constant as when one side of mesh moves apart other moves closer.
C	1979	Silicone elastomer, collagen coated	static	membrane clamped in a square frame with adjustable side to apply strain	biaxial	simple apparatus, inexpensive, range of strain possible
D	1979	Millipore filter	static	used split millipore filter to apply strain	uniaxial	inexpensive
E	1980	Silicone rubber sealant	cyclic	substrate anchored at one end in tissue culture flask, the other end attached to nylon cord which passed through cap to an eccentric mounted on motorized shaft	uniaxial	simple
F	1982	Elastin polycarbonate silicone, collagen coated	continuous	substrate attached to two parallel rollers driven by a synchronous motor with speed transmission	uniaxial	simple
G	1983	collagen on wire mesh	static	The wire mesh embedded in the collagen is deformed by applying spring forceps to the peripheral meshes at points equidistant from one corner	uniaxial	Inexpensive
H	1984	None, the culture itself formed an elastic membrane	cyclic	Modified incubator with a motor driven shaft running through it. Eccentric cams and linkages moved one end of clamped membrane	Uniaxial	Could stretch four sets of cultures at once

Table 10-1. *(Continued).*

Model	Year/ (Ref)	Substrate/ Frame	Static/ Cyclic/ Continuous	Mechanism of Deformation	Direction of Force	Advantages
I	1985	Silicone rubber	Cyclic/ static	Vacuum applied underneath flexible bottom well. Precursor to flexercell system	Radial	Compatible with conventional cell culture equipment
J	1986	Silastic-plasma and collagen coated	Cyclic	Synchronous motor drives an eccentric disk formed to imitate the diameter changes in the aorta	Uniaxial	Relatively simple, multiple strain regimens possible
K	1988	FLEXRI flexible bottom culture plate	Cyclic strain	Flexercell instrument	Radial	Computer controlled computer records strain applied, Easy to use, multiple strain regimens possible. Multiple plates can be stretched at the same time

moveable stage and moved in a horizontal plane. A pushing movement of the motor moved the stage and attached bars away from the motor, stretching the substratum on which cells were grown; a pulling motion relaxed the substratum.

The second method (Figure 10-3B) employs a Flexercell strain unit (Banes *et al.*, 1985; Banes *et al.*, 1990) which consists of a computer, a control module which regulates a solenoid-valve controlled vacuum pump, baseplate and gasket. Six-well plates with silastic membrane bottoms are mounted on the baseplate. The vacuum generated by the pump causes a deformation of the silastic membrane. The amount of vacuum determines the degree of strain applied across the membrane.

Both models of cyclic strain apply the same premise that endothelial cells grown on the deformable membranes experience the same degree of strain as the membrane since the cells are firmly attached. A major difference however between the methods is the fact that the first unit applies a uniform amount of uniaxial strain to the membrane while the Flexercell apparatus produces a heterogeneous strain pattern. The disadvantage of this is that the amount of strain applied to the membrane has to be averaged by a mathematical analysis of the strain pattern across the membrane. However there is a benefit which derives from the presence of a high and low strain region within the same membrane: one can compare high versus low strain effects on cells grown on the same membrane where all other factors are identical.

Shear stress and cyclic strain have been shown to cause a variety of effects in endothelial cells including morphological changes (such as cytoskeletal changes) (Vandenburgh and Kaufman, 1979; Buck, 1980; Dewey *et al.*, 1981; Davies *et al.*, 1984; Eskin *et al.*, 1984; Sumpio *et al.*, 1988; Terracio *et al.*, 1988; Iba and Sumpio, 1991), modulation of proliferation and migration (Davies *et al.*, 1984; Eskin *et al.*, 1984; Sumpio *et al.*, 1987; Li *et al.*, 1994; Yano *et al.*, 1996; Yano *et al.*, 1996; Yano *et al.*, 1997), and induction of

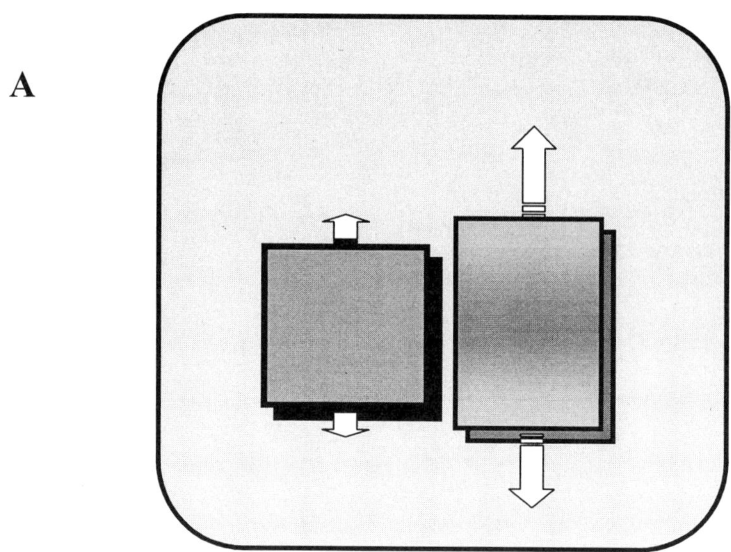

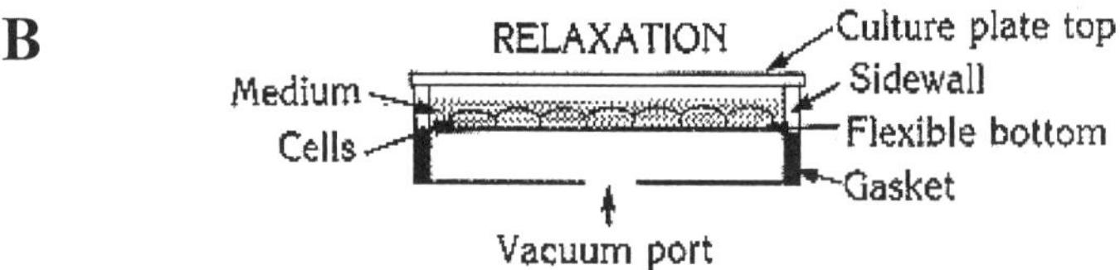

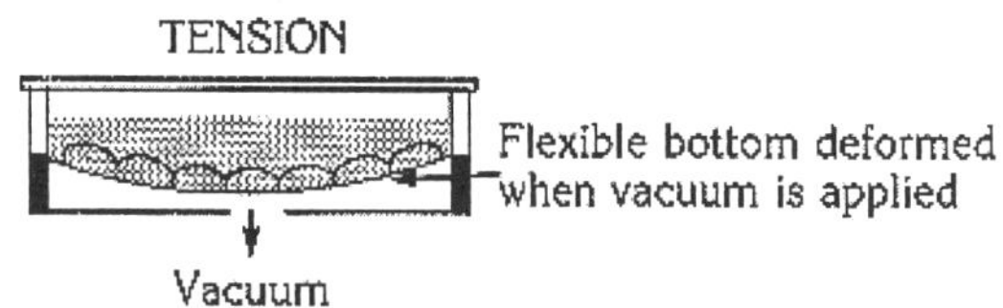

Figure 10-3. *In vitro* methods for studying cyclic strain.

A. A four-phase linear activator stepper motor was mounted in a horizontal position and attached to a moveable linear positioning translation stage. Three pairs of aluminum bars are positioned on top of the translation stage to which the culture wells (4 per pair) were attached by stainless steel screws. The rear bars stayed fixed in position while the front bars were attached to the moveable stage and moved in a horizontal plane. A pushing movement of the motor moved the stage and attached bars away from the motor, stretching the substratum on which cells are grown; a pulling motion relaxed the substratum.

B. Flexercell strain unit consists of a computer, a control module which regulates a solenoid-valve controlled vacuum pump, baseplate and gasket. Six-well plates with silastic membrane bottoms are mounted on the baseplate. The vacuum generated by the pump causes a deformation of the silastic membrane. The amount of vacuum determines the degree of strain applied across the membrane.

gene expression tPA (Diamond *et al.*, 1989; Iba *et al.*, 1991; Iba and Sumpio, 1992; Sumpio *et al.*, 1997), PDGF (Mitsumata *et al.*, 1993; Resnick *et al.*, 1993; Resnick *et al.*, 1994; Resnick and Gimbrone, 1995; Sumpio *et al.*, 1998), endothelin (Sumpio and Widmann, 1990; Sharefkin *et al.*, 1991; Kuchan and Frangos, 1993) among others. We will however be concentrating on the effects on NOS expression.

IN VITRO EFFECTS OF MECHANICAL DEFORMATION ON NOS EXPRESSION

Shear stress induced expression of NOS

Diamond and coworkers studied the effect of laminar shear stress on eNOS expression in cultured human umbilical vein EC (HUVEC) and bovine aortic EC (BAEC (Ranjan, 1995)). Using the parallel-plate flow chamber, cultured confluent HUVEC and BAEC were subjected to steady laminar shear stress of 4 to 25 dyn/cm^2. The cells were pretreated for 24 hr with dexamethasone to inhibit the expression of iNOS which reportedly is expressed in EC. There was a 2-fold and almost 3-fold increase in NOS protein levels in BAEC and HUVEC, respectively, that had been exposed to 25 dyne/cm^2 for 6 hr compared to control cells. Low shear stress of 4 dyne/ cm^2 did not cause a significant induction of NOS protein in either BAEC or HUVEC. In the presence of inhibitors of PKC (H-7) and NOS (L-NAME), there was no significant inhibition of shear stress induced induction of NOS protein. They demonstrated a 3-fold increase in NOS steady state mRNA transcript in cells exposed to low (4 dyne/cm) and high (25 dyne/cm) laminar shear stress for 6 hr compared to control (Figure 10-4A). This induction remained elevated after 12 hr of exposure. They concluded that the shear stress induction of NOS was neither PKC- nor cGMP-dependent. They also showed that although shear stress induced endothelin-1 secretion appears to be dependent on NO production, L-NAME while blocking the shear stress induced suppression of endothelin-1, did not suppress NOS induction by shear stress.

Using the parallel plate model of shear stress, Nishida and coworkers (Nishida *et al.*, 1992) subjected cultured bovine aortic endothelial cells to 24 h of laminar shear stress. They demonstrated an increase in eNOS mRNA steady-state transcript levels in cells exposed to shear stress as compared to cells grown in static conditions. They also showed by Western blot analysis that the same regimen of shear stress increased expression of eNOS protein. In contrast, bovine aortic endothelial cells exposed to TNF-α (which is known to increase expression of inducible NOS) demonstrated a down-regulation in eNOS mRNA transcription. This study among others, also supports the evidence that eNOS should not be regarded as a constitutive enzyme.

In a more recent study from the same group of investigators, the characteristics of eNOS expression by shear stress were examined (Uematsu, 1995). Bovine aortic EC and human aortic EC were subjected to 3–24 hrs of laminar shear stress. By 3 hrs, there was a maximal 3 fold increase in eNOS mRNA from bovine EC. The increase was sustained for up to 24 hrs of laminar shear stress (Figure 10-4B). A similar set of results was also obtained for human aortic EC. Administration of Actinomycin-D blocked the increased mRNA induction by shear stress, however it had no effect on the basal expression of eNOS mRNA. The PKC inhibitor calphostin C also had no effect on the increased induction of eNOS in bovine aortic EC. They were also able to show that shear stress caused an increase in

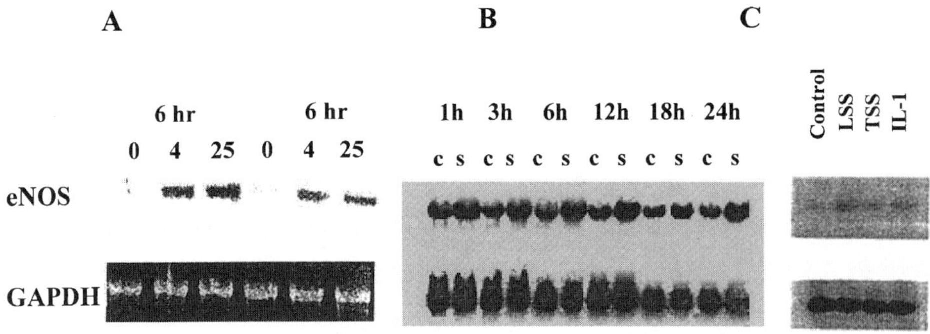

Figure 10-4. Exposure of endothelial cells to shear stress results in an enhanced NOS expression.
A. Bovine aortic EC and human aortic EC subjected to laminar shear stress using a parallel plate perfusion chamber demonstrating a 2-fold and almost 3-fold increase in NOS protein levels in BAEC (left three lanes) and HUVEC (right three lanes), respectively, that had been exposed to 25 dyne/cm^2 for 6 hr compared to control cells. Low shear stress of 4 dyne/cm^2 did not cause a significant induction of NOS protein in either BAEC or HUVEC.
B. Bovine aortic EC were subj ccted to 3–24 hrs of laminar shear stress. By 3 hrs, there was a maximal 3 fold increase in ENOS MRNA from bovine EC. The increase was sustained for up to 24 hrs of laminar shear stress. C = control, s = shear stress.
C. The effect of laminar shear stress (LSS) and turbulent shear stress (TSS) on endothelial cell genes from human umbilical vein EC, using differential display and quantitative reverse transcriptase PCR. There was a 3 fold increase in ENOS gene expression as measured by quantitative PCR. Positive control using interleukin-1 (IL-1 is shown on the right lane.

baseline production of nitrites by endothelial cells. The implications of this study were to demonstrate that shear stress could be a regulatory stimulus for the production of eNOS. They postulate that this could be the mechanism by which sustained bouts of exercise bestow a beneficial effect on the circulatory system. We know that exercise conditions skeletal muscle — this conditioning may be a response to elevated levels of eNOS, which enhances the vasodilator capacity in exercising muscle.

A study by Topper and colleagues recently examined the effect of laminar shear stress on endothelial cell genes from human umbilical vein EC, using the powerful technique of differential display and quantitative reverse transcriptase PCR (Topper, 1996). The cells were subjected to physiological levels of laminar shear stress (10 dyne/cm^2). They were able to identify three genes that were consistently upregulated by shear stress, namely the cyclooxygenase-2 (COX-2), superoxide dismutase, and eNOS genes. The COX-2 and superoxide dismutase genes showed early responses to shear stress (within 1 to 6 hr), while the response in the eNOS gene was demonstrated only after 24 hrs of shear stress. (There was a 3 fold increase in eNOS gene expression as measured by quantitative PCR, Figure 10-4C).

Cyclic Strain Upregulation of Nitric Oxide

In work from our laboratory, Awolesi *et al.* (1994) demonstrated the effect of cyclic strain on the production and gene expression of NOS protein (Awolesi *et al.*, 1994, 1995). This study not only examined the effect of cyclic strain on EC, but also the effect of different

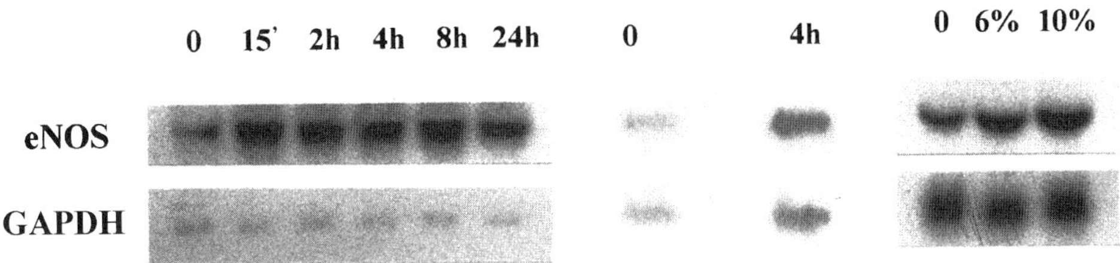

Figure 10-5. Exposure of endothelial cells to cyclic strain results in enhanced NOS expression.
A. Northern blot for ENOS of EC exposed to 10% average strain (–20 kPa vacuum) at 60 cycles/min. GAPDH is the constitutive control for loading. There was increased expression of the ENOS gene which was significantly greater than that seen with EC subjected to 6% average strain.
B. Nuclear run-off assays demonstrating that there was significant induction of new ENOS transcripts in nuclei isolated from EC exposed to 4 h of 10% average strain compared with nuclei from control stationary cells (0). The almost 3 fold increase in ENOS transcription was specific because there was only a minimal induction of GAPDH transcripts observed with cyclic strain.
C. Northern blot of EC exposed to either 6% or 10% average strain at 60 cycles/min for 24 h. to directly compare the effects of low versus high strain on ENOS gene expression. Both strain regimens increased ENOS gene expression 2 fold but the increase with 6% average strain was lower than that seen in EC exposed to 10% average strain.

degrees and duration of strain. Analysis of eNOS gene expression compared with GAPDH expression showed a 3 fold increase in bovine aortic EC exposed to strain compared with control (stationary) cells. To determine whether this increase in eNOS gene expression was dependent on the amplitude of strain, EC were exposed to 6% average strain at 60 cycles min for various times. Figure 10-5A demonstrates that this level of strain also increased the expression of the eNOS gene but to a lesser extent than that seen with EC subjected to 10% average strain. The increase in gene expression was not evident until the cells were exposed to strain. Furthermore, nuclear run-off assays demonstrates that there was significant induction of new eNOS transcripts in nuclei isolated from EC exposed to 4 h of 10% average strain compared with nuclei from control stationary cells (Figure 10-5B). The almost 3 fold increase in eNOS transcription was specific because there was only a minimal induction of GAPDH transcripts observed with cyclic strain. To directly compare the effects of low versus high strain on eNOS gene expression, the same bovine aortic EC cells were exposed to either 6 or 10% average strain at 60 cycles/min for 24 h followed by Northem blot analysis. Both strain regimens increased eNOS gene expression 2 fold but the increase with 6% average strain was lower than that seen in EC exposed to 10% average strain (Figure 10-5C). To ascertain the significance of increased eNOS gene expression, eNOS protein levels were examined by Western blot analysis in bovine aortic EC that were subjected to 6 or 10% average strain for 24 h. With both strain regimens there was an increase in eNOS protein after 24 h compared with stationary controls. Moreover, EC subjected to greater strain had a higher level (2.3-fold) of eNOS protein (Figure 10-6A).

As mentioned previously, in this model of cyclic strain the strain pattern across the stretch membrane is inhomogeneous. Typically, cells seeded in the periphery of the membrane experience maximum strain (24% at 20 kPa of vacuum deformation) while cells at the very center of the membrane experience minimum strain (0%). To examine if differential patterns of eNOS protein expression occurred in areas of high versus low strain, eNOS was localized by immunohistochemical staining (Figure 6B). Bovine aortic EC at the

A

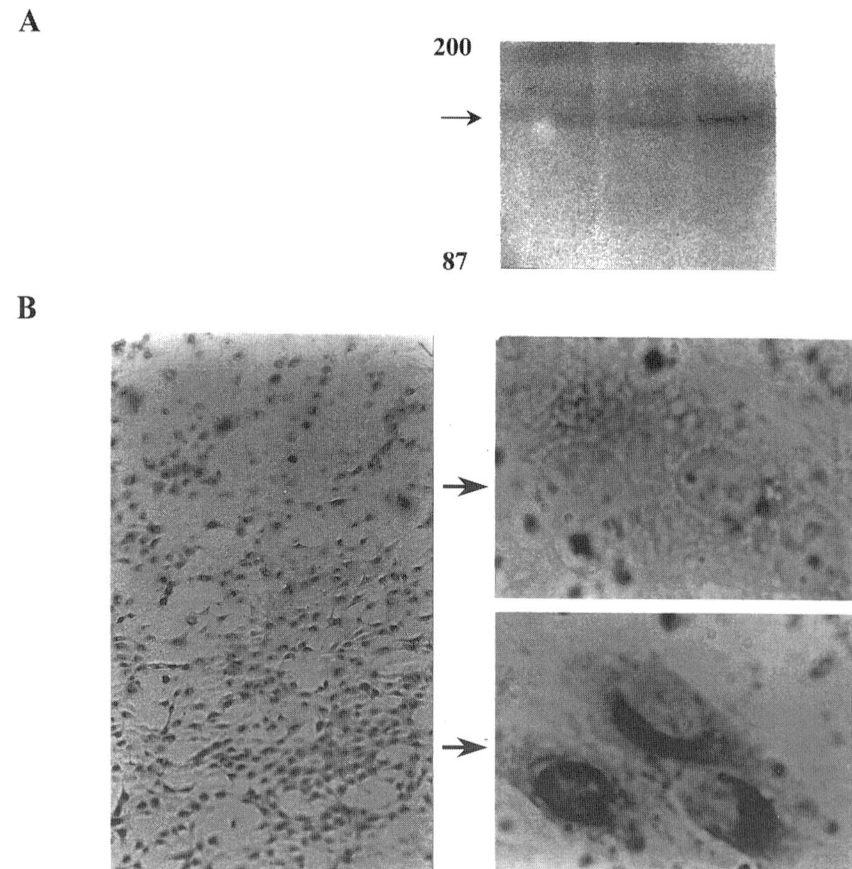

B

Figure 10-6. A. Western blot analysis for ENOS protein in bovine aortic EC subjected to cyclic. EC were subjected to 6 (middle lane) or 10% (right lane) average strain for 24 h. With both strain regimens there was an increase in ENOS protein after 24 h compared with stationary controls (left lane). Moreover, EC subjected to greater strain had a higher level (2.3-fold) of eNOS protein.
B. Immunohistochemical staining for ENOS protein. Bovine aortic EC at the periphery of the membrane (top panel) (7–24% strain) showed more intense staining compared with bovine aortic EC at the center of the membrane (bottom panel) (0–7% strain). High power photomicrograph revealed an intense perinuclear distribution of the punctate ENOS staining and a lesser stain which appears to be arranged around intracellular vacuoles. Unstretched bovine aortic EC showed very minimal staining.

periphery of the membrane (7–24% strain) showed more intense staining compared with bovine aortic EC at the center of the membrane (0–7% strain). High power photomicrograph revealed an intense perinuclear distribution of the punctate eNOS staining and a lesser stain which appears to be arranged around intracellular vacuoles. Unstretched bovine aortic EC showed very minimal staining.

Our studies demonstrate that cyclic strain (at two different strain regimens) increases eNOS gene expression in bovine aortic EC. Direct comparison of the two strain regimens showed that there was a differential expression of eNOS gene with 10% average strain, inducing a more rapid and robust increase in eNOS gene transcription compared with 6%

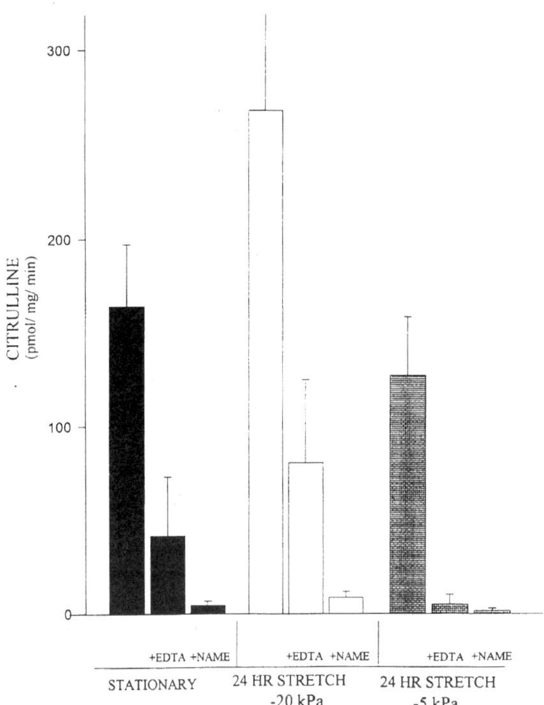

Figure 10-7. Assessment of ENOS activity in EC exposed to cyclic strain on membranes stretched with –2O kPa vacuum at 60 cycles/min for 24 hours. There was a strain-dependent increase (61%) in ENOS activity by the L-citrulline assay shown above and a 4.75-fold increase in nitrite accumulation (Greiss assay, data not shown). Stationary controls and EC exposed to –5 kPa of vacuum were not significantly different. Results are mean ± SE, *p < 0.05 compared to stationary controls.

average strain. The functional effect of this increased eNOS expression is confirmed by a previous report (Awolesi *et al.*, 1994) which demonstrated a strain-dependent increase (61%) in eNOS activity by the L-citrulline assay and a 4.75-fold increase in nitrite accumulation (Greiss assay) in bovine aortic EC exposed to 24 h of cyclic strain (Figure 10-7). These data corroborate the previous report that shear stress increases eNOS gene and protein, and both studies are consistent with recent *in vivo* studies showing that hemodynamic forces can alter endothelium-dependent relaxations and eNOS. Taken together, our studies and the findings of others suggest that eNOS and ultimately NO production can be regulated by changes in pulsatile flow.

DIFFERENCES BETWEEN SHEAR STRESS AND CYCLIC STRAIN INDUCTION OF NOS

While studies on shear stress and cyclic strain induction of NOS show a 3 to 4-fold increase in both NOS protein and mRNA steady state transcript, there are important differences in the mechanism of the response (Figure 10-8). The shear stress responses tend to be faster (Diamond *et al.*, 1989). Ranjan *et al.* (1995) showed a 3 fold increase by 6 hr while the

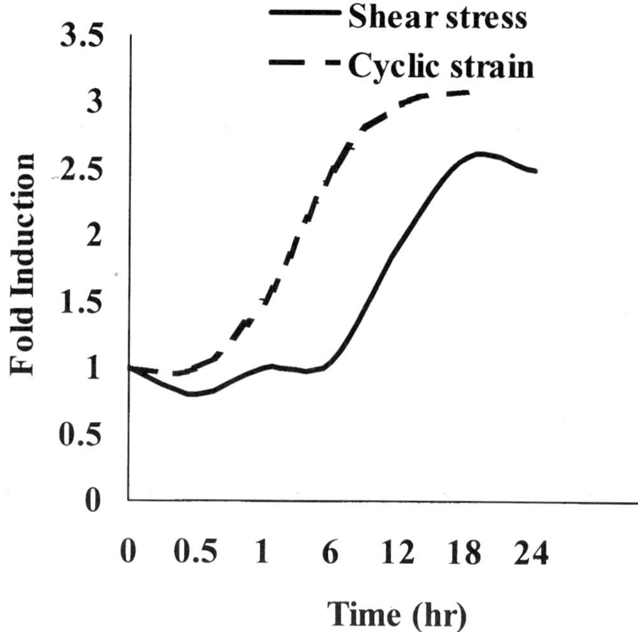

Figure 10-8. Temporal and quantitative comparison of shear stress and cyclic strain induction of NOS.

maximal effect from Awolesi's study (1995) was 24 hr. The response was less dependent on the degree of shear stress as opposed to cyclic strain where low strain (6% average strain) produced a smaller induction (1.95) compared to a 2.2 fold increase with 10% average strain. The induction of NOS protein was similarly less with low strain compared to high strain regions. The protein induction was similarly over as longer time frame (24 hr) compared to shear stress. Most importantly however the mechanism of induction also seems to differ. While both shear stress and cyclic strain induce NOS , L-NAME blocks the cyclic strain induction of NOS while shear stress induction is unaffected by L-NAME. In addition, cyclic strain does not suppress endothelin-1 induction (Sumpio and Widmann, 1990) while shear stress does (Kuchan and Frangos, 1993).

Why are there differences between the induction between shear stress and cyclic strain? There is no clear answer to the question yet however it is clear that the pathway by which they modulate their effects differ. Work on other vasoactive mediators such as the PDGF has revealed a shear stress response element (SSRE), present in the PDGF promoter genes (Resnick *et al.*, 1993). The NOS gene has been postulated to contain a similar SSRE site (Sessa *et al.*, 1992; Marsden *et al.*, 1993) and there is ongoing work to determine if there is a corresponding "cyclic strain" response element present also.

TRANSCRIPTIONAL REGULATION OF NOS

Endothelial NOS expression may be potentially regulated at several points in the protein synthesis pathway (Figure 10-9). eNOS rnRNA levels may be changed by alterations in

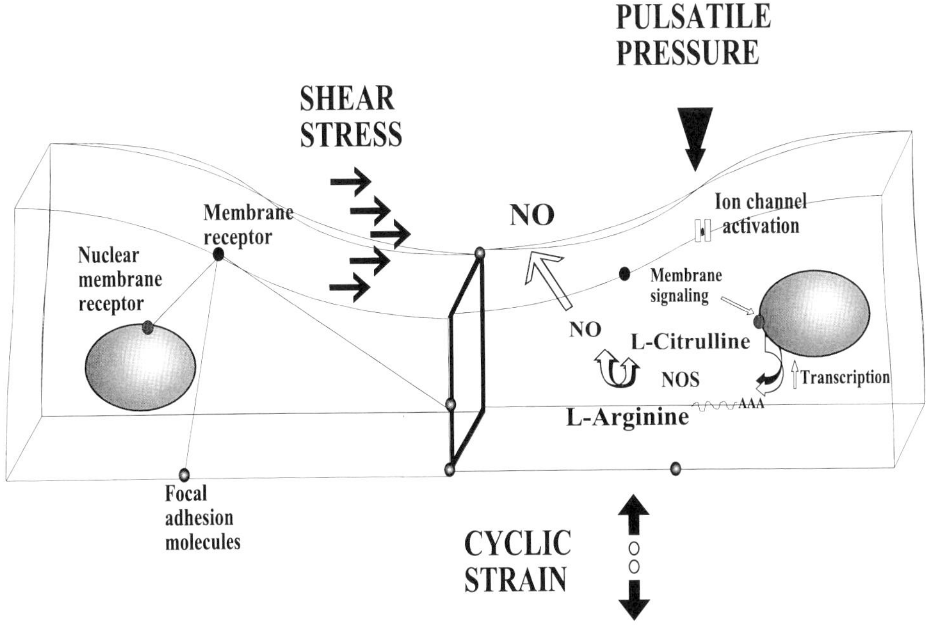

Figure 10-9. Regulation of endothelial NOS expression can occur at several points in the protein synthesis pathway.

mRNA transcription or by posttranscriptional modifications which change the rate of mRNA degradation, and finally eNOS protein levels may be controlled by changes in protein stability. Previous studies in our laboratory have demonstrated that cyclic strain alters the second messenger metabolic pathways which may potentially play a role in the regulation of eNOS. For example, cyclic strain leads to activation of protein kinase A/ adenylate cyclase (Cohen *et al.*, 1997), thereby increasing intracellular cAMP which is also accompanied by an increased binding of the nuclear factor, cAMP responsive element (CRE) binding protein (Du *et al.*, 1995; Sumpio *et al.*, 1997). Other investigators have shown that increased intracellular cAMP levels may play a role in the regulation of NO production (Graier *et al.*, 1992). Interestingly, the promoter region of the human eNOS gene has a CRE binding site (Marsden *et al.*, 1993; Robinson *et al.*, 1994) and while there is no direct evidence to implicate CRE in the regulation of eNOS in bovine EC, we can postulate that CRE may play a role in the regulation of this gene. Furthermore, we have shown recently that strain activates nuclear factors which bind to AP-1 (Sumpio *et al.*, 1994; Du *et al.*, 1995) and the shear stress responsive element (Resnick *et al.*, 1994; Sumpio *et al.*, 1998). Shear stress responsive element and AP-1 binding sites have also been demonstrated in the human eNOS promoter (Sessa *et al.*, 1992; Marsden *et al.*, 1993; Robinson *et al.*, 1994), therefore, cyclic strain-induced binding of these nuclear factors to the eNOS promoter may mediate amplification of the mRNA in EC exposed to strain.

The promoter region of the eNOS gene has been described and cloned for bovine EC (Sessa *et al.*, 1992; Marsden *et al.*, 1993; Robinson *et al.*, 1994). There are many transcription factors present in the region which attest to the facility for cytokine induced transcription of eNOS. Other transcription factors have been described in the eNOS promoter, among which are AP-1 sites, CRE sites and Sp-1 sites. Sessa and coworkers have demonstrated by promoter deletion and mutant studies the importance of an Sp-1 site (–103) in the basal activity of the core eNOS promoter (Zhang *et al.*, 1995). They have also noted the presence of a shear stress response element, GAGACC, responsible for the induction of PDGF activity by laminar shear stress. Whether this binding site is necessary for the stimulation of NOS transcription by hemodynamic forces in not known. At present there are ongoing studies to define the presence or absence of a distinct element which may account for the cyclic strain induced upregulation of eNOS activity and gene expression. Studies in our laboratory investigating the effect of cyclic strain on the regulation of the eNOS promoter have demonstrated a modest upregulation of promoter activity that can not be wholly ascribed to any the known transcription factors, nor can it be explained by an upregulation of the shear stress responsive element.

THE ROLE OF NOS AND HEMODYNAMIC FORCES IN ANGIOGENESIS

Angiogenesis is the process whereby new vessels are formed from pre-existing vasculature (as opposed to vasculogenesis which is the *de novo* formation of vessels) (Folkman, 1982; Folkman and Shing, 1992). The process involves the formation of capillary buds from mobilized endothelial cells, and subsequent canalization of these buds to form patent vessels. The process leads to neovascularization, which can be both physiological and pathological as seen in solid tumors. There is increasing evidence that nitric oxide is involved and perhaps necessary in the process of angiogenesis.

Early work on angiogenesis in embryo revealed that the developmental changes in blood flow were important in modulating early development of the circulatory system. Thoma (1893) observed that the parts of the early vasculature that had the highest flow rates became conduit vessels spawning the hypothesis that flow over the endothelium was the necessary stimulus in these observations. In corollary, when the right third aortic arch (which becomes the adult aortic arch was occluded), the left third arch became the adult arch. These early observations have been borne out by more recent studies examining the role of NO in the modulation of angiogenesis.

The process of capillary migration and proliferation which leads to neovascularization is posited to be mediated by NO. Corroborative evidence for this is provided by the observation that various vasodilators that are NO donors such as sodium nitroprusside, or glyceryl trinitrate promote the growth and proliferation of capillary endothelial cells *in vitro* (Ziche *et al.*, 1993). This response is a dose dependent one. Furthermore it has been observed that NO synthase inhibitors such as L-NMMA, abolished the proliferation and migration seen with sodium nitroprusside, indicating that NO production from constitutive NOS (and not iNOS) is involved in the process of angiogenesis. D-NMMA was ineffective in reverting the block while the provision of L-arginine enhanced the process. This indicates that NO mediated proliferation and migration is a competitive process in that the production of new NO lead to a reversal of the inhibition caused by L-NMMA or L-NAME.

An important study on the role of NO in the mediation of angiogenesis was carried out by Ziche *et al.* (1993) in which they evaluated the effect of NO donors, such as sodium nitroprusside, or stimulators of NO endogenous production, such as substance P (SP), in the angiogenic process. SP is known to cause vasodilatation and increased vascular permeability. The vasorelaxant action is mediated by the NK_1 receptor. Ziche and coworkers had previously shown that SP activates cyclic GMP. They had shown in previous experiments that SP promotes angiogenesis *in vivo*, while also promoting endothelial cell proliferation and migration *in vitro*. Their hypothesis from all the evidence at hand was that the action of SP was NO mediated. Using both an *in vivo* model (corneal implants) and an *in vitro* model of capillary endothelial cells grown in culture, they stimulated angiogenesis with [Sar^9]-SP-sulfone, which is a stable and selective agonist for the NK_1 receptor, and PGE_1. They created two adjacent pockets in the cornea of rabbits, and in one pocket, they placed either SP or PGE_1 and in the other, they placed sodium nitroprusside or control. SP and PGE_1 in the absence of sodium nitroprusside led to 50% corneal neovascularization. When SNP was added to the adjacent pocket however, there was an augmentation of the angiogenic effect of either SP or PGE_1 with a 100% neovascularization of the corneal, while with control there was no added angiogenic effect over and above that observed with either SP or PGE_1 alone (Ziche *et al.*, 1995). When the animals had been fed L-NAME for 1 week prior to the placement of the corneal implants, the angiogenic effect of SP or PGE_1 was totally abolished. When the animals were treated with D-NAME, or a lower dose of L-NAME, there was no modification of the angiogenic effect. These *in vivo* findings were substantiated by *in vitro* studies on corneal endothelial cell migration (using the Boyden chamber) and proliferation (using cell count and thymidine incorporation). NO synthase inhibition led to drastic decreases in both cell migration and proliferation, that were reversed by the addition of L-arginine to the media.

The above data has also been reproduced by Leibovich *et al.* (1994) in human monocytes where they showed that NO played a similar role in monocyte induced angiogenesis. However in a study by Pipili-Synetos *et al.* (1993), they showed the opposite effect. They examined the involvement of nitric oxide (NO) in the regulation of angiogenesis using the *in vivo* system of the chorioallantoic membrane (CAM) of the chick embryo and in the matrigel tube formation assay. Their experiments demonstrated that sodium nitroprusside (SNP) which releases NO spontaneously, caused a dose-dependent inhibition of angiogenesis in the CAM *in vivo* and reversed completely the angiogenic effects of alpha-thrombin and phorbolester. In the matrigel tube formation assay, which is another *in vitro* assay of angiogenesis, both SNP and the cell permeable cyclic GMP analogue, Br-cGMP reduced tube formation. Furthermore, when NO synthase was inhibited with L-NMMA and L-NAME they demonstrated a stimulation of angiogenesis in the CAM *in vivo*, in a dose-dependent fashion. D-NMMA and D-NAME on the other hand had no effect on angiogenesis in this system. The effect of L-NNMA and L-NAME was inhibited by L-Arginine. Inhibition of inducible NO synthase with low dose dexamethasone also produced a stimulation of angiogenesis in the CAM, whereas at a higher concentration, it inhibited this process. These results seem to contradict the studies by Ziche and Leibovich, indicating that both the constitutive and the inducible NO synthase may contribute to the NO-mediated inhibition of angiogenesis. However, it is important to note that the Pipili-Synetos experiments were done on chick embryonic tissue while the previous studies were on adult fully differentiated tissue — this may account for the difference in results.

CONCLUSION

The mechanisms involved in the responses to cyclic strain and shear stress observed in eNOS gene expression have yet to be clearly elucidated. However, there is enough evidence to confirm that there is de novo synthesis of mRNA and that the effect is not a result of increased mRNA instability. Ongoing are studies designed at examining the eNOS promoter. Although certain elements in the promoter have been shown to be important in the basal expression of eNOS, there has not been any convincing evidence of a shear stress or cyclic strain responsive element in the promoter sequence.

Recent studies have sought to understand the mechanisms involved in the activity of the eNOS protein. It has been shown that there is posttranslational modification of the eNOS protein, which may affect its bioavailability and activity (Garcia-Cardena *et al.*, 1996; Garcia-Cardena *et al.*, 1996). The eNOS gene is known to contain a sequence that allows for the myristoylation of the protein. The significance of this myristoylation site has been examined are recent studies looking at how eNOS expression alters eNOS protein activity. NOS has been shown to be a membrane bound protein with most NOS activity present in the particulate fraction of cell preparations (Sessa *et al.*, 1995). In the amino acid sequence is a sequence associated with a rare posttranslational modification known as myristolation. The significance of this sequence is that myristolation allows the protein to be membrane associated. Site directed mutagenesis of this sequence leads to loss of the membrane association with most of the eNOS being present in the cytosol fraction. The myristolation of eNOS may be important in the regulation of its activity. Phosphorylation of eNOS has also been described and may be another mechanism important in the regulation of eNOS activity.

An understanding of the role of endothelial mediators such as nitric oxide in the pathogenesis of many disease processes such as hypertension, arteriosclerosis, diabetic microangiopathy and certain bleeding diathesis may hold the key to future therapies. It is therefore imperative that we understand what factors control the regulation of these mediators. Hemodynamic forces have been shown to induce production of these mediators at the molecular level as documented in some of the studies reported in this discourse, however more work needs to be done to understand the specifics of how these forces may control the production of endothelial mediators such as nitric oxide.

REFERENCES

Awolesi, M.A. *et al.* (1994) Cyclic strain increases endothelial nitric oxide synthase. *Surgery*, **116**, 439–444.

Awolesi, M.A. *et al.* (1995) Cyclic strain upregulates nitric oxide syntase in cultured bovine aortic endothelial cells. *J. Clin. Invest.*, **96**, 1449–1454.

Banes, A.J. *et al.* (1985) A New Vacuum-operated Stress-providing Instrument that Applies Static or Variable Duration Cyclic Tension or Compression to Cells *In Vitro*. *Journal of Cell Science*, **75**, 35–42.

Banes, A.J. *et al.* (1990) Culturing cells in a mechanically active environment. *Am. Biotech. Lab.*, 13–22.

Buck, R.C. (1980) Reorientation response of cells to repeated stretch and recoil of the substratum. *Exp. Cell Res.*, **127**, 470–474.

Cohen, C.R. *et al.* (1997) Activation of adenylate cyclase, cAMP, PKA pathway in endothelial cells exposed to cyclic strain. *Exp. Cell Res.*, **231**, 184–189.

Davies, P.F. *et al.* (1984) Influence of hemodynamics forces on vascular endothelial function. *In vivo* studies shear stress and pinocytosis in bovine airtic cells. *Journal of Clinical Investigation*, **73**, 1121–1129.

Delp, M. *et al.* (1993) Exercise training alters endothelium-dependent vasoreactivity of rat abdominal aorta. *J. Applied Physiol.*, **753**, 1354–63.

Deng, S.X. *et al.* (1994) New experiments on shear modulus of elasticity of arteries. *American Journal of Physiology*, **266**, H1–10.

Dewey, C.F. *et al.* (1981) The dynamic response of vascular endothelial cells to fluid shear stress. *J. Biochem. Eng.*, **103**, 177–185.

Diamond, S.L. *et al.* (1989) Fluid flow stimulates tissue plasminogen activator secretion by cultured human endothelial cells. *Science*, **243**, 1483–1485.

Du, W. *et al.* (1995) Cyclic Strain Causes Heterogeneous Induction of Transcription Factors, AP-1, CRE binding protein and NF-κB, in Endothelial Cells: Species and Vascular Bed Diversity. *Journal of Biomechanics*, **28**, 1485–1491.

Ehsani, A. *et al.* (1981) Effects of 12 months of intense exercise training on ischemic segment depression in patients with coronary disease. *Circulation*, **64**, 1116–1124.

Eskin, S.G. *et al.* (1984) Response of cultured endothelial cells to steady flow. *Microvascular Research*, **28**, 87–94.

Folkman, J. (1982) Angiogenesis: initiation and control. *Ann. NY Acad. Sci.*, **401**, 212–227.

Folkman, J. and Y. Shing (1992) Angiogenesis. *J. Biol. Chem.*, **267**(16), 10931–10934.

Frangos, J.A. *et al.* (1985) Flow effects on prostacyclin production in cultured human endothelial cells. *Science*, **227**, 1477–1479.

Fung, Y.C. and S.Q. Liu (1995) Determination of the mechanical properties of the different layers of blood vessels *in vivo*. *Proceedings of the National Academy of Sciences of the United States of America*, **92**, 2169–2173.

Garcia-Cardena, G. *et al.* (1996) Endothelial nitric oxide synthase is regulated by tyrosine phosphorylation and interacts with caveolin-1. *Journal of Biological Chemistry*, **271**, 27237–27240.

Garcia-Cardena, G. *et al.* (1996) Targeting of nitric oxide synthase to endothelial cell caveolae via palmitoylation: implications for nitric oxide signaling. *Proceedings of the National Academy of Sciences of the United States of America*, **93**, 6448–6453.

Glagov, S. *et al.* (1992) Micro-architecture and composition of artery walls, relationship to location, diameter and the distribution of mechanical stress. *Journal of Hypertension Supplement*, **10**, S101–104.

Glagov, S. *et al.* (1988) Hemodynamics and atherosclerosis. Insights and perspectives gained from studies of human arteries. *Archives of Pathology & Laboratory Medicine*, **112**, 1018–1031.

Graier, W.F. *et al.* (1992) Increases in endothelial cyclic AMP levels amplify agonist-induced formation of endothelium-derived relaxing factor (EDRF). *Biochem. J.*, **288**, 345–349.

Han, H.C. and Y.C. Fung (1995) Longitudinal strain of canine and porcine aortas. *Journal of Biomechanics*, **28**, 637–641.

Hellenius, M.L. *et al.* (1993) Diet and exercise are equally effective in reducing risk for cardiovascular disease. Results of a randomized controlled study in men with slightly to moderately raised cardiovascular risk factors. *Atherosclerosis*, **1031**, 31–91.

Hiatt, W.R. *et al.* (1990) Benefits of exercise conditions for patients with peripheral arterial disease. *Circulation*, **81**, 602–609.

Iba, T. *et al.* (1991) Effect of Cyclic Stretch on Endothelial Cells From Different Vascular Beds. *Circulatory Shock*, **35**, 193–198.

Iba, T. and B.E. Sumpio (1991) Morphologic evaluation of human endothelial cells subjected to repetitive stretch *in vitro*. *In Vitro*. In Press.

Iba, T. and Sumpio, B.E. (1992) Tissue plasminogen activator expression in endothelial cells exposed to cyclic strain *in vitro*. *Cell Transplantation*, **1**, 43–50.

Kassab, G.S. and Fung, Y.C. (1995) The pattern of coronary arteriolar bifurcations and the uniform shear hypothesis. *Annals of Biomedical Engineering*, **23**, 13–20.

Koller, A. and Huang, A. (1994) Impaired nitric oxide-mediated flow-induced dilation in arterioles of spontaneously hypertensive rats. *Circulation Research*, **74**, 416–421.

Koller, A. and Huang, A. (1995) Shear stress-induced dilation is attenuated in skeletal muscle arterioles of hypertensive rats. *Hypertension*, **25**, 758–763.

Kuchan, M. and Frangos, J. (1993) Shear Stress Regulates endothelin-1 Release via Protein Kinase C and cGMP in Cultured Endothelial Cells. *Am. J. Physiol.*, **264**, H150–H155.

Langille, L. (1993) Remodeling of developing and mature arteries: endothelium, smooth muscle, and matrix. *J. Cardiovasc. Pharm.*, **21**(Suppl. 1), S11–S17.

Leaf, C.D. *et al.* (1990) Nitrate biosynthesis in rats, ferrets and humans. Precursor studies with L-arginine. *Carcinogenesis*, **115**, 855–8.

Leibovich, S. *et al.* (1994) Production of angiogenic activity by human monocytes requires an L-arginine/nitric oxide-synthase dependent effector mechanism. *Proc. Natl. Acad. Sci. USA*, **91**, 4190–4194.

Li, G. *et al.* (1994) Cyclic strain stimulates endothelial cell proliferation, Characterization of strain requirements. *Endothelium*, **22**, 177–181.

Liu, S.Q. and Fung, Y.C. (1993) Material coefficients of the strain energy function of pulmonary arteries in normal and cigarette smoke-exposed rats. *Journal of Biomechanics*, **26**, 1261–1269.

Marsden, P. *et al.* (1993) Structure and chromosomal localization of the human constitutive endothelial nitric oxide synthase gene. *J. Biol. Chem.*, **268**, 17478–17488.

Miller, V.M. and Burnett, J.C.Jr. (1992) Modulation of NO and endothelin by chronic increases in blood flow in canine femoral arteries. *Am. J. Physiol.*, **263**, H103–108.

Miller, V.M. and Vanhoutte, P.M. (1988) Enhanced release of endothelium-derived factor(s) by chronic increases in blood flow. *Am. J. Physiol.*, **255**, H446–451.

Mitsumata, M. *et al.* (1993) Fluid shear stress stimulates platelet-derived growth factor expression in endothelial cells. *American Journal of Physiology*, **265**, H3–8.

Nadaud, S. *et al.* (1996) Sustained increase in aortic endothelial nitric oxide synthase expression *in vivo* in a model of chronic high blood flow. *Circ. Res.*, **79**, 857–863.

Nakamura, M. *et al.* (1992) The efficacy of aerobic exercise therapy on hypertensive patients with mild cardiac complications. *Ann. Acad. Med.*, **211**, 38–41.

Nishida, K. *et al.* (1992) Molecular cloning and chracterization of the constitutive bovine endothelial cell nitric oxide synthase. *J. Clin. Invest.*, **90**, 2092–2093.

Ojha, M. (1993) Spatial and temporal variations of wall shear stress within an end-to-side arterial anastomosis model. *J. Biomech.*, **26**, 1377–1388.

Ojha, M. (1994) Wall shear stress temporal gradient and anastomotic intimal hyperplasia. *Circ. Res.*, **74**, 1227–1231.

Paffenbarger, R.S. *et al.* (1983) Physical activity and incidence of hypertension in college alumni. *Am. J. Epidemiol.*, **117**, 245–257.

Pipili-Synetos, E. *et al.* (1993) Nitric oxide is involved in the regulation of angiogenesis. *Br. J. Pharmacology*, **108**, 855–857.

Poehlman, E.T. *et al.* (1992) Resting energy metabolism and cardiovascular disease risk in resistance-trained and aerobically trained males. *Metabolism: Clinical & Experimental*, **41**, 1351–60.

Posner, J.D. *et al.* (1990) Effects of exercise training in the elderly on the occurrence and time to onset of cardiovascular diagnoses. *J. Am. Geriatrics Soc.*, **383**, 205–10.

Ranjan, V., Xiao, Z. and Diamond, S.L. (1995) Constitutive NOS expression in cultured endothelial cells is elevated by fluid shear stress. *Am. J. Physiol.*, **269**, H550–555.

Remuzzi, A. *et al.* (1984) Orientation of endothelial cells in shear fields *in vitro*. *Biorheology*, **21**, 617–630.

Resnick, N. *et al.* (1993) Platelet-derived growth factor B chain promoter contains a cis-acting fluid shear-stress-responsive element. *Proc. Natl. Acad. Sci. USA*, **90**(16), 7908.

Resnick, N. and M.A. Gimbrone, Jr. (1995) Hemodynamic forces are complex regulators of endothelial gene expression. *Faseb J.*, **9**(10), 874–82.

Resnick, N. *et al.* (1994) Endothelial Regulation by Biomechanical Forces. *Proceedings of Atherosclerosis*.

Rinaldi, G. and Bohr, D. (1989) Endothelium-mediated spontaneous response in aortic rings of deoxycorticos-terone acetate-hypertensive rats. *Hypertension*, **13**, 256–261.

Robinson, L. *et al.* (1994) Isolation and chromosomal localization of the human endothelial nitric oxide synthase gene. *Genomics*, **19**, 350–357.

Salzar, R.S. *et al.* (1995) Pressure-induced mechanical stress in the carotid artery bifurcation: a possible correlation to atherosclerosis. *Journal of Biomechanics*, **28**, 1333–1340.

Sessa, W.C., Pritchard, K., Seyedi, N., Wang, J. and Hintze, T.H. (1994) Chronic exercise in dogs increases coronary vascular nitric oxide production and endothelial cell nitric oxide synthase gene expression. *Circ. Res.*, **74**, 349–353.

Sessa, W.C. *et al.* (1995) The Golgi association of endothelial nitric oxide synthase is necessary for the efficient synthesis of nitric oxide. *Journal of Biological Chemistry*, **270**, 17641–17644.

Sessa, W.C. *et al.* (1992) Molecular cloning and expression of a cDNA encoding endothelial cell nitric oxide synthase. *J. Biol. Chem.*, **267**(22), 15274–6.

Sharefkin, J.B. *et al.* (1991) Fluid flow decreases preproendothelin mRNA levels and suppresses endothelin-1 peptide release in cultured human endothelial cells. *Journal of Vascular Surgery*, **14**, 1–9.

Sumpio, B. *et al.* (1994) Exposure of endothelial cells to cyclic strain induces c-fos, fos B and c-jun but not jun B or jun D and increases the transcription factor AP-1. *Endothelium*, **2**, 149–156.

Sumpio, B.E., Ed. (1993) *Hemodynamic Forces and Vascular Cell Biology.* Austin, Texas, RG Landes Publisher.

Sumpio, B.E. *et al.* (1988) Alterations in aortic endothelial cell morphology and cytoskeletal protein synthesis during cyclic tensional deformation. *J. Vasc. Surg.*, **7**, 130–138.

Sumpio, B.E. *et al.* (1987) Mechanical stress stimulates aortic endothelial cells to proliferate. *J. Vasc. Surg.*, **6**, 252–256.

Sumpio, B.E. *et al.* (1997) Regulation of tPA in endothelial cells exposed to cyclic strain: role of CRE, AP-2, and SSRE binding sites. *American Journal of Physiology*, **273** (5 Pt 1), C1441–8.

Sumpio, B.E. *et al.* (1998) Regulation of PDGF-B in endothelial cells exposed to cyclic strain. *Arteriosclerosis, Thrombosis & Vascular Biology*, **18**(3), 349–55.

Sumpio, B.E. and Widmann, M.D. (1990) Enhanced production of an endothelium derived contracting factor by endothelial cells subjected to pulsatile stretch. *Surgery*, **108**, 277–282.

Terracio, L. *et al.* (1988) Effects of cyclic mechanical stimulation of the cellular components of the heart: *in vitro*. *In vitro cellular and developmental biology*, **24**(1), 53–58.

Thomas, R. (1893) *Untersuchagen uberdie histogenese und histomechanik des gefassystems*. Stuttgart, Enke.

Thubrikar, M.J. and Robicsek, F. (1995) Pressure-induced arterial wall stress and atherosclerosis. *Annals of Thoracic Surgery*, **59**, 1594–1603.

Topper, J.N., Cai, J., Falb, D. and Gimbrone, M.A., Jr. (1996) Identification of vascular endothelial genes differentially responsive to fluid mechanical stimuli: cyclooxygenase-2, manganese superoxide dismutase, and endothelial cell nitric oxide synthase are selectively up-regulated by steady laminar shear stress. *Proc. Natl. Acad. Sci. USA*, **93**, 10417–10422.

Uematsu, M., Ohara, Y., Navas, J.P., Nishida, K., Murphy, T.J., Alexander, R.W. *et al.* (1995) Regulation of endothelial cell nitric oxide synthase mRNA expression by shear stress. *Am. J. Physiol.*, **269**, C1371–1378.

Vandenburgh, H. and S. Kaufman (1979) *In vitro* Model for Stretch-Induced Hypertrophy of Skeletal Muscle. *Science*, **203**, 265–268.

Xie, J. *et al.* (1995) Bending of blood vessel wall: stress-strain laws of the intima-media and adventitial layers. *Journal of Biomechanical Engineering*, **117**, 136–145.

Yano, Y. *et al.* (1996) Tyrosine phosphorylation of pp125FAK and paxillin in aortic endothelial cells induced by mechanical strain. *American Journal of Physiology*, **271** (2 Pt 1), C635–49.

Yano, Y. *et al.* (1997) Cyclic strain induces reorganization of integrin $\alpha5\beta1$ and $\alpha2\beta1$ in human umbilical vein endothelial cells. *Journal of Cellular Biochemistry*, **64**(3), 505–13.

Yano, Y. *et al.* (1996) Involvement of rho p21 in cyclic strain-induced tyrosine phosphorylation of focal adhesion kinase (pp125FAK), morphological changes and migration of endothelial cells. *Bioch. Biophys. Res. Comm.*, **224**, 508–515.

Zhang, R. *et al.* (1995) Functional Analysis of the human endothelial nitric oxide synthase promoter, SP1 and GATA factors are necessary for basal transcription in endothelial cells. *J. Biol. Chem.*, **270**, 153260–152366.

Ziche, M. *et al.* (1993) Nitric Oxide promotes DNA synthesis and cyclic GMP formation in endothelial cells from post-capillary venules. *Biochem. Biophys. Res. Commun.*, **192**, 1198–1203.

Ziche, M. *et al.* (1995) Nitric oxide modulates angiogenesis elicited by prostaglandin E1 in rabbit cornea. *Advances in Prostaglandin, Thromboxane, & Leukotriene Research*, **23**, 495–497.

11 Nitric Oxide Synthase Regulation by 17β-Estradiol

Katalin Kauser and Gabor M. Rubanyi

Department of Cardiovascular Research, Berlex Biosciences, Richmond, CA, USA

The vasculoprotective effects of 17β-estradiol are well documented, although the exact mechanism(s) of action of the female sexual steroid hormone on the cardiovascular system is not entirely understood. Regulation of nitric oxide production by the different nitric oxide synthase isoenzymes may contribute to the observed benefits of hormone replacement therapy. This chapter will discuss the possible mechanisms of the regulation of nitric oxide production by 17β-estradiol by reviewing the current literature and summarizing experimental data generated in our laboratory.

Key words: 17β-estradiol, vasculoprotection, nitric oxide, nitric oxide synthase, atherosclerosis

INTRODUCTION

Nitric oxide (NO) is a multifunctional molecule as it has been indicated by previous chapters. NO is synthesized by nitric oxide synthase-II (NOS-II; inducible NOS) or NOS-I (brain constitutive NOS) in large quantities to mediate cytotoxic effects as part of inflammatory reactions, or in smaller amounts by the endothelial isoform, NOS-III, to exert antioxidant, antiinflammatory and antithrombotic effects and to mediate vasodilation of blood vessels.

Cardioprotection by the female sexual steroid hormone, estrogen, has been widely documented by epidemiological as well as more recent clinical studies and numerous experimental data (Isles *et al.*, 1992; Stampfer *et al.*, 1991; Farhat *et al.*, 1996; Kauser and Rubanyi, 1997/a, 1997/b). Despite of all the supporting evidence the exact mechanism of the cardiovascular protective effect of 17β-estradiol is not known. NO, produced by NOS-III, seems to be an ideal mediator of the beneficial cardiovascular effect of 17β-estradiol due to its multiple vasculoprotective actions.

All three isoforms of NOS have been shown to be regulated by 17β-estradiol. The endothelial isoform (NOS-III) has been the most extensively investigated both *in vivo* and *in vitro*. Besides NOS-III, the regulation of the neuronal (NOS-I) and the inducible (NOS-II) isoenzymes have also been shown by estrogen treatment *in vivo* (Weiner *et al.*, 1994/b; Kauser *et al.*, 1997). Inhibition of NOS-II has been studied *in vitro* in our laboratory using cytokine treated isolated rat aortic rings (Kauser *et al.*, 1994).

In this chapter we will review data indicating the upregulation of NOS-III by 17β-estradiol and discuss the experiments, which demonstrated the inhibition of NOS-II by the ovarian sex steroid hormone.

Correspondence: Katalin Kauser, MD, PhD, Berlex Biosciences, Cardiovascular Department, 15049 San Pablo Avenue, Richmond, CA 94804-0099, Ph: (510) 669-4282; Fax: (510) 669-4246; E-mail: Katalin_Kauser@Berlex.com.

UPREGULATION OF ENDOTHELIAL NO PRODUCTION BY 17β-ESTRADIOL

Gender Difference and the Effect of 17β-Estradiol on Endothelial Function

The possible influence of sexual steroid hormones on NO synthesis is indicated by studies demonstrating gender difference in endothelium-dependent modulation of vascular tone (Maddox *et al.*, 1987; Stallone *et al.*, 1991; Hayashi *et al.*, 1992; Kauser and Rubanyi, 1995/a). These studies showed that the vascular endothelium of females is capable to produce larger amount of NO. Our own studies utilizing a superfusion bioassay system demonstrated the first time that more NO is released from the perfused aortas of female rats than that of the males (Figure 11-1) (Kauser and Rubanyi, 1994/b). The hypothesis that the greater production of NO is due to the stimulating effect of the female sexual steroid hormone was supported by studies investigating the effect of ovariectomy and/or 17β-estradiol treatment on endothelium-dependent vascular responses (Gisclard *et al.*, 1988; Hayashi *et al.*, 1992; Kauser and Rubanyi, 1995/b). The original observation by Gisclard *et al.* (1988) described increased endothelium-dependent relaxation in isolated femoral arteries of rabbits after four days treatment with 17β-estradiol. Similar potentiation of endothelial NO-mediated responses were found in coronary arteries isolated from 17β-estradiol treated monkeys (Williams *at al.*, 1990). Hayashi *et al.* (1992) demonstrated that basal release of NO from female rabbits were greater than that of released from male and ovariectomized female rabbits.

These experimental studies have been confirmed by clinical investigations, demonstrating attenuation of abnormal vasomotor responses by administration of ethinyl-estradiol (Reis *et al.*, 1994) or physiological doses of 17β-estradiol (Gilligan *et al.*, 1994). A two year follow up study reported increases in circulating NO levels in postmenopausal women on hormone replacement therapy compared to women not taking estrogen (Rosselli *at al.*, 1995). More recently increases in basal NO release have been shown in response to 17β-estradiol treatment in the forearm circulation of perimenopausal women (Sudhir *et al.*, 1996).

Potential Mechanisms Involved in Increased Endothelial NO Level by 17β-Estradiol

Regulation of NOS-III gene expression

One of the possible mechanisms by which 17β-estradiol can enhance NO synthesis is via increasing the amount of the NOS-III enzyme. Data in several species suggest that 17β-estradiol modulation of NO availability is due to up-regulation of NOS-III gene expression (Weiner *et al.*, 1994/b; Goetz *et al.*, 1994). Pregnancy has been associated with a four-fold increase in NOS-III activity in the guinea pig uterine artery (Weiner *at al.*, 1994/a). Similarly increases of NOS-III mRNA has been demonstrated in pregnant rat aortas using semiquantitative polymerase chain reaction (PCR) (Goetz *et al.*, 1994). Besides pregnancy, chronic treatment with 17β-estradiol also increased Ca^{2+}-dependent NO-activity in female guinea pigs (Weiner *et al.*, 1994/b).

Regulation of the NOS-III gene expression by 17β-estradiol has been extensively studied in cultured endothelial cells (Hishikawa *et al.*, 1995; Hayashi *et al.*, 1995; MacRitchie

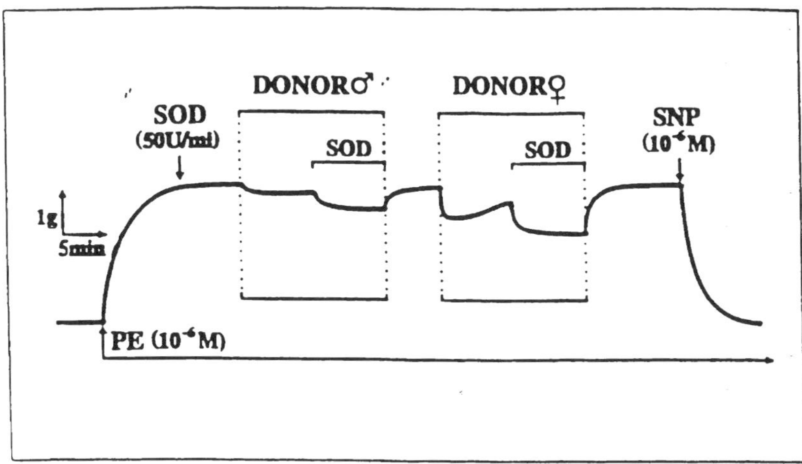

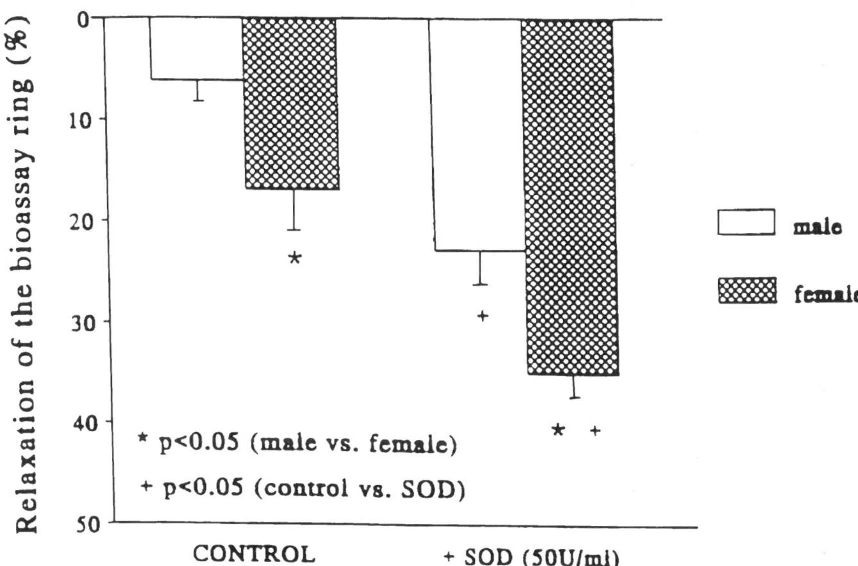

Figure 11–1. Gender difference in bioassayable endothelium-derived NO generation in rat aorta. The upper panel illustrates an original tracing of a typical bioassay experiment. Perfusate from the donor segment isolated from female Wistar rats (DONOR ♀) evoked greater relaxation of a denuded rat thoracic aortic ring, than the perfusate from the male donor (DONOR ♂). Superoxide dismutase (SOD; 50 U/ml) further enhanced the nitric oxide mediated relaxation in both cases, but did not affect the gender difference. The lower panel demonstrates the mean values of 8 experiments. Open bars represent responses (mean ± SEM) of the de-endothelialized bioassay rings to perfusate from male donor segments and the cross-hatched bars illustrate the responses evoked by the perfusate from the female donors. Relaxation of the bioassay ring is expressed as a percent inhibition of the phenylephrine induced contraction. Asterisk indicate significantly (*p < 0.05) greater relaxation of the same assay ring in response to the perfusate from the female donor segments. Reproduced with the permission of the American Physiological Society (from Ref. Kauser and Rubanyi, 1994/a).

et al., 1997). In early passages of human aortic and ovine fetal pulmonary endothelial cells, physiological concentrations of 17β-estradiol caused a significant increase in NO production parallel to increased NOS-III protein levels (Hishikawa *et al.*, 1995; MacRitchie *et al.*, 1997). In other experiments it has been demonstrated that bovine aortic endothelial cells in culture are able to synthesize estrogens and contain functional estrogen receptors (Bayard *et al.*, 1995). The transactivating capability of these receptors has also been reported using transiently transfected cultured bovine endothelial cells with a plasmid containing the luciferase reporter gene with an estrogen response element (ERE) as a transcriptional enhancer (Bayard *et al.*, 1995). The finding of the two half palindromic sites of ERE on the human NOS-III gene supports a potential receptor-mediated effect of estrogen on NOS-III gene expression in endothelial cells (Venema *et al.*, 1994).

Regulation of NOS-III enzyme activity

Increased synthesis of NO can be a result of elevated enzyme activity. Activity of NOS-III can be augmented by increased availability of the substrate L-arginine or cofactors, such as tetrahydrobiopterin (BH_4), calmodulin or intracellular calcium level. Studies have demonstrated that the administration of exogenous L-arginine, the physiological precursor of NO production, restored endothelium-dependent relaxation in hypercholesterolemic humans (Creager *et al.*, 1992) and decreased aortic lesion formation in cholesterol fed rabbits (Cooke *et al.*, 1992). These findings indicate that under certain pathological conditions substrate availability of NOS-III may be limited and supplementing L-arginine can provide vasculoprotective benefit. Results of recent report by Hayashi *et al.* (1995) argue against the possibility that sexual steroid hormones would regulate the level of circulating L-arginine.

BH_4, a cofactor of NO synthesis by all NOS enzymes, has been reported recently to restore NO-mediated vasodilation in hypercholesterolemic patients (Stroes *et al.*, 1997). BH_4 synthesis is shown to be a rate limiting step for NO production by cultured human endothelial cells (Schmidt *et al.*, 1992; Werner-Felmayer *et al.*, 1993; Rosenkranz-Weiss *et al.*, 1994). It is feasible, that BH_4 synthesis can be regulated by 17β-estradiol via modulation of the level and/or activity of the GTP-cyclohydrolase enzyme, which is responsible for the intracellular level of BH_4 (Hattori and Gross, 1993). However, there is no data available describing the effect of 17β-estradiol on BH_4 synthesis.

The probability that 17β-estradiol may modulate the effect of NOS-III activity via interaction with calmodulin is also conceivable (Bouhoute and Leclercq, 1995). The influence of 17β-estradiol on intracellular calmodulin synthesis has been described in rabbit myometrium (Matsui *et al.*, 1983), but it has not yet been investigated in vascular endothelial cells.

Protection of NO by 17β-estradiol

Another way of regulating the level of bioavailable NO is to interfere with its degradation. Increased amount of guanylate cyclase has been measured by ethinyl estradiol treated bovine aortic endothelial cells without seeing any increase in NOS-III mRNA and protein amount or enzyme activity (Arnal *et al.*, 1996). Measurement of superoxide anion radical production from the same cells revealed that estrogen treatment dramatically reduced superoxide generation in the cultured bovine endothelial cells. Superoxide anion is a potent inactivator of NO (Rubanyi and Vanhoutte, 1986). Decreasing the level of superoxide anions could lead to increased amount of bioavailable NO without changing it's synthetic rate. Besides regulating superoxide production of endothelial cells, estrogens also have antioxidant properties (Subbiah *et al.*, 1993; Ruiz-Larrea *et al.*, 1995). They have been shown to inhibit copper-induced as well as cell-mediated oxidative modification of LDL (Maziere *et al.*, 1991) and they are also potent inhibitors of HDL oxidation (Banka, 1996). It is conceivable that the antioxidant properties of 17β-estradiol may contribute to the increased bioactivity of NO in the presence of estrogen, although 17β-estradiol does not directly scavenge superoxide anion and it does not prolong the half-life of endothelium-derived NO in superfusion bioassay experiments (unpublished observations from our laboratory).

Role of the Estrogen Receptor in the Regulation of NOS-III

The classical mechanism of estrogen's action on different target organs is mediated by the activation of estrogen receptors (ER). The presence of functional ER has been demonstrated in endothelial cells (Bayard *et al.*, 1995; Hayashi *et al.*, 1995; Kim-Schulze *et al.*, 1996; Venkov *et al.*, 1996; MacRitchie *et al.*, 1997). Dose-dependent, saturable increase in NOS-III mRNA and protein as well as enzyme activity in response to 17β-estradiol was inhibited by the ER antagonists, ICI 182780 and tamoxifen in human umbilical endothelial cell and bovine aortic endothelial cell in culture (Bayard *et al.*, 1995; Hayashi *et al.*, 1995; MacRitchie *et al.*, 1997), suggesting that the ER is involved in the regulation of NOS-III by 17β-estradiol.

We have recently studied the role of ER in the modulation of the release of endothelium derived NO in a homozygous ERα mutant mouse (ERKO) (Rubanyi *et al.*, 1997), in which the ERα gene was disrupted resulting in the lack of expression of functional ER (Lubahn *et al.*, 1993). We found that the lack of the expression of ERα lead to significant impairment in basal NO release (Figure 11-2), without affecting the amount of NOS-III enzyme. This finding demonstrated that ERα is probably involved in the regulation of NOS-III enzyme activity, by affecting the level of free cytosolic calcium and/or participating in posttranslational modifications of NOS-III resulting in increased enzyme activity.

Since the construction of the ERKO animal, lacking the functional ERα gene, a second estrogen receptor, ERβ has been identified (Kuiper *et al.*, 1996). The mRNA expression of this novel human estrogen receptor is higher in cultured human endothelial cells (personal communications by Dr. Zajchowski of Berlex Biosciences) than ERα. ERβ mediated transactivation by 17β-estradiol was demonstrated in transiently transfected CHO cells (Mosselman *et al.*, 1996). The potential role of ERβ in the regulation of NOS-III remains to be elucidated.

Role of Increased Endothelial NO in the Anti-Atherosclerotic Effect of 17β-Estradiol

Endothelial cell injury or endothelial dysfunction is considered to play an important role in the development of atherosclerotic plaques (Rubanyi, 1993; Busse and Fleming, 1996). Endothelial dysfunction is manifested mainly in impaired NO production. The role of NO

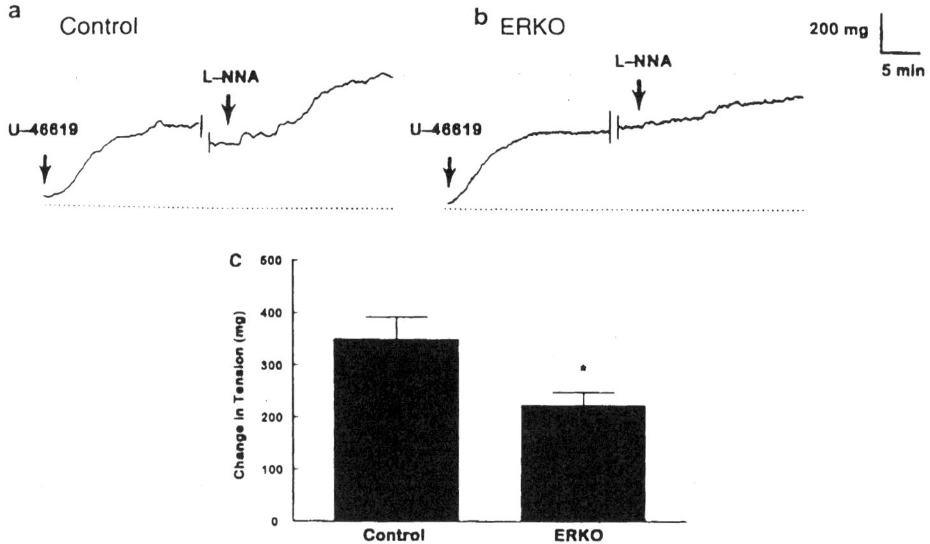

Figure 11–2. Decreased basal endothelial NO production in estrogen receptor deficient (ERKO) mouse aorta. Endothelium-dependent contraction of the thoracic aorta to N^G-nitro-L-arginine (L-NA) was significantly (*p < 0.05) diminished in rings isolated from ERKO mice compared to age matched wild type (WT) controls. The upper two panels of the graph illustrate original tracing of the organ chamber experiments and the lower panel demonstrates bar graph of the average data (mean ± SEM) obtained from n = 5 animals. The bar graphs represent total contraction of the isolated mouse thoracic aortic rings to the thromboxane receptor agonist U-46619 and L-NA. Reproduced with the permission of the American Society of Clinical Investigation (from Ref. Rubanyi *et al.*, 1997).

in mediating vasculoprotection has been demonstrated by achieving acceleration of atherosclerosis in hypercholesterolemic rabbits by chronic inhibition of NO formation (Cayatte *et al.*, 1994; Naruse *et al.*, 1994). Enhanced acetylcholine-induced, endothelium-dependent relaxation of coronary arteries in cholesterol fed cynomolgus monkeys following 17β-estradiol treatment (Williams *et al.*, 1990) suggested, that the effect of estrogen may be mediated by increased NO production.

Recently we investigated the role of NO in mediating the protective effect of 17β-estradiol on vascular lesion development in ovariectomized, hypercholesterolemic rabbits. Rabbits were treated with either 17β-estradiol or N^ω-nitro-L-arginine methylester (L-NAME) and their combination for eight weeks. Diminished aortic plaque formation in the estrogen treated animals was accompanied by enhanced endothelium-dependent responses of isolated aortic rings, suggesting that the anti-atherosclerotic effect of 17β-estradiol could be mediated, at least in part, by increased NO formation of the aortic endothelium (Figure 11-3). Chronic, systemic administration of L-NAME to 17β-estradiol-treated hypercholesterolemic rabbits significantly attenuated the vasculoprotective effect of estrogen, but did not completely prevent the reduction in plaque/surface ratio evoked by 17β-estradiol (Figure 11-3). Our results demonstrated a strong correlation between the elevated NO

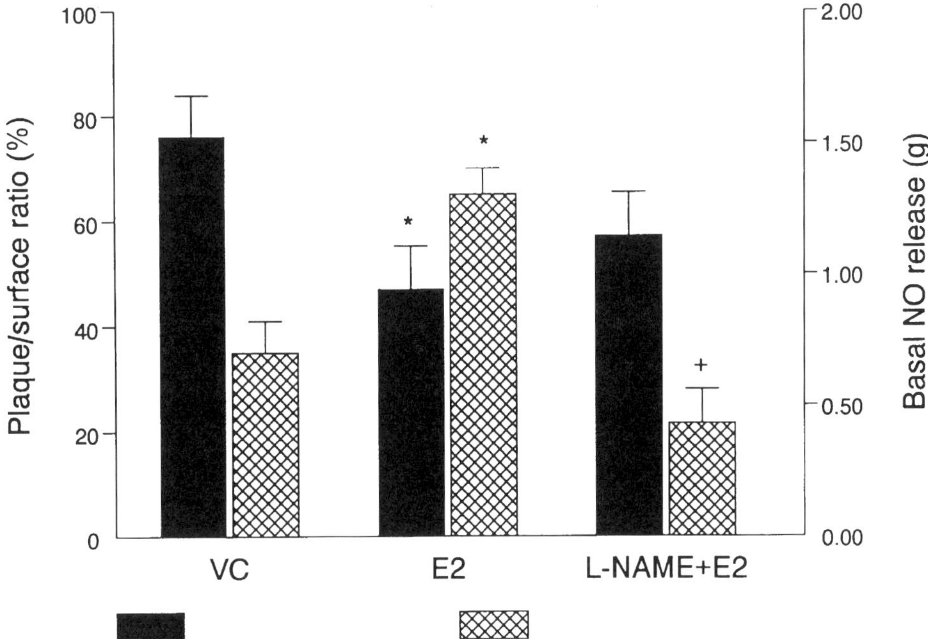

Figure 11–3. Effect of chronic NOS inhibition on the anti-atherosclerotic effect of 17β-estradiol in hypercholesterolemic rabbits.
Filled bars illustrate plaque/surface area ratio (left Y axis) of thoracic aortae isolated from control, 17β-estradiol treated and 17β-estradiol and L-NAME treated cholesterol fed rabbits. 6–week treatment of the rabbits with 17β-estradiol resulted in a significant ($p < 0.05$) reduction in the plaque/surface area, which was diminished by chronic NOS inhibition. Endothelium-dependent contraction to L-NA is represented by cross-hatched bars (right Y axis). 17β-estradiol significantly ($p < 0.05$) augmented the endothelium-mediated contraction to L-NA (estimation of basal NO production), whereas L-NAME treatment was accompanied by significantly ($+p < 0.05$) attenuated basal NO production. In each group n = 8 rabbits were studied. Bars represent mean ± SEM plaque/surface area (left Y axis) and NO mediated contraction (right Y axis). VC = vehicle control.

production and vasculoprotective effect of estrogen in this model, however the anti-atherosclerotic effect of 17β-estradiol persisted after partial inhibition of endothelial NO production. Our findings support the role of NO in vasculoprotection, although it also suggests the contribution of other potential mechanisms mediating the effect of estrogen, which are able to sustain significant protection against atherosclerotic plaque development despite of the significant reduction in NO.

In agreement with our study it has been reported recently that long-term inhibition of nitric oxide synthesis by L-NAME significantly reduced the anti-atherogenic effect of 17β-estradiol and levormeloxifene, a partial estrogen receptor agonist (Holm *et al.*, 1997). In this study the rabbits were balloon-injured in the middle thoracic aorta and their plasma cholesterol was controlled at a desired level, resulting in the demonstration that the lipid-independent beneficial effect of estrogens on atherosclerosis progression is mediated at least in part by endothelium-derived nitric oxide.

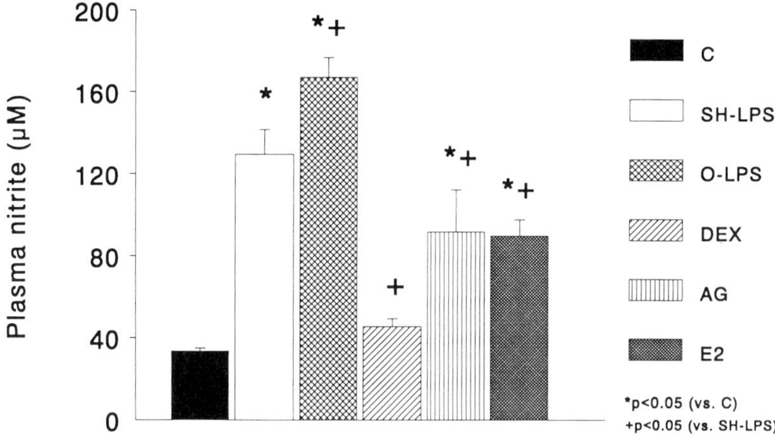

Figure 11–4. Plasma nitrite level in sham operated and ovariectomized female rats. Animals were divided into 6 groups (n = 8 in each): C = control; SH-LPS = sham operated female+LPS injection; O-LPS = ovariectomized female + LPS injection; DEX = ovariectomized female+LPS+dexamethasone injection 1h prior LPS; AG = ovariectomized female + LPS + aminoguanidine 1h after LPS injection; E2 = ovariectomized female + LPS + 17β-estradiol injection 48h before LPS. LPS injection resulted in a four-to five fold increase in plasma nitrite concentration compared to the control group. Nitrite levels of the plasma was determined by the Griess reaction. Nitrite level was significantly higher in the O-LPS compared to the SH-LPS group. Dexamethasone completely prevented nitrite increase (+p < 0.05). 17β-estradiol treatment (E2) resulted in a similar degree of significant inhibition (+p < 0.05) in nitrite formation than that of the effect of aminoguanidine (AG) compared to the O-LPS group. Reproduced with the permission of the American Physiological Society (from Ref. Kauser *et al.*, 1997/c).

EXCESSIVE NO PRODUCTION OF NOS-II IS INHIBITED BY 17β-ESTRADIOL

The ovarian sex steroid hormone, 17β-estradiol is one of the few physiological regulators of NOS-III (see above), however little is known about the effect of estrogens on excessive NO production by NOS-II.

Expression of NOS-II and excessive production of nitrite/nitrate by vascular smooth muscle cells mediates endothelium-independent suppression of blood vessel contractility during septic shock contributing to the severe hypotension (Thiemermann and Vane, 1990).

In a recent study we found that physiological substitution doses of 17β-estradiol reduced nitrite production in lipopolysaccharide (LPS)-treated ovariectomized rats (Kauser *et al.*, 1997) (Figure 11-4). Ovariectomized and sham operated female rats were injected with LPS and excessive NO generation was estimated by measuring plasma nitrite levels, 12 hours after LPS injection. LPS treatment of the rats resulted in a 4 to 5-fold increase of circulating plasma nitrite level. The plasma level was significantly higher in the ovariec-tomized animals compared to sham operated females, suggesting that endogenous ovarian sex steroid hormones may suppress excessive NO production by NOS-II in this model. The LPS-induced plasma nitrite increase was prevented by dexamethasone. 17β-estradiol treatment of ovariectomized rats significantly attenuated LPS-induced elevation of plasma nitrite, which was comparable to the effect of aminoguanidine. Our finding suggests that

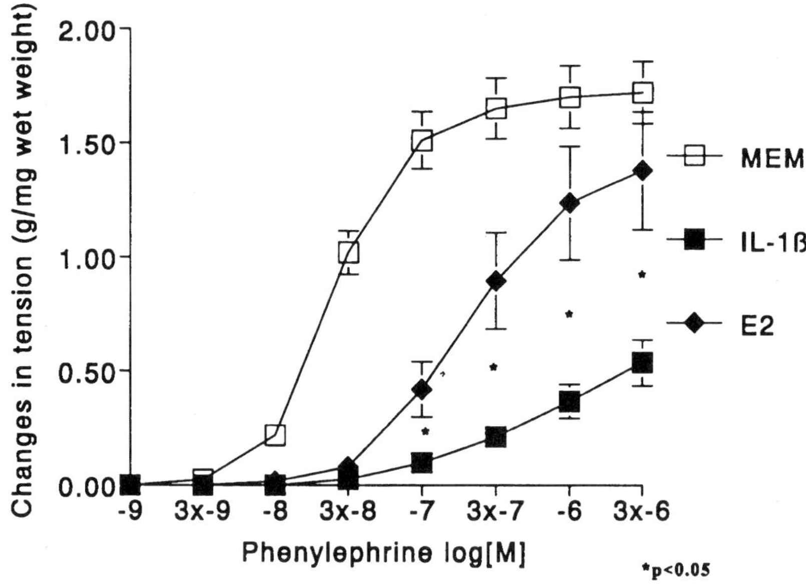

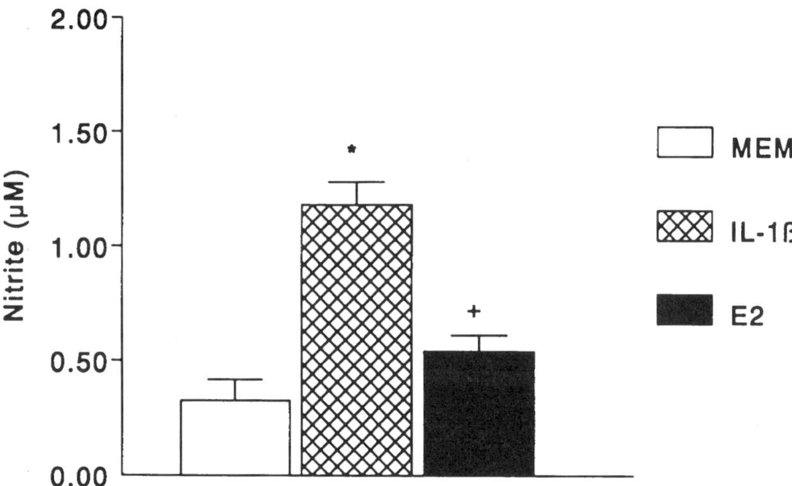

Figure 11–5. Effect of 17β-estradiol on IL-1β-induced suppression of contractility and nitrite production by isolated rat thoracic aortic rings. The upper panel of the figure illustrates dose-response curves to phenylephrine in control (open square), IL-1β (filled square) and IL-1β + 17β-estradiol (filled diamond) incubated de-endothelialized rat aortic rings. IL-1β significantly attenuated smooth muscle contraction. Treatment with 17β-estradiol significantly (*p < 0.05) reversed the inhibition of contractility by IL-1β. The lower panel demonstrates nitrite accumulation in the incubation media of the vessels. IL-1β induced significant (*p < 0.05) nitrite production (cross-hatched bar) by the aortic rings in the organ culture, which was significantly inhibited (+p < 0.05) by 17β-estradiol treatment of the cytokine incubated rings (filled bar).

estrogens may be beneficial in pathological conditions associated with excessive NO generation.

The effect of estrogen has been investigated earlier on the spontaneous NO production in cultured hepatocytes (Pittner and Spitzer, 1993) and rat peritoneal macrophages (Chao *et al.*, 1994), however its effect on cytokine-induced NOS-II activity and expression has not yet been studied. We designed an *in vitro* experiment to investigate the effect of 17β-estradiol on interleukin-1β (IL-1β)-induced nitrite production and suppressed contractile reactivity in isolated rat aortic rings. We found, that similar to the selective NOS-II inhibitor, aminoguanidine, co-incubation of the aortic rings with the cytokines and 17β-estradiol significantly reduced nitrite accumulation and increased contractile reactivity to phenylephrine (Kauser *et al.*, 1994). These results show, that in contrast to augmentation of NOS-III expression/activity, 17β-estradiol decreases IL-1β-induced excessive nitrite generation and restores contractile responsiveness in isolated rat aortic rings (Figure 11-5).

The presence of ER has been demonstrated in the rat aorta (Lin *et al.*, 1986; Knauthe *et al.*, 1996). It is possible, that the effect of 17β-estradiol is mediated by the activation of ER, since another ligand of ER, the non-steroidal, partial ER agonist, 4-OH-tamoxifen, also inhibited nitrite production and attenuated the suppression of phenylephrine contraction (Kauser *et al.*, 1998).

Inhibition of excessive NO production has been shown to provide some therapeutic benefits (Nava *et al.*, 1992), however non-specific inhibition of NOSes in general can be detrimental (Nava *et al.*, 1991; Harbrecht *et al.*, 1992), emphasizing the need for isoenzyme-specific NOS inhibition. The properties, that 17β-estradiol is able to upregulate NOS-III in the rat aorta (Goetz *et al.*, 1994; Kauser and Rubanyi, 1995) where it can also inhibit NOS-II activity, make this ovarian sex steroid hormone an ideal candidate for NOS isoenzyme regulation in pathological conditions, where the therapeutic goal is inhibition of NOS-II expression/activity without reduction in endothelial NO production.

SUMMARY

Due to its anti-atherogenic properties, NO is an ideal mediator of the vasculoprotective effect of 17β-estradiol. In this chapter we reviewed experimental evidence which supports the hypothesis that NO, at least in part, contributes to vasculoprotection by 17β-estradiol. This question however cannot be answered unambiguously based on the available data. The mechanism(s) by which 17β-estradiol increases bioavailable NO is not entirely understood. The role of the estrogen receptor has been demonstrated, but it is not clear whether the regulation of NOS-III is taking place at the transcriptional, translational or post-translational level. Inhibition of NOS-II and concomitant increase in NOS-III activity make 17β-estradiol an ideal agent in treatment of pathological conditions associated with excessive amount of NO production by NOS-II.

REFERENCES

Arnal, J.F., Clamens, S., Pechet, C., Negre-Salvayre, A., Allera., C., Girolami, J.-P. *et al.* (1996) Ethinylestradiol does not enhance the expression of nitric oxide synthase in bovine endothelial cells but increases the release of bioactive nitric oxide by inhibiting superoxide anion production. *Proceedings of the National Academy of Sciences of the USA*, **93**, 4108–4113.

Banka, C.L. (1996) High density lipoprotein and lipoprotein oxidation. *Current Opinions in Lipidology*, **7**, 139–142.

Bayard, F., Clamens, S., Delson, G., Blaes, N., Maret A. and Faye, J.-C. (1995) Oestrogen synthesis, oestrogen metabolism and functional oestrogen receptors in bovine aortic endothelial cells. In *Non-Reproductive Actions of Sex Steroids*, Ciba Foundation Symposium, 191, pp. 122–138, Chichester: John Wiley and Sons.

Bouhoute, A. and Leclercq, G. (1995) Modulation of estradiol and DNA binding to estrogen receptor upon association with calmodulin. *Biochemical and Biophysical Research Communications*, **208**, 748–755.

Brandes, R.P. and Mugge, A. (1996) Gender differences in the generation of superoxide anions in the rat aorta. *Life Sciences*, **60**, 391–396.

Busse, R. and Fleming, I. (1996) Endothelial dysfunction in atherosclerosis. *Journal of Vascular Research*, **33**, 181–194.

Cayatte, A.J., Palacino, J.J., Horten, K. and Cohen R.A. (1994) Chronic inhibition of nitric oxide production accelerates neointima formation and impairs endothelial function in hypercholesterolemic rabbits. *Arteriosclerosis and Thrombosis*, **14**, 753–759.

Chao, T.-Ch., VanAlten, P. J. and Walter, R. J. (1994) Steroid sex hormones and macrophage function: Modulation of reactive oxygen intermediates and nitrite release. *American Journal of Reproductive Immunology*, **32**, 43–52.

Cooke, J.P., Singer, A.H., Tsao, P., Zera, P., Rowan, R.A. and Billingham M.E.(1992) Antiatherogenic effects of L-arginine in the hypercholesterolemic rabbit. *The Journal of Clinical Investigation*, **90**, 168–1172.

Creager, M.A., Girerd, X.J., Gallagher, S.H., Coleman, S., Dzau, V.J. and Cooke, J.P. (1992) L-arginine improves endothelium-dependent vasodilation in hypercholesterolemic humans. *The Journal of Clinical Investigation*, **90**, 1248–1253.

Farhat, M.Y., Lavigne, M.C. and Ramwell, P.W. (1996) The vascular protective effects of estrogen. *The FASEB Journal*, **10**, 615–624.

Gilligan, D.M., Quyyumi, A.A., Cannon III, R.O., Johnson, G.B. and Schenke, W.H. (1994) Effects of physiological levels of estrogen on coronary vasomotor function in postmenopausal women. *Circulation*, **89**, 2545–2551.

Gisclard, V., Miller, V.M. and Vanhoutte, P.M. (1988) Effect of 17β-estradiol on endothelium-dependent responses in the rabbit. *Journal of Pharmacological and Experimental Therapeutics*, **244**, 19–22.

Goetz, R.M., Morano, I., Calvoni, T., Studer, R. and Holtz, J. (1994) Increased expression of endothelial constitutive nitric oxide synthase in rat aorta during pregnancy. *Biochemical Biophysical Research Communication*, **205**, 905–910.

Harbrecht, B.G., Billiar, T.R. and Stadler, J. (1992) Inhibition of nitric oxide synthesis during endotoxemia promotes intrahepatic thrombosis and an oxygen radical mediated hepatic injury. *Journal of Leukocyte Biology*, **52**, 390–394.

Hattori, Y. and Gross, S.S. (1993) GTP-cyclohydrolase I mRNA is induced by LPS in vascular smooth muscle: characterization, sequence and relationship to nitric oxide synthase. *Biochemical Biophysical Research Communication*, **195**, 435–441.

Hayashi, T., Fukuto, J.M., Ignarro, L.J. and Chaudhuri, G. (1992) Basal release of nitric oxide from aortic rings is greater in female rabbits than in male rabbits: Implications for atherosclerosis. *Proceedings of National Academy of Science, USA*, **89**,11259–11263.

Hayashi, T., Fukuto, J.M., Ignarro, L.J. and Chaudhuri, G. (1995) Gender differences in atherosclerosis: possible role of nitric oxide. *Journal of Cardiovascular Pharmacology*, **26**, 792–802.

Hayashi, T.K., Yamada, K., Esaki, T., Kuzuya, M., Satake, S., Ishikawa, T., Hidaka, H. and Iguchi, A. (1995) Estrogen increases endothelial nitric oxide by a receptor-mediated system. *Biochemical Biophysical Research Communication*, **214**, 847–855.

Hishikawa, K., Makaki, T., Marumo, T., Suzuki, H., Kato, R. and Saruta, T. (1995) Up-regulation of nitric oxide synthase by estradiol in human aortic endothelial cells. *FEBS Letter*, **36**, 291–293.

Holm, P., Korsgaard, N., Shalmi, M., Andersen, H.L., Hougaard, P. and Skouby, S.O. (1997) Significant reduction of the antiatherogenic effect of estrogen by long-term inhibition of nitric oxide synthesis in cholesterol-clamped rabbits. *Journal of Clinical Investigation*, **100**, 821–828.

Isles, C.G., Hole, D.J., Hawthorne, V.M. and Lever, A.F. (1992) Relation between coronary risk and coronary mortality in women of the Refew and Paisely survey: comparison with men *Lancet*, **339**, 702–706.

Kauser, K and Rubanyi, G.M. (1994/b) 17β-estradiol and endothelial nitric oxide synthase. *Endothelium*, **2**, 203–208.

Kauser, K. and Rubanyi, G.M. (1994/a) Gender difference in bioassayable endothelium-derived nitric oxide release from isolated rat aortae. *American Journal of Physiology*, **267**, H2311–H2317.

Kauser, K. and Rubanyi, G.M. (1995/a) Gender difference in endothelial dysfunction in the aorta of spontaneously hypertensive rats. *Hypertension*, **25**, 517–523.

Kauser, K. and Rubanyi, G.M. (1995/b) Effect of 17β-estradiol on endothelial dysfunction in the aorta of spontaneously hypertensive rats. *Endothelium*, **25**, 517–523.

Kauser, K. and Rubanyi, G.M. (1997/a) Vasculoprotection by estrogen contributes to gender difference in cardiovascular diseases; potential mechanism and role of endothelium. In: Rubanyi, G.M. and Dzau, V.J., eds.. *The Endothelium in Clinical Practice*, pp. 439–467. New York, NY: Marcel Dekker, Inc.

Kauser, K. and Rubanyi, G.M. (1997/b) Potential cellular signaling mechanisms mediating upregulation of endothelial nitric oxide production by estrogen. *Journal of Vascular Research*, **34**, 229–236.

Kauser, K., Sonnenberg, D., Henrichmann, F. and Rubanyi, G.M. (1994) 17β-estradiol inhibits IL-1β-induced suppression of contractile reactivity in isolated rat thoracic aorta. *Circulation*, **90**, 1–577.

Kauser, K., Sonnenberg, D., Tse, J. and Rubanyi, G.M. (1997) 17β-Estradiol attenuates endotoxin-induced excessive nitric oxide production in ovariectomized rats *in vivo*. *American Journal of Physiology*, **273**, H506–H509.

Kauser, K., Sonnenberg, D., Diel, P. and Rubanyi, G.M. (1998) 17β-Estradiol inhibits cytokine-induced nitric oxide production in isolated rat aorta. *British Journal of Pharmacology*, **123**, 1089–1096.

Kim-Schulze, S., McGowan, K.A., Hubchak, S.C., Cid, M.C., Martin, M.B., Kleinman, H.K. *et al.* (1996) Expression of an estrogen receptor by human coronary artery and umbilical vein endothelial cells. *Circulation*, **94**, 1402–1407.

Knauthe, R., Diel, P., Hegele-Hartung, C.H., Engelhaupt, A. and Fritzemeier, K.-H. (1996) Sexual dimorphism of steroid hormone receptor messenger ribonucleic acid expression and hormonal regulation in rat vascular tissue. *Endocrinology*, **137**, 3220–3227.

Kuiper, G.G.J., Enmark, E., Pelto-Huikko, M., Nilsson, S. and Gustafsson, J.-A. (1996) Closing of a novel estrogen receptor expressed in rat prostate and ovary. *Proceedings of the National Academy of Sciences of the USA*, **9**, 5925–5030.

Lin, A.L., Shain, S.A. and Gonzales, R. (1986). Sexual dimorphism characterizes steroid hormone modulation of rat aortic steroid hormone receptors. *Endocrinology*, **119**, 296–302.

Lubahn, D.B., Moyer, J.S., Golding, T.S., Course, J.F., Korach, K.S. and Smithies, O. (1993) Alteration of reproductive function but not prenatal sexual development after insertional disruption of the mouse estrogen receptor gene. *Proceedings of the National Academy of Sciences of the USA*, **90**, 11162–11166.

MacRitchie, A.N., Jun, S.S., Chen, Z., German, Z., Yuhanna, I.S., Sherman, T.S. and Shaul P.W. (1997) Estrogen upregulates endothelial nitric oxide synthase gene expression in fetal pulmonary artery endothelium. *Circulation Research*, **81**, 355–362.

Maddox, Y.T., Falcon, J.G., Ridinger, M., Cunard, C.M. and Ramwell, P.W. (1987) Endothelium-dependent gender differences in the response of the rat aorta. *Journal of Pharmacological and Experimental Therapeutics*, **240**, 392–395.

Matsui, K., Higashi, K., Fukunga, K., Miyazaki, K., Maeyama, M. and Miyamoto, E. (1983) Hormone treatments and pregnancy alter myosin light chain kinase and calmodulin levels in rabbit myometrium. *Journal of Endocrinology*, **97**, 11–16.

Maziere, C., Ronveaus, M.F., Salmon, S., Santus, R. and Mazier, J.C. (1991) Estrogens inhibit copper and cell-mediated modification of low density lipoprotein. *Atherosclerosis*, **89**, 175–82.

Mosselman, S., Polman, J. and Dijkema, R. (1996) ER-β: identification and characterization of a novel human estrogen receptor. *FEBS Letters*, **392**, 49–53.

Naruse, K.M., Shimizu, M., Muramatsu, M., Toki, Y., Miyazaki, Y., Okurama, K., Hashimoto, H. and Ito, T. (1994) Long-term inhibition of NO synthesis promotes atherosclerosis in the hypercholesterolemic rabbit thoracic aorta. *Arteriosclerosis and Thrombosis*, **14**, 746–752.

Nava, E., Palmer, R.M.J. and Moncada, S. (1991) Inhibition of nitric oxide synthesis in septic shock: how much is beneficial? *Lancet*, **338**, 1555–1557.

Nava, E., Palmer, R.M.J. and Moncada, S. (1992) The role of nitric oxide in endotoxic shock: effects of N[G]-methyl-L-arginine. *Journal of Cardiovascular Pharmacology*, **20**(Suppl. 12), S132–134.

Pittner, R.A. and Spitzer, J.A. (1993) Steroid hormones inhibit induction of spontaneous nitric oxide production in cultured hepatocytes without changes in arginase activity or urea production. *PSEBM*, **202**, 499–504.

Reis, S.E., Gloth, S.T., Blumenthal, R.S., Resar, J.R., Zacur, H.A., Gerstenblith, G. and Brinker, J.A. (1994) Ethinyl estradiol acutely attenuates abnormal coronary vasomotor responses to acetylcholine in postmenopausal women. *Circulation*, **89**, 52–60.

Rosenkranz-Weiss, P., Sessa, W.C., Milstien, S., Kaufman, S., Watson, C.A. and Pober, J.S. (1994) Regulation of nitric oxide synthesis by proinflammatory cytokines in human umbilical vein endothelial cells. Elevations in tetrahydrobiopterin levels enhance endothelial nitric oxide synthase specific activity. *The Journal of Clinical Investigation*, **93**, 2236–2243.

Rosselli, M., Imthurn, B., Keller, P.J., Jackson, E.K. and Dubey, R.K. (1995) Circulating nitric oxide (nitrite/nitrate) levels in postmenopausal women substituted with 17β-estradiol and norethisterone acetate. *Hypertension*, **25**, part 2, 848–853.

Rubanyi, G.M. (1993) The role of endothelium in cardiovascular homeostasis and diseases. *Journal of Cardiovascular Research*, **22**(Suppl.4), S1–S14.

Rubanyi, G.M. and Vanhoutte P.M. (1986) Oxygen-derived free radicals, endothelium and responsiveness of vascular smooth muscle. *American Journal of Physiology*, **250**, H815–H821.

Rubanyi, G.M., Freay, A.D., Kauser, K., Sukovich, D., Burton, G., Lubahn, D.B., Couse, J.F., Curtis, S.W. and Korach, K.S. (1997) Vascular estrogen receptors and endothelium-derived nitric oxide production in the mouse aorta gender difference and effect of estrogen receptor gene disruption. *The Journal of Clinical Investigation*, **99**, 2429–2437.

Ruiz-Larrea, B., Leal, A., Martin, C., Martinez, R. and Lacort, M. (1995) Effects of estrogens on the redox chemistry of iron: A possible mechanism of the antioxidant action of estrogens. *Steroids*, **60**, 780–783.

Schmidt, K., Werner, E.R., Mayer, B., Wachter, H. and Kukovetz, W.R. (1992) Tetrahydrobiopterin-dependent formation of endothelium-derived relaxing factor (nitric oxide) in aortic endothelial cells. *Biochemical Journal*, **281**, 297–300.

Stallone, J.N., Crofton, J.T. and Share, L. (1991) Sexual dimorphism in vasopressin-induced contraction of rat aorta. *American Journal of Physiology*, **260**, H453–458.

Stampfer, M.J., Colditz, G.A., Willett, W.C., Manson, J.E., Rosner, B., Speizer, F.E. and Hennekens, C.H. (1991) Postmenopausal estrogen therapy and cardiovascular disease: ten-year follow-up from the Nurses' Health Study. *New England Journal of Medicine*, **325**, 756–762.

Stroes, E., Kastelein, J., Cosentino, F., Erkelens, W., Wever, R., Koomans, H., Luscher, T. and Rabelink, T. (1997) Tetrahydrobiopterin restores endothelial function in hypercholesterolemia. *The Journal of Clinical Investigation*, **99**, 41–46.

Subbiah, M.T.R., Kessel, B., Agrawal, M., Rajan, R., Ablanalp, W. and Rymaszewski, Z. (1993) Antioxidant potential of specific estrogens on lipid peroxidation. *Journal of Clinical Endocrinology and Metabolism*, **77**, 1095–1097.

Sudhir, K., Jennings, G.L., Funder, J.W. and Komesaroff, P.A. (1996) Estrogen enhances basal nitric oxide release in the forearm vasculature in perimenopausal women. *Hypertension*, **28**, 330–334.

Thiemermann, C. and Vane, J.R. (1990) Inhibition of nitric oxide synthesis reduces the hypotension induced by bacterial lipopolysaccharides in the rat *in vivo*. *European Journal of Pharmacology*, **182**, 591–595.

Venema, R.C., Nishida, K., Alexander, R.W., Harrison, D.G. and Murphy, T.J. (1994) Organization of the bovine gene encoding the endothelial nitric oxide synthase. *Biochemical Biophysical Acta*, **1218**, 413–420.

Venkov, C.D., Rankin, A.B. and Vaughan, D.E. (1996) Identification of authentic estrogen receptor in cultured endothelial cells. A potential mechanism for steroid hormone regulation of endothelial function. *Circulation*, **94**, 727–733.

Weiner, C.P., Knowles, R.G. and Moncada S. (1994/a) Induction of nitric oxide synthases early in pregnancy. *American Journal of Obstetrics and Gynecology*, **171**(3), 838–843.

Weiner, C.P., Lizasoain, I., Baylis, S.A., Knowles, R.G., Charles, I.G. and Moncada, S. (1994/b) Induction of calcium-dependent nitric oxide synthases by sex hormones. *Proceedings of the National Academy of Sciences of the USA*, **91**(11), 5212–5216.

Werner-Felmayer, G., Werner, E.R., Fuchs, D., Hausen, A., Reibnegger, G., Schmidt, K., Weiss, G. and Wachter, H. (1993) Pteridine biosynthesis in human endothelial cells. Impact on nitric oxide-mediated formation of cyclic GMP. *The Journal of Biological Chemistry*, **268**, 1842–1846.

Williams, J.K., Adams, M.R. and Klopfenstein, H.S. (1990) Estrogen modulates responses of atherosclerotic coronary arteries. *Circulation*, **81**, 1680–1687.

12 Mice Deficient in eNOS and nNOS Isoforms

Paul L. Huang

Cardiovascular Research Centre, Massachusetts General Hospital-East, Charlestown, USA

Targeted disruption of NOS genes offers a genetic approach that complements pharmacologic approaches to study the roles of each NOS isoform. It circumvents specificity problems of NOS inhibitors and reveals the effects of deleting each NOS isoform on physiologic processes in intact animals. In this chapter, we review the rationale, methodology, and limitations of gene knockout approaches and discuss the phenotypes of mice deficient in neuronal NOS and endothelial NOS. These knockout mice have been useful to dissect apart the complexities of the nitric oxide signaling system and demonstrate molecular and physiologic compensations brought into play in the absence of individual NOS isoforms.

Key words: Nitric oxide, nitric oxide synthase, targeted disruption, gene knockout, animal models, homologous recombination.

GENETIC APPROACH TO NITRIC OXIDE SYNTHASES

Rationale For Genetic Approach

Nitric oxide synthase (NOS) enzymes form nitric oxide from one of the terminal guanidino nitrogens of the substrate L-arginine and convert the rest of the molecule to citrulline. The three major isoforms of NOS are encoded by separate genes: type I, or neuronal NOS (nNOS), type II, or inducible NOS (iNOS), and type III, or endothelial NOS (eNOS), although there is substantial overlap in expression patterns (Marletta, 1994; Nathan and Xie, 1994). Many individual cell types contain more than one NOS isoform and most tissues, because they contain nerves and blood vessels, contain more than one isoform. Pharmacologic approaches are limited by the lack of specificity of NOS inhibitors that affect multiple NOS isoforms, for example L-nitro arginine, L-N-arginine-methyl ester (L-NAME), or L-N-monomethyl arginine (L-NMMA).

A genetic approach, to disrupt or knockout the genes for each NOS isoform, circumvents problems with specificity of NOS inhibitors. Furthermore, it allows the study of the roles of each NOS isoform on physiologic processes in the context of intact animals. In addition, it gives insights into possible developmental roles for NO and parallel processes that may compensate for the absence of each NOS isoform. Our group has generated neuronal NOS knockout mice (Huang *et al.*, 1993) and endothelial NOS knockout mice (Huang *et al.*, 1995). MacMicking (1995), Wei (1995), and Laubach (1995) have knocked out the inducible NOS gene. The major phenotypes of the nNOS and eNOS mutant mice are shown in Table 12-1.

Gene Knockout Technology

The generation of knockout mice is made possible by two important advances; the first is the use of homologous recombination in mammalian cells. Targeting vectors can be

Table 12-1. Phenotypes of nNOS and eNOS knockout mice

NOS Isoform	*Phenotype*
Neuronal NOS (type I)	Pyloric stenosis, absent slow IJP in gastric smooth muscle
	Resistance to focal and global cerebral ischemia
	Increased aggressive and sexual behavior
	Preserved hippocampal LTP
	Reduced airway responsiveness to methacholine
	Reduced baseline renal renin content
	Non-NO mediated compensation in cerebrovascular response to hypercarbia and whisker stimulation, nociception, and MAC for isoflurane anesthesia
Endothelial NOS (type III)	Absence of EDRF activity in aorta
	Elevated mean arterial blood pressure (hypertension)
	Increased susceptibility to global ischemia
	Increased cardiac contractility responses to adrenergic agonists
	Exaggerated neointimal formation in response to vessel injury
	Preserved hippocampal LTP
	Neuronal NO dependent compensation in pial arteriolar response to acetylcholine
eNOS/nNOS double knockout	Reduced hippocampal LTP (stratum radiatum)
	Preserved LTP (stratum oriens)

designed so that homologous recombination between the vector and a native gene results in the deletion of a targeted segment, replacement of a targeted segment, or insertion of new sequences into a targeted segment. Targeting vectors generally contain several kilobases of homologous DNA that flank the targeted segment, to allow the vector to align with the region of the gene to be modified. Homologous recombination between a foreign DNA molecule like the targeting vector and the native genome is a rare event in mammalian cells, but cells in which targeting has occurred can be selected by the incorporation of selectable markers like antibiotic resistance into the targeting vector. The second important advance is the ability to grow embryonic stem (ES) cells in culture. Homologous recombination in ES cells results in genetically modified cells that can theoretically contribute to any tissues in the developing embryo. Recent methodology enables the culture of ES cells under conditions that allow them to remain pluripotent so the ES cells can transmit the genetic modification through the germ line to the next generation.

The process of generating gene knockouts has been described in detail in recent monographs (Sedivy and Joyner, 1992). The steps include designing and generating the targeting construct, introducing the construct into ES cells (generally by electroporation), selecting the ES cells that have undergone appropriate homologous recombination, and introducing these modified ES cells into embryos (either by injection into blastocysts or by aggregation into morulae). The resulting animal is chimeric since it contains cells from the modified ES cells and cells from the host embryo. Because the goal is to obtain mice homozygous for the genetic modification, chimeric animals are bred to check if there is germ-line transmission of the genetic modification. If there is, offspring that carry the modification are heterozygous for the mutation, and breeding two such animals will give rise to offspring of which one quarter will be homozygous for the mutation, one quarter will be wild-type, and one half will be heterozygous.

Most genes, including the NOS genes, are too large to delete in their entirety, so a region must be selected for targeting. The targeted region (which exon or exons are deleted or modified), and the type of modification (deletion, insertion, substitution) affects the outcome of the gene knockout, as well as interpretations of experiments using the animals. To date, reported NOS gene knockout mice have been generated by deletion of key exons. One approach is to target the transcriptional or translational start site, so that the resulting targeted gene is incapable for generating any protein. However, there is potential for the use of alternative transcriptional or translational start sites. Another approach is to target key regions of the molecule, so that the altered gene encodes a protein that, even if made, is non functional. This approach carries the risk of a dominant phenotype in heterozygous mice if the aberrant form of the protein itself has effects. Substitutions or insertions have been performed in other systems, but so far have not been reported for the NOS genes.

Another parameter that affects the phenotype of the knockout mouse is the genetic background of the animals. This is influenced by the origin of the ES cells used and by the breeding scheme to generate the animals. This is discussed in more detail below.

Limitations To Gene Knockout Approaches

Developmental effects

The absence of the target gene throughout development may affect the development in significant ways. The most extreme examples of this are the embryonic lethal phenotypes of many gene knockouts with clear cut developmental abnormalities, for example endothelin-1 (Kurihara *et al.*, 1994) and neuregulin (Meyer and Birchmeier, 1995). However, more subtle effects are possible, particularly for genes that are expressed and play a role during embryonic development. These range from the upregulation of other genes that can substitute for the disrupted one, as in the case of MyoD and Myf-5 (Rudnicki *et al.*, 1992), or physiologic compensation by parallel pathways, as observed for hypercapnia or whisker response in nNOS mutant mice (see below), to abnormal anatomy or circuitry. Disruption of genes that exert effects at several stages of development can lead to very complex phenotypes.

The nNOS isoform is expressed widely throughout the developing brain (Bredt and Snyder, 1994). Expression levels drop precipitously after birth, and in adult mice, only 1–2% of cortical neurons express nNOS. The functions of nNOS during brain development are unknown. NO has been proposed as a mediator of synaptic plasticity in long-term potentiation and during brain development. So far, nNOS and eNOS mutant mice do not demonstrate differences in neuroanatomy at the light microscopic level. The extent to which developmental compensation may affect phenotype in the mutant animals is not known. For example, behavioral abnormalities noted in the nNOS mutant mice (Nelson *et al.*, 1995) may be due to subtle changes in neural circuitry not reflected in detectable anatomic changes. Furthermore, nNOS and eNOS knockout mice show evidence of physiologic compensation, as discussed below.

Tissue specificity

Traditional gene knockout approaches reveal the function of each NOS isoform, but they do not address the contribution of different cell types to the phenotype. The nNOS isoform is expressed not only in neurons, but also in pancreatic cells and skeletal muscle (Schmidt *et al.*, 1992; Kobzik *et al.*, 1994). The eNOS isoform is expressed not only in endothelial cells, but also in smooth muscle cells and neurons (Koide *et al.*, 1994; Loesch *et al.*, 1994). Cell specific gene knockout will be required to separate the effects of different cell types in generating NO.

Genetic background

The genetic background of mouse strain has a significant impact on the phenotype. Most traditional gene knockout work is done using ES cells from SV129 mice because they maintain pluripotency and contribute to the germ line with higher efficiency than cells from other strains. Mice that result from mating of SV129 mice with other strains like C57BL/6 have genetic backgrounds that are a combination of their parental backgrounds. The contribution of genetic background to phenotype of knockout mice has received much attention (Gerlai, 1996). Phenotypes that differ between SV129 and C57BL/6 raise the most problems, since the proper wild-type control mice to compare with mutant mice is not clear. Even wild-type littermate controls from heterozygous matings are genetically different from the homozygous knockout animals, not only at the disrupted gene, but at other loci on the same chromosome.

Several potential solutions have been considered to address these concerns (Crawley, 1996; Crusio, 1996; Gerlai, 1996; Lathe, 1996).

1. Crossing the chimeric mouse derived from SV129 ES cells back to the SV129 background gives animals that are pure SV129 except for the mutation. Conceptually this is a valid approach, but it is limited in practice by abnormalities inherent in the SV129 mouse strain. SV129 mice are neuroanatomically abnormal and lack a corpus callosum. They are severely impaired in spatial learning tasks and behaviorally passive. They are susceptible to teratomas and other tumors, which may explain why the SV129 strain is such a good source of pluripotent ES cells.
2. Mutant mice of combined genetic background can be backcrossed to a standard inbred strain like C57BL/6. Although some investigators advocate backcrossing for 20 generations, calculations reveal that even after this many generations, the amount of linked SV129 DNA that cosegregates with the mutation is likely to still contain 300 genes or more.
3. Another approach is to create a mouse where the targeted gene is intact but tagged, so that one can select for the same region that is associated with the null allele. However, both mutant and control populations are continually drifting, and the generation number and degree of backcrossing have to be matched between the control mice and the mutant mice. Controls of one generation cannot be used for valid comparison with mutant mice of a different generation.
4. Replacement of the disrupted gene in the mutant animals to rescue a phenotype provides strong evidence that the observed phenotype is due to the missing gene. However, it would be tedious to perform this verification for each experimental study done using the mutant mice.
5. Finally, generation of knockout animals using ES cells from a standard pure inbred strain like C57BL/6 would obviate many difficulties. Wild-type C57BL/6 mice would clearly be the proper controls for the mutant animals, and variability between strains would not be an issue. Such work is currently in progress for the nNOS and eNOS genes.

NEURONAL NOS MUTANT MICE

Generation Of nNOS Mutant Mice And Molecular Analysis

In, 1993, our group reported the generation of neuronal NOS knockout mice by targeting the first translated exon of the nNOS gene, including the protein initiation codon ATG (Huang *et al.*, 1993). We used J1 ES cells of SV129 origin, and crossed the chimeric mice with C57BL/6 mice. The nNOS knockout mice are viable and fertile. They do not express nNOS, as detected by Western blot analysis, NADPH diaphorase histochemical staining, or diaphorase formazan staining by electron microscopy (Huang *et al.*, 1993; Darius *et al.*, 1995).

Despite the absence of native nNOS protein, we found a low level of NOS catalytic activity in the brain of nNOS mutant mice with a different distribution than is present in wild-type mice (Huang *et al.*, 1993). The cortex and the caudate-putamen contain the most residual activity, about 5% of wild-type levels, whereas the cerebellum and the superior colliculus contain the least activity, about 0.2% of wild-type levels. Some residual NOS activity is due to the presence of other isoforms in the brain, like endothelial NOS (Dinerman *et al.*, 1994; O'Dell *et al.*, 1994), as confirmed by quantitative autoradiography (Hara *et al.*, 1996). Some may be due to the existence of low level natural splicing variants that remain in the mutant animals (Brenman *et al.*, 1996). These splice variants utilize an alternative transcriptional start site and are present in wild-type mice as well, so they do not account for the phenotype of the mutant mice. Furthermore, the deleted exon in the nNOS knockout mice encodes a PDZ domain that is critical to the membrane association of the protein, so the residual splice variants are not fully functional, despite possessing catalytic activity.

There is substantial evidence that NO production is abolished in the brain of nNOS mutant animals. Nitric oxide radical complexes cannot be detected by electron spin resonance techniques either at baseline or following ischemia in nNOS mutant mice (Mullins *et al.*, 1995). Levels of downstream effectors like cGMP are diminished in the brain of nNOS mutant mice at baseline as compared to wild-type mice, and nNOS knockout mice show no increase following ischemia, despite robust increases in wild-type mice (Huang *et al.*, 1994). Pharmacologic evidence includes the absence of L-NA dependent activities like cerebrovascular response to hypercarbia or whisker stimulation (Irikura *et al.*, 1995; Ma *et al.*, 1996).

Double labeling studies show that neurons that normally express nNOS are present and their locations and cell numbers are unaffected in nNOS knockout mice despite absence of nNOS expression (Huang *et al.*, 1993). The nNOS mutant mice do not show gross neuroanatomic abnormalities, and the cell numbers of each layer of the hippocampus and the cerebellum are normal.

Pyloric Stenosis

The nNOS mutant mice have variably enlarged stomachs, which are often several times normal size (Huang *et al.*, 1993). Some animals die suddenly and are found at post-mortem examination to have gross gastric outlet obstruction. On histologic examination, the pyloric sphincter is hypertrophied and nNOS immunostaining normally present in the myenteric plexus of wild-type mice is absent in the nNOS knockout mice. This phenotype resembles the human disorder infantile hypertrophic pyloric stenosis, and nNOS knockout mice serve as the first animal model for this disease. In fact, Vanderwinden et al reported that the myenteric plexus of children with pyloric stenosis do not stain using NADPH diaphorase (Vanderwinden *et al.*, 1992), indicating that a defect in nNOS expression may be related to disease pathogenesis. Because NADPH diaphorase staining elsewhere in such children is normal, the human disorder is not due to nNOS gene knockout, but rather to an abnormality in nNOS gene expression.

nNOS mutant mice have been useful to clarify the molecular details of non-adrenergic, non-cholinergic neurotransmission in the gastrointestinal tract. Electrical field stimulation of gastric fundus circular smooth muscle results in membrane hyperpolarization, a response that is sensitive to tetrodotoxin and is therefore mediated by substances released from nerves. This membrane hyperpolarization response is called an inhibitory junction potential (IJP). While it had been known that pharmacologic inhibition of NOS affects IJP, it was not clear whether NO serves as a true antegrade neurotransmitter, or as a mediator produced in post-junctional smooth muscle cells by the action of other putative inhibitory neurotransmitters like ATP or vasoactive intestinal peptide (VIP). Both possibilities have precedent, since NO is made by both neurons and smooth muscle cells.

Studies in wild-type and nNOS knockout mice establish that the IJP in mouse stomach muscle consists of overlapping fast and slow components (Mashimo *et al.*, 1996). The fast component can be blocked by ATP receptor desensitization by α-β methylene ATP, blockade of P2 purinergic receptors by reactive blue 2, and blockade of calcium activated potassium channels by apamin. The slow component can be blocked by L-NA and other NOS inhibitors, and by VIP antagonists. In wild-type mice, blockade of one IJP component leaves the other IJP component, and simultaneous blockade of both with L-NA and methyl-ATP obliterates the IJP response. In nNOS knockout mice, electrophysiologic tracings show only fast IJP, which is not affected by inhibitors of NOS or VIP. The response can be totally abolished with apamin, indicating that only fast IJP, and not slow IJP, is present in the nNOS knockout mice. Thus, the nNOS isoform mediates slow IJP in concert with VIP.

To examine whether NO acts downstream from VIP (e.g. VIP causes NO release), or vice versa (NO causes VIP release), exogenous VIP was applied to the smooth muscle. VIP produces hyperpolarization of wild-type muscle strips which is blocked by pretreatment with L-NA, suggesting that VIP stimulates activity of a preexisting NOS isoform. VIP fails to produce any response in nNOS knockout mice, indicating that NO acts downstream from VIP, and not vice versa. VIP-induced hyperpolarization is sensitive to ω-conotoxin GVIA, which blocks N-type calcium channels on nerve terminals, but it is not sensitive to nifedipine, which blocks voltage-dependent L-type calcium channels involved in calcium entry into smooth muscle cells. Thus, nNOS knockout mice establish unequivocally that VIP acts pre-junctionally to stimulate nNOS activity.

Airway Responsiveness

NO is present in expired gas of humans and animals but its enzymatic source is not known. All three NOS isoforms are present in tracheobronchial tissue: nNOS in submucosal airway nerve terminals, eNOS in epithelial basal cells, and iNOS in tracheal and bronchial columnar epithelial cells. NO could potentially modulate airway function in several ways. Blockade of NO synthesis leads to enhanced airway responsiveness, suggesting that NO may normally inhibit airway tone (Belvisi *et al.*, 1992; Bai and Bramley, 1993). In contrast, patients with asthma have elevated levels of NO in their exhaled air, and these levels decrease with steroid treatment and clinical improvement, suggesting that NO may also be involved in airway narrowing (Kharitonov *et al.*, 1994; Massaro *et al.*, 1996).

The levels of expired NO are reduced in nNOS knockout mice to 60% of wild-type levels, suggesting that 40% of expired NO is derived from the nNOS isoform (G.T. DeSanctis and J.M. Drazen, personal communication). This assumes that there are no secondary changes in eNOS or iNOS activity or distribution in the nNOS knockout mice, assumptions that may be validated by complementary measurements in eNOS knockout mice. The nNOS knockout mice also demonstrate less airway responsiveness to methacholine, indicating that the nNOS isoform contributes to hyperresponsiveness in the mouse. The possibility that eNOS or iNOS have the opposite effect, to suppress these responses, are being addressed by studies in appropriate knockout mice.

Role In Renin Release

NO has been implicated in the regulation of renin release by *in vivo* and *in vitro* studies. The nNOS isoform is expressed at high levels in the macula densa (Briggs *et al.*, 1993), whereas eNOS is expressed in the glomerulus and endothelial cells (Ujiie *et al.*, 1994). The activity of renal nNOS, but not eNOS, is increased by a low salt diet. As with other biological processes, there is evidence for multiple biological effects of NO. In isolated cultured JG cells, NO donors inhibit renin release, and NO inhibitors increase renin release (Beierwaltes and Carretero, 1992). In contrast, NO stimulates renin release in isolated kidney preparations (Scholz and Kurtz, 1993). Taken together, it appears that renin secretion is affected differently by NO derived from at least two sources: tubular or macula densa nNOS, versus afferent arteriole eNOS. Harding *et al.* have addressed this (1997) by studying the renal renin content and response to low salt in nNOS knockout mice. The nNOS mutant mice show a 45% reduction in renal renin content, indicating that the nNOS isoform is involved in establishing the baseline level of renin release. The nNOS knockout mice are, however, able to increase renal renin content in response to low sodium diet, demonstrating that this response does not require nNOS. The induction of COX-2 appears important for this process in both wild-type and nNOS mutant mice.

Aggressive Behavior

Nelson *et al.* (1995) made the observation that male neuronal NOS knockout mice are more aggressive than wild-type littermates. They defined aggressive encounters as offensive attacks, biting, wrestling, and chasing. In an intruder-resident model, when a wild-type male mouse (the intruder) is placed into a cage with a resident mouse, the latency time to the first attack is the same for wild-type mice and nNOS knockout mice. However, nNOS knockout mouse residents display three to four times as many aggressive encounters, and far fewer submissive postures. Among groups of male mice housed together, nNOS knockout mice show a latency time to attack that is one fifth that of wild-type mice, and have twice as many aggressive encounters. Male nNOS knockout mice also fail to show the normal reduction in the number of attempts in mounting female mice that are not in estrus, leading to the suggestion that nNOS may play a role in sexual as well as aggressive behavior. No detectable sensory or motor abnormalities, differences in strength and agility, or indications of increased anxiety in open field tests accounted for these behaviors. These findings raise the possibility that nNOS may play a role in complex behavior, although it is also possible that secondary effects or compensatory mechanisms may contribute to the behavioral phenotype.

ENDOTHELIAL NOS MUTANT MICE

EDRF And Vascular Tone

In 1995, our group reported the generation of eNOS knockout mice by deleting the exons that encode the NADPH ribose and adenine binding sites in J1 ES cells (Huang *et al.*, 1995). The chimeric mice were mated with C57BL/6 mice. The resulting eNOS knockout mice are viable and fertile, despite absence of detectable eNOS mRNA, protein, or enzymatic activity. Aortic rings from wild-type mice show a dose-dependent relaxation in response to acetylcholine, which is blocked by L-NA. These experiments are similar to the ones performed by Furchgott and Zawadzki in their original description of EDRF (Furchgott and Zawadzki, 1980). Aortic rings from eNOS mutant mice, on the other hand, do not respond to acetylcholine. They do relax upon treatment with sodium nitroprusside or papaverine, indicating that the vascular smooth muscle is capable of relaxation. These experiments provide the first genetic evidence that the eNOS isoform is responsible for EDRF activity.

Hypertension And Blood Pressure Response

Because of its role as an endogenous vasodilator, NO has been postulated to regulate blood pressure. The eNOS knockout mice are indeed hypertensive, with mean arterial blood pressures that are about 30% higher than wild-type littermates under anesthesia or awake, confirming the importance of NO to blood pressure regulation (Huang *et al.*, 1995). It is not clear why other mechanisms involved in blood pressure control do not compensate fully for the absence of eNOS. For example, one might expect that the levels of other vasodilators and vasoconstrictors could be altered to return the blood pressure to normal. It is possible that the renin-angiotensin system and the autonomic nervous system evolved primarily as a defense against hypotension, and diminution of their activity is a poor buffer against hypertension. It is also possible that eNOS may be involved in establishing the baroreceptor setpoint.

These experiments are subject to the possible confounding factor of genetic background. While some mouse strains have a single renin gene, as do humans and other species, other mouse strains, including the SV129 strain, have a gene duplication resulting in a second renin gene expressed in the submandibular gland. However, the blood pressure of anesthetized C57BL/6 and SV129 wild-type mice, as well as F1 crosses and wild-type littermates of mutant animals are all similar, so at least in this situation, the phenotype is most likely due to eNOS gene deletion.

The eNOS knockout mice reveal a possible opposing role for nNOS in the regulation of blood pressure. Treatment of wild-type animals with L-NA increases blood pressure, consistent with the contribution of basal production of NO to vascular tone. Treatment of eNOS knockout mice with L-NA on the other hand, decreases blood pressure. This effect is blocked by L-arginine, and is not seen with D-NA, suggesting that it is due to inhibition of residual isoforms of NOS present in the eNOS mutant mice. Neuronal NOS mutant mice have blood pressures similar to wild-type littermates, but they have a tendency toward hypotension under anesthesia (Irikura *et al.*, 1995). A role for nNOS in maintenance of blood pressure would explain both the hypotensive tendency in nNOS mutant mice and the abnormal response to L-NA in the eNOS mutant mice.

Cardiac Contractility And Relaxation

The role of NO in cardiac function is debated, although in heart failure, cardiac allograft rejection and sepsis, NO derived from the iNOS isoform is speculated to diminish contractility (Brady *et al.*, 1992; Yang *et al.*, 1994; Hare *et al.*, 1995). The enzymatic source of NO under physiologic conditions and in disease states is unclear, since all three isoforms are expressed in the heart. We measured arterial pressure, left ventricular pressure, and heart rate at baseline and in response to stimulation by the β-adrenergic agonist isoproterenol in eNOS knockout mice. To assay systolic contractile function, we measured the first derivative of left ventricular pressure with respect to time (dP/dt_{max}). For diastolic function, we determined τ, the time constant of relaxation.

Wild-type littermates of eNOS mutant mice, wild-type C57BL/6 mice and wild-type SV129 mice had indistinguishable blood pressure, heart rate and contractility, indicating that genetic variation does not affect these parameters. The eNOS mutant mice are hypertensive compared with wild-type mice, but have comparable heart rates. The LV dP/dt_{max} values of eNOS mutant mice were not significantly different from those of wild-type mice, but eNOS knockout mice exhibit a much greater increase in dP/dt_{max} in response to increasing doses of isoproterenol. Other hemodynamic effects of isoproterenol, on heart rate, mean arterial pressure, and left-ventricular end-diastolic pressure (preload) were indistinguishable between the two groups of animals. Acute pharmacologic inhibition of all NOS isoforms with L-NA also enhances the contractility response of wild-type mice, and results in a pattern very similar to the eNOS knockout mice. These results indicate that the eNOS isoform normally buffers the contractility response to β-adrenergic agonists. It is conceivable that myocardial changes in response to chronic hypertension could alter the acute contractility responses, but this seems unlikely to explain the results because other models of chronic hypertension in mice such as aortic banding impair, rather than augment, ventricular contractility.

Ventricular relaxation is reflected by the time constant of ventricular relaxation (τ), the time required for the LV pressure to decline to 1/e of its initial value during isovolumetric relaxation. τ is the same in eNOS mutant mice as in wild-type mice at baseline and varies little following treatment with isoproterenol. L-NA however, markedly increases the time required for LV relaxation in wild-type mice, whereas it has no effect in eNOS mutant mice. These effects are not due to increased afterload, since increasing blood pressure of wild-type mice by phenylephrine does not affect τ. Isoproterenol lowers τ in a dose dependent manner in L-NA treated wild-type mice.

These results indicate that NO plays a role in the normal regulation of systolic contractility in response to beta adrenergic agonists as well as diastolic relaxation, and that the eNOS isoform mediates these responses. The observations also suggest several important differences between the effects of NO on systole and diastole, in that eNOS appears essential to the physiologic regulation of systolic function, but not diastolic relaxation.

Response To Vascular Injury And Atherosclerosis

Nitric oxide has physiologic effects that may normally prevent atherosclerosis, including suppression of smooth muscle proliferation (Mooradian *et al.*, 1995), inhibition of platelet aggregation and adhesion (Radomski *et al.*, 1991), and inhibition of leukocyte activation and adhesion (Bath, 1993; Lefer and Ma, 1993). A reduction in endothelial NO would be expected to diminish these normally protective effects, and thereby predispose to atherosclerosis.

In fact, patients with atherosclerosis, hypertension, diabetes mellitus, and familial hypercholesterolemia demonstrate reduced endothelial NO levels (Freiman *et al.*, 1986; Linder *et al.*, 1990; Flavahan, 1992). Considerable controversy remains as to whether decreased endothelial NO production results from decreased production by eNOS, as seen with oxidized LDL (Liao *et al.*, 1995) or tumor necrosis factor α (Yoshizumi *et al.*, 1993), or from increased destruction of NO in the vessel wall (Mugge *et al.*, 1991), or both. In any case, eNOS knockout mice serve as an important animal model for these conditions, and reveal the consequences of diminished endothelial NO levels on vascular injury or atherogenesis.

We applied a model of vessel injury to eNOS knockout mice. In this model, the endothelium is left intact and a cuff is placed around the vessel so that signals from the adventitia stimulate formation of a neointimal layer in a predictable manner. Quantitative morphometry was used to measure the volumes of the lumen, intima, and media. Wild-type mice, which have no detectable intima at baseline, develop neointima within 14 days of cuff injury, with an intima to media (I/M) volume ratio of 0.29 for males and 0.18 for females. The eNOS mutant animals developed substantially more neointima, with an I/M volume ratio of 0.73 (males) and 0.43 (females). Thus, absence of eNOS expression alone is sufficient for an increased proliferative response to vessel injury. These results support the importance of endothelial dysfunction in the pathophysiology of intimal hyperplasia in atherogenesis.

CEREBROVASCULATURE

Differing Roles Of nNOS And eNOS In Cerebral Ischemia

The nNOS and eNOS knockout mice have been useful to define the sometimes opposing roles of multiple NOS isoforms in cerebral ischemia. Brain NO levels rise dramatically following ischemia (Malinski *et al.*, 1993), but its source, purpose, and effects are unclear. On the one hand, NO may mediate neurotoxicity by peroxynitrite anion formation, activation of poly-ADP ribose synthase, or binding to iron-sulfur complexes and heme. On the other hand, NO may serve a protective function by vasodilation, improving blood flow to ischemic regions, or by effects on platelets and leukocytes. In some studies, NOS inhibitors limit the amount of neuronal damage after ischemia, although in others they it had no effect, and in still others, they had a detrimental effect (Iadecola *et al.*, 1994).

One possible explanation for these discordant results is that nNOS may contribute to tissue damage following cerebral ischemia whereas eNOS may be protective. Using a focal ischemia model of stroke due to middle cerebral artery occlusion, nNOS knockout mice demonstrate significantly reduced infarct size by TTC staining than wild-type, with a 38% reduction over SV129 and C57BL/6 wild-type mice (Huang *et al.*, 1994). The functional outcome was also significantly improved. There were no neuroanatomic or vascular differences between the nNOS mutant mice and wild-type mice. Measurements of rCBF by laser Doppler flow probe demonstrate that the filament used to occlude the middle cerebral artery causes the same reduction in rCBF in both the core ischemic area as well as the penumbra in both wild-type and nNOS mutant mice, so differences in blood flow do not account for the reduction in infarct size. These results support the notion that nNOS contributes to neuronal damage after focal ischemia. Treatment of nNOS mutant mice with L-NA, which inhibits remaining isoforms of NOS including eNOS, caused larger strokes, consistent with the protective role of eNOS. The nNOS mutant mice show similar protection using a model of transient focal ischemia followed by reperfusion (Hara *et al.*, 1996) and a model of global ischemia (Panahian *et al.*, 1996).

The eNOS mutant mice, on the other hand, had larger infarcts following permanent MCAO than either C57BL/6 or SV129 wild-type controls, even when their blood pressure was reduced to normotensive levels for the duration of the experiment (Huang *et al.*, 1996). These results confirm that eNOS plays a protective role, and its absence following cerebral ischemia is detrimental. The hemodynamic effect of middle cerebral artery occlusion, as judged by laser Doppler flowmetry, is greater in the eNOS mutant mice than in wild-type. Lo *et al.* (1996) mapped blood flow kinetics by functional CT scanning using the contrast agent iodohexol. Each pixel in a coronal view was mapped as normal flow (normal kinetics), core infarct (absent flow), or penumbra (flow present, but kinetics abnormal). The total area of abnormality (infarct plus penumbra) was the same in SV129 mice as in eNOS mutant mice, but the size of the core infarct area was greater in the eNOS mutant mice and the ischemic penumbra was correspondingly smaller. These results demonstrate that the larger infarcts in the eNOS mutant mice are due to absence of the normal protective effects of eNOS which serve to preserve cerebral blood flow.

Regulation of Cerebrovascular Blood Flow and Physiologic Compensation

Roy and Sherrington (1890) first proposed that local neuronal activity in the brain is coupled to relative cerebral blood flow. Possible mediators include products of neuronal metabolism (H+ and adenosine), ions released following neuronal activation, or neurotransmitters released adjacent to blood vessels. Because it diffuses freely across cell membranes, has a short biological half-life, and causes vasodilation, NO is an attractive candidate for coupling cerebral blood flow to brain metabolism in response to external stimuli. Two such processes are the response to hypercapnia, and the response to whisker stimulation.

Cerebrovascular blood flow normally increases in response to moderate hypercapnia, such as breathing 5%CO_2. This response can be partially blocked by topical L-NA in a closed cranial window preparation, indicating that NO is involved in this response. Surprisingly, Irikura *et al.* (1995) found that the cerebrovascular response to 5% CO_2 in nNOS mutant mice is indistinguishable from wild-type mice, despite absence of the nNOS isoform. However, topical L-NA has no effect in the nNOS mutant mice. Thus, the response of nNOS mutant mice is not due to eNOS or another NOS isoform. This is an example of compensation in knockout animals by additional mechanisms to vasodilate cerebral blood vessels. These same mediators participate in wild-type animals as well, since NOS inhibition blocks the rCBF response only over a very narrow range of $PaCO_2$.

A similar example of physiologic compensation in the nNOS mutant mice is found in the blood flow response to whisker stimulation. The mouse somatosensory cortex is organized in discrete cortical barrels that correspond to single contralateral whiskers. Whisker stimulation at 2–3 Hz for 60 seconds results in a predictable increase in blood flow overlying the associated cortical barrel. This response can be measured by laser Doppler flowmetry, and is widely used as a model of coupling neuronal activity with cerebral blood flow. In wild-type mice, topical L-NA blocks the rCBF response to whisker stimulation, implicating NO in the response. Ma *et al.* (1996) showed that nNOS mutant mice demonstrate quantitatively normal cerebrovascular responses to whisker stimulation, but unlike wild-type mice, the response is not sensitive to L-NA. Endothelium-dependent relaxation, as assessed by pial dilation to acetylcholine, is the same in wild-type and nNOS mutant mice. These results indicate that endothelial NOS does not mediate the whisker response in nNOS mutant mice, and that the effect of L-NA in wild-type mice is due to nNOS inhibition. In addition, they suggest that NO-independent mechanisms couple rCBF with metabolism in nNOS mutant mice, again demonstrating physiologic compensation in the knockout mice. In separate studies, Ayata *et al.* (1996) found that eNOS mutant mice have a response indistinguishable from wild-type, including sensitivity to L-NA, confirming that nNOS mediates blood flow responses in the cortical barrel fields.

Taken together, these studies demonstrate that the nNOS isoform, not the eNOS isoform, mediates the cerebral blood flow changes in response to hypercapnia and whisker stimulation. Furthermore, nNOS-derived NO acts in parallel with other vasodilators, since NOS inhibition attenuates but does not obliterate these responses. In the chronic absence of nNOS expression, as in the nNOS knockout mice, redundant vasodilatory mechanisms compensate, so the response is preserved and no longer L-NA sensitive. Additional examples of compensation in nNOS mutant mice include nociception (Crosby *et al.*, 1995) and minimum alveolar concentration and righting reflex ED_{50} for isoflurane anesthesia (Ichinose *et al.*, 1995). The common feature of these responses is dependence on NO in wild-type animals, as demonstrated by sensitivity to NOS inhibitors, and a preserved effect in nNOS mutant mice that is not sensitive to NOS inhibitors.

Cerebral Pial Vessel Dilation and Evidence that nNOS May Compensate for Absence of eNOS

Meng *et al.* (1996) found another form of compensation, in which one NOS isoform substitutes for another. Pial arterioles dilate in response to acetylcholine in a closed cranial window preparation. In wild-type mice, as in other species, this response is blocked by superfusion with atropine or L-NA. The expected source of NO in pial vasodilation is eNOS, and this response is not normally affected by tetrodotoxin. In nNOS mutant mice as well as eNOS mutant mice, acetylcholine dilates pial arterioles to the same extent as in wild-type animals, and L-NA attenuates this response. In contrast to wild-type animals or nNOS mutant mice, the response in eNOS mutant mice is blocked by tetrodotoxin. These results suggest that in the eNOS mutant mice, acetylcholine-induced arteriolar dilation is still mediated by NOS; the sensitivity to tetrodotoxin suggests a possible role for the neuronal isoform. It is possible that additional mediators, like endothelium-derived hyperpolarizing factor, carbon monoxide generated by heme oxygenase, adenosine, or products of the cyclooxygenase pathway are involved as well. These results suggest that in certain situations, the nNOS isoform is able to compensate for the absence of the eNOS isoform. This type of alternative isoform compensation is unusual, and contrasts with compensation for lack of nNOS activity by non-NO vasodilatory mediators seen in hypercapnia or whisker response.

LONG TERM POTENTIATION

Long-term potentiation (LTP) in the hippocampus represents a persistent increase in synaptic strength thought to be involved in certain forms of learning and memory. Although LTP involves events in the post-synaptic cell, it also involves increase in transmitter release from the pre-synaptic cell. NO has been postulated as a retrograde messenger, a signal originating in a post-synaptic cell that feeds back to a pre-synaptic cell to stimulate changes necessary for LTP. Evidence for a role for NO includes the inhibition of LTP by NOS inhibitors or scavengers, and activity-dependent enhancement of synaptic transmission by NO generating compounds (Bohme *et al.*, 1991; Schuman and Madison, 1991; Haley *et al.*, 1992; O'Dell *et al.*, 1994).

Because nNOS accounts for over 95% of NOS catalytic activity in the brain, one might expect that the neuronal isoform is involved in LTP. However, O'Dell *et al.* (1994) found that nNOS mutant mice demonstrate preserved LTP. LTP is blocked by NOS inhibitors in the nNOS knockout mice, just as it is in wild-type mice, raising the possibility that another NOS isoform may be responsible. In support of this, immunocytochemical studies localized eNOS to the CA1 region (Dinerman *et al.*, 1994; O'Dell *et al.*, 1994), leading the speculation that the eNOS isoform may mediate LTP instead of the nNOS isoform. However, eNOS knockout mice, like nNOS knockout mice, have normal LTP. The answer to this apparent paradox came when Son *et al.* (1996) examined LTP in double nNOS/eNOS knockout mice made by a breeding scheme involving eNOS knockout mice and nNOS knockout mice. One type of LTP (stratum radiatum) was significantly reduced in the double nNOS/eNOS knockout mice, whereas another type of LTP (stratum oriens) was unaffected. These results provided the first genetic evidence that NO is involved in stratum radiatum LTP. Furthermore, they suggest that nNOS and eNOS can compensate for each other in single knockout mice, a situation similar to the pial vessel dilation response to acetylcholine.

SUMMARY

Targeted disruption of NOS genes offers a complementary approach to the use of pharmacologic inhibitors of NOS. NOS knockout mice have been useful to study the role of each individual NOS isoform and the effects of its deletion in intact animals. The neuronal NOS mutant mice serve as a model for the human disorder hypertrophic pyloric stenosis. They also demonstrate the contribution of the neuronal isoform to toxicity following cerebral ischemia, and suggest that selective blockade of nNOS but not eNOS may be a useful approach in the treatment of stroke. Endothelial NOS knockout mice establish the role of eNOS in generating EDRF. They are hypertensive and demonstrate increased cardiac contractility and increased neointimal proliferation in response to vessel injury. The eNOS knockout mice serve as useful animal models for human conditions of hypertension, atherosclerosis, diabetes, and aging, in which endothelial production of NO is diminished.

The phenotypes of the nNOS and eNOS mutant mice demonstrate the existence and importance of physiologic compensation for the absence of the deleted genes. Study of the knockout animals may reveal pathways in parallel with NO signaling that may be explored in hopes of restoring responses in human conditions where NO production is diminished.

Limitations of NOS gene knockouts include the possibility of developmental abnormalities that can affect phenotype, and variability in phenotype due to genetic background effects. These two areas have been largely unexplored to date, but will be important in interpreting results from NOS knockout mice. The advent of inducible and tissue-specific gene knockouts, so far achieved in other systems, promises to address these issues and extend the power of the genetic approach.

REFERENCES

Ayata, C., Ma, J., Meng, W., Huang, P. and Moskowitz, M.A. (1996) L-NA sensitive rCBF augmentation during vibrissal stimulation in type III nitric oxide synthase mutant mice. *J. Cereb. Blood Flow Metab.*, **16**, 539–41.

Bai, T.R. and Bramley, A.M. (1993) Effect of an inhibitor of nitric oxide synthase on neural relaxation of human bronchi. *Am. J. Physiol.*, **264**, L425–30.

Bath, P.M.W. (1993) The effect of nitric oxide donating vasodilators on monocytic chemotaxis and intracellular cGMP concentrations *in vitro. Eur. J. Clin. Pharmacol.*, **45**, 53–8.

Beierwaltes, W.B. and Carretero, O.A. (1992) Nonprostanoid endothelium-derived factors inhibit renin release. *Hypertension*, **19** (suppl. II), 1168–73.

Belvisi, M.G., Stretton, C.D., Yacoub, M. and Barnes, P.J. (1992) Nitric oxide as an endogenous modulator of cholinergic neurotransmission in guinea pig airways. *Eur. J. Pharmacol.*, **198**, 219–21.

Bohme, G.A., Bon, C., Stutzman, J.M., Doble, A. and Blanchard, J.C. (1991) Possible involvement of nitric oxide in long-term potentiation. *Eur. J. Pharmacol.*, **199**, 379–81.

Brady, A.J.B., Poole-Wilson, P.A., Harding, S.E. and Warren, J.B. (1992) Nitric oxide production within cardiac myocytes reduces their contractility in endotoxemia. *Am. J. Physiol.*, **263**, H1963–H6.

Bredt, D.S. and Snyder, S.H. (1994) Transient nitric oxide synthase neurons in embryonic cerebral cortical plate, sensory ganglia, and olfactory epithelium. *Neuron*, **13**(2), 301–13.

Brenman, J.E., Chao, D.S., Gee, S.H., McGee, A.W., Craven, S.E., Santillano, D.R. *et al.* (1996) Interaction of nitric oxide synthase with the post synaptic density protein PSD-95 and α1-syntrophin mediated by PDZ domain. *Cell*, **84**, 757–67.

Briggs, J.P., Wang, W., He, X., Brosius, F.C. and Schnermann, J. (1993) Expression of cerebellar cNOS in kidney cortex; characterization by direct PCR sequencing. *J. Am. Soc. Nephrol.*, **3**, 11.

Crawley, J.N. (1996) Unusual behavioral phenotypes of inbred mouse strains. *Trends Neurosci.*, **19**, 181–2.

Crosby, G., Marota, J.J.A. and Huang, P.L. (1995) Intact nociception-induced neuroplasticity in transgenic mice deficient in neuronal nitric oxide synthase. *Neuroscience*, **69**, 1013–7.

Crusio, W.E. (1996) Gene-targeting studies: new methods, old problems. Trends *Neurosci.*, **19**, 186–7.

Darius, S., Wolf, G., Huang, P.L. and Fishman, M.C. (1995) Localization of NADPH-diaphorase/nitric oxide synthase in the rat retina: an electron microscopic study. *Brain Res.*, **690**, 231–5.

Dinerman, J.L., Dawson, T.M., Schell, M.J., Snowman, A. and Snyder, S.H. (1994) Endothelial nitric oxide synthase localized to hippocampal pyramidal cells: implications for synaptic plasticity. *Proc. Natl. Acad. Sci. USA*, **91**, 4214–8.

Flavahan, N.A. (1992) Atherosclerosis or lipoprotein induced endothelial dysfunction: potential mechanisms underlying reduction in EDRF/nitric oxide activity. *Circulation*, **85**, 1927–38.

Freiman, P.C., Mitchell, G.C., Heistad, D.D., Armstrong, M.L. and Harrison, D.G. (1986) Atherosclerosis impairs endothelium-dependent vascular relaxation to acetylcholine and thrombin in primates. *Circ. Res.*, **58**, 783–9.

Furchgott, R.F. and Zawadzki, J.V. (1980) The obligatory role of endothelial cells in the relaxation of arterial smooth muscle by acetylcholine. *Nature*, **288**, 373–6.

Gerlai, R. (1996) Gene targeting in neuroscience: the systematic approach. *Trends Neurosci.*, **19**, 188–9.

Gerlai, R. (1996) Gene targeting studies of mammalian behavior: is it the mutation or the background genotype? *Trends Neurosci.*, **19**, 177–81.

Haley, J.E., Wilcox, G.L. and Chapman, P.F. (1992) The role of nitric oxide in hippocampal long-term potentiation. *Neuron*, **8**, 211–6.

Hara, H., Huang, P.L., Panahian, N., Fishman, M.C. and Moskowitz, M.A. (1996) Reduced brain edema and infarction volume in mice lacking the neuronal isoform of nitric oxide synthase after transient MCA occlusion. *J. Cereb. Blood Flow Metab.*, **16**, 605–11.

Hara, H., Waeber, C., Huang, P.L., Fujii, M., Fishman, M.C. and Moskowitz, M.A. (1996) Brain distribution of nitric oxide synthase in neuronal or endothelial nitric oxide synthase mutant mice using nitro-arginine autoradiography. *Neuroscience*, **75**, 881–90.

Harding, P., Sigmon, D.H., Alfie, M.E., Huang, P.L., Fishman, M.C., Beierwaltes, W.H. *et al.* (1997) Cyclooxygenase-2 mediates increased renal content induced by low-sodium diet. *Hypertension*, **29**, 297–302.

Hare, J.M., Loh, E., Creager, M.A. and Colucci, W.S. (1995) Nitric oxide inhibits the positive inotropic response to beta-adrenergic stimulation in humans with left ventricular dysfunction. *Circulation*, **92**, 2198–203.

Huang, P.L., Dawson, T.M., Bredt, D.S., Snyder, S.H. and Fishman, M.C. (1993) Targeted disruption of the neuronal nitric oxide synthase gene. *Cell*, **75**(7), 1273–86.

Huang, P.L., Huang, Z., Mashimo, H., Bloch, K.D., Moskowitz, M.A., Bevan, J.A. *et al.* (1995) Hypertension in mice lacking the gene for endothelial nitric oxide synthase. *Nature*, **377**(6546), 239–42.

Huang, Z., Huang, P.L., Ma, J., Meng, W., Ayata, C., Fishman, M.C. et al. (1996) Enlarged infarcts in endothelial nitric oxide synthase knockout mice are attenuated by nitro-L-arginine. *J. Cereb. Blood Flow Metab.*, **16**, 981–7.

Huang, Z., Huang, P.L., Panahian, N., Dalkara, T., Fishman, M.C. and Moskowitz, M.A. (1994) Effects of cerebral ischemia in mice deficient in neuronal nitric oxide synthase. *Science*, **265**(5180), 1883–5.

Iadecola, C., Pelligrino, D.A., Moskowitz, M.A. and Lassen, N.A. (1994) Nitric oxide synthase inhibition and cerebrovascular regulation. *J. Cereb. Bllod Flow Metab.*, **14**, 175–92.

Ichinose, F., Huang, P.L. and Zapol, W.M. (1995) Effects of targeted neuronal nitric oxide synthase gene disruption and nitroG-L-arginine methylester on the threshold for isoflurane anesthesia. *Anesthesiol.*, **83**(1), 101–8.

Irikura, K., Huang, P.L., Ma, J., Lee, W.S., Dalkara, T., Fishman, M.C. *et al.* (1995) Cerebrovascular alterations in mice lacking neuronal nitric oxide synthase gene expression. *Proc. Natl. Acad. Sci. USA*, **92**(15), 6823–7.

Kharitonov, S.A., Yates, D., Robbins, R.A., Logan-Sinclair, R., Shinebourne, E.A. and Barnes, P.J. (1994) Increased nit ric oxide in exhaled air of asthmatic patients. *Lancet*, **343**, 133–5.

Kobzik, L., Reid, M.B., Bredt, D.S. and Stamler, J.S. (1994) Nitric oxide in skeletal muscle. *Nature*, **372**(6506), 546–8.

Koide, M., Kawahara, Y., Tsuda, T., Nakayama, I. and Yokoyama, M. (1994) Expression of nitric oxide synthase by cytokines in vascular smooth muscle cells. *Hypertension*, **23**(1 Suppl), I45–8.

Kurihara, Y., Kurihara, H., Suzuki, H., Kodama, T., Maemura, K., Nagai, R. *et al.* (1994) Elevated blood pressure and craniofacial abnormalities in mice deficient in endothelin-1. *Nature*, **368**(6473), 703–10.

Lathe, R. (1996) Mice, gene targeting and behaviour: more than just genetic background. *Trends Neurosci.*, **19**, 183–6.

Laubach, V.E., Shesely, E.G., Smithies, O. and Sherman, P.A. (1995) Mice lacking inducible nitric oxide synthase are not resistant to lipopolysaccaride-induced death. *Proc. Natl. Acad. Sci. USA*, **92**, 10688–92.

Lefer, A.M. and Ma, X. (1993) Decreased basal nitric oxide release in hypercholesterolemia increases neutrophil adherence to rabbit coronary artery endothelium. *Arterioscler. Thromb.*, **13**, 771–6.

Liao, J.K., Shin, W.S., Lee, W.Y. and Clark, S.L. (1995) Oxidized low density lipoprotein decreases the expression of endothelial nitric oxide synthase. *J. Biol. Chem.*, **270**, 319–24.

Linder, L., Kiowski, W., Buhler, F.R. and T.F., L. (1990) Indirect evidence for release of endothelium-derived relaxing factor in human forearm circulation *in vivo*: blunted response in essential hypertension. *Circulation*, **81**, 1762–7.

Lo, E., Hara, H., Rogowska, J., Trocha, M., Pierce, A.R., Huang, P.L. *et al.* (1996) Temporal correlation mapping analysis of the hemodynamic penumbra in mutant mice deficient in endothelial nitric oxide synthase gene expression. *Stroke*, **27**, 1381–5.

Loesch, A., Belai, A. and Burnstock, G. (1994) An ultrastructural study of NADPH-diaphorase and nitric oxide synthase in the perivascular nerves and vascular endothelium of the rat basilar artery. *J. Neurocytol.*, **23**, 49–59.

Ma, J., Ayata, C., Huang, P.L., Fishman, M.C. and Moskowitz, M.A. (1996) Regional cerebral blood flow response to vibrissal stimulation in mice lacking type I NOS gene expression. *Am. J. Physiol.*, **271**, H1717–9.

MacMicking, J.D., Nathan, C., Hom, G., Chartrain, N., Fletcher, D.S., Trumbauer, M. *et al.* (1995) Altered responses to bacterial infection and endotoxic shock in mice lacking inducible nitric oxide synthase. *Cell*, **81**(4), 641–50.

Malinski, T., Bailey, F., Zhang, Z.G. and Chopp, M. (1993) Nitric oxide measured by a porphyrinic microsensor in rat brain after transient middle cerebral artery occlusion. *J. Cereb. Blood Flow Metab.*, **13**(3), 355–8.

Marletta, M.A. (1994) Nitric oxide synthase: aspects concerning structure and catalysis. *Cell*, **78**(6), 927–30.

Mashimo, H., He, X.D., Huang, P.L., Fishman, M.C. and Goyal, R.K. (1996) Neuronal constitutive nitric oxide synthase is involved in murine enteric inhibitory neurotransmission. *J. Clin. Invest.*, **98**, 8–13.

Massaro, A.F., Mehta, S., Lillly, C.M., Kobzik, L., Reilly, J.J. and Drazen, J.M. (1996) Elevated nitric oxide concentrations in lower airway gas of asthmatic subjects. *Am. J. Respir. Crit. Care Med.*, **153**, 1510–4.

Meng, W., Ma, J., Ayata, C., Hara, H., Huang, P.L., Fishman, M.C. *et al.* (1996) Acetylcholine dilates pial arterioles in endothelial and neuronal nitric oxide synthase knockout mice by nitric oxide-dependent mechanisms. *Am. J. Physiol.*, **271**, H1145–H50.

Meyer, D. and Birchmeier, C. (1995) Multiple essential functions of neuregulin in development. *Nature*, **378**(6555), 386–90.

Mooradian, D.L., Hutsell, T.C. and Keefer, L.K. (1995) Nitric oxide (NO) donor molecules: effect of NO release rate on vascular smooth muscle cell proliferation *in vitro*. *J. Cardiovasc. Pharmacol.*, **25**(4), 674–8.

Mugge, A., Elwell, J.H., Peterson, T.E. and Harrison, D.G. (1991) Release of intact endothelium-derived relaxing factor depends on endothelial superoxide dismutase activity. *Am. J. Physiol.*, **260**, 219–25.

Mullins, M.E., Sondheimer, N.J., Huang, Z., Singel, D.J., Huang, P.L., Fishman, M.C. *et al.* (1995) *The Biology of Nitric Oxide*, Part 5. 4th International Meeting on the Biology of Nitric Oxide, Amelia Island, Florida, USA, Portland Press.

Nathan, C. and Xie, Q.W. (1994) Nitric oxide synthases, roles, tolls, and controls. *Cell*, **78**(6), 915–8.

Nelson, R.J., Duman, G.E., Huang, P.L., Fishman, M.C., Dawson, V.L., Dawson, T.M. *et al.* (1995) Behavioural abnormalities in male mice lacking neuronal nitric oxide synthase. *Nature*, **378**, 383–6.

O'Dell, T.J., Huang, P.L., Dawson, T.M., Dinerman, J.L., Snyder, S.H., Kandel, E.R. *et al.* (1994) Endothelial NOS and the blockade of LTP by NOS inhibitors in mice lacking neuronal NOS. *Science*, **265**(5171), 542–6.

Panahian, N., Yoshida, T., Huang, P.L., Hedley-Whyte, E.T., Fishman, M. and Moskowitz, M.A. (1996) Attenuated hippocampal damage after global cerebral ischemia in knock-out mice deficient in neuronal nitric oxide synthase. *Neuroscience*, **72**, 343–54.

Radomski, M.W., Palmer, R.M. and Moncada, S. (1991) Modulation of platelet aggregation by an L-arginine-nitric oxide pathway. *Trends in Pharmacological Sciences*, **12**(3), 87–8.

Roy, C.W. and Sherrington, C.S. (1890) On the regulation of the blood supply of the brain. *J. Physiol.*, **11**, 85–108.

Rudnicki, M.A., Braun, T., Hinuma, S. and Jaenisch, R. (1992) Inactivation of MyoD in mice leads to up-regulation of the myogenic HLH gene Myf-5 and results in apparently normal development. *Cell*, **71**, 383–90.

Schmidt, H.H.H.W., Warner, T.D., Ishii, K., Sheng, H. and Murad, F. (1992) Insulin secretion from pancreatic B cells caused by L-arginine-derived nitrogen oxides. *Science*, **255**, 721.

Scholz, H. and Kurtz, A. (1993) Involvement of endothelium-derived relaxing factor in the pressure control of renin secretion from isolated perfused kidney. *J. Clin. Invest.*, **91**, 1088–94.

Schuman, E.M. and Madison, D.V. (1991) A requirement for the intercellular messenger nitric oxide in long-term potentiation. *Science*, **254**, 1503–6.

Sedivy, J. and Joyner, A. (1992) *Gene Targeting*. New York, W. H. Freeman and Company.

Son, H., Hawkins, R.D., Martin, K., Kiebler, M., Huang, P.L., Fishman, M.C. *et al.* (1996) Long-term potentiation is reduced in mice that are doubly mutant in endothelial and neuronal nitric oxide synthase. *Cell*, **87**, 1015–23.

Ujiie, K., Yuen, J., Hogarth, L., Dazinger, R. and Star, R.A. (1994) Localization and regulation of endothelial NO synthase mRNA expression in rat kidney. *Am. J. Physiol.*, **267**, F296–F302.

Vanderwinden, J.M., Mailleux, P., Schiffmann, S.N., Venderhaeghen, J.J. and De Laet, M.H. (1992) Nitric oxide synthase activity in infantile hypertrophic pyloric stenosis. *N. Engl. J. Med.*, **327**, 511–5.

Wei, X.Q., Charles, I.G., Smith, A., Ure, J., Feng, G.J., Huang, F.P. *et al.* (1995) Altered immune responses in mice lacking inducible nitric oxide synthase. *Nature*, **375**(6530), 408–11.

Yang, X., Chowdhury, N., Cai, B., Brett, J., Marboe, C., Sciacca, R.R. *et al.* (1994) Induction of myocardial nitric oxide synthase by cardiac allograft rejection. *J. Clin. Invest.*, **1994**, 714–21.

Yoshizumi, M., Perrella, M.A., Burnett, J.C. and Lee, M.E. (1993) Tumor necrosis factor downregulates an endothelial nitric oxide synthase mRNA by shortening its half-life. *Circ. Res.*, **73**, 205–9.

Index

α1-syntrophin, 52
α-granules, 237
β_2 integrins, 308, 312
β-thromboglobulin, 240
β-VLDL, 64
γ-IRE, 44
·O_2-, 101, 160
·OH, 160
1-(2-aminoethyl)imidazole, 514
13-hydroxyoctadecadienoic acid
   (13-hode), 231
1400W, 329
^{15}N-L-arginine, 7, 21
17β-estradiol:
   anti-atherosclerotic effect of, 199
   physiological substitution doses of, 202
   protection of NO by, 198
   regulation of NOS-III enzyme activity
     by, 198
   regulation of NOS-III gene
     expression by, 196
   role of increased endothelial NO, 199
   upregulation of endothelial NO
     production by, 196
   vasculoprotective effects of, 195
1-phenylimidazoles, 514
2-amino-5,6-dihydro-4H-1,3-thiazine, 512
2-aminopicoline, 512
2-aminopyridine, 512
2-aminopyridines, 512
2-aminothiazole, 512
2-aminothiazoline, 512
2-deoxyglucose, 133
2-iminobiotin, 514
3-aminobenzamide, 88
3-bromo-7-nitroindazole, 515
4-amino-BH_4, 506
7-nitroindazole, 515
A23187, 2
Acetaminophen, 11
Acetylcholine, 1
Acid-activatable inhibitory factor
   (BRPIF), 9
Acidosis, 435

Aconitase, 433, 435
Activated endothelium, 100
Activated factor V (Va), 232
Activated factor X (Xa), 232
Activated macrophages, 8
Activated PMNs, 306, 316
Acute arterial occlusion, 284
Acute bronchopneumonia, 53
Acute cardiovascular events, 284
Acute exercise, 174
Acute lung injury, 482
Acute rejection, 581
Acute renal failure, 336
Acute respiratory distress syndrome
   (ARDS), 269, 483
Acyclic amidines, 513
Adeno-associated virus, 556
Adenosine, 456, 457
Adenovirus, 556, 574, 579
Adhesion molecule:
   expression, 311
Adhesion molecules, 287, 310, 312
Adjuvant-induced arthritis, 329, 414
ADP ribosylation, 436
ADP, 2
Adrenal glands, 51
Adult cardiac disease, 479
Adventitia, 558, 562
aFGF, 131
Air pollution:
   role of NO, 455
Alveolar macrophages, 60
Amidine-containing inhibitors, 510
Amidines, 506
Amino acid hydroxylases, 508
Aminogaunidine, 159, 328, 378, 506
Amyl nitrite, 456, 457, 458
Androgens, 366
Angeli's salt, 463
Angina, 285
Angiogenesis, 13, 471
Angiogenesis:
   the role of NOS and hemodynamic
     forces, 188

Angioplasty, 560
Angiotensin II, 271, 362, 370
Angiotensin-converting enzyme inhibitors, 290
Animal models, 290
Anorexic agents, 269
Anoxia, 3
Antibiotics, 456
Antihypertensives, 456, 459
Anti-IFN-γ antibodies, 43
Anti-inflammatory effects, 106
Antioxidant defense pathways, 62
Antioxidants, 70, 108, 113, 290, 383
Antiprogestins, 359
Antisense, 43
Antithrombin III, 234
Antithrombotic processes, 241
AP endonucleases, 72
AP-1, 45, 100, 104, 134, 136, 188, 287, 430
Apoptosis, 75, 155, 292, 398, 405, 419, 431
Arachidonic acid, 3, 156
ARDS, 484
Arginase, 340, 382
Arteriosclerosis, 555
Arteriovenous anastomosis, 173
Arteriovenous fistula, 174
Arthritis, 14, 397, 413
Arylimidazoles, 506
Aspirin, 243
Assymetric dimethylarginine (ADMA), 242
Astrocytes, 379, 381, 387
Asymmetric dimethylarginine (ADMA), 294, 295
Atherectomy, 285
Atherogenesis, 100, 286, 290, 312
Atherosclerosis, 121, 173, 240, 294, 569, 578, 581
Atherosclerosis: progression, 290
Atherosclerotic lesions, 53
Atherosclerotic plaques, 199
Atherothrombosis, 240
ATP, 2, 69
Atrial natriuretic peptide, 289
Autoimmune diseases, 384, 405, 413
Autooxidation, 60
Azide, 133, 461

B lymphocyte, 43
Balloon angioplasty, 240, 285, 289, 569
Basic fibroblast growth factor (bFGF), 124, 129, 131, 429, 569
Bcl-2, 417
BH$_4$, 18, 28, 198, 507, 577
BH$_4$ binding antagonists, 506
BH$_4$ synthesis inhibitors, 506
Bioavailable NO, 199
Biopterin, 328
Bis-amidines, 513
Bleeding diathesis, 229
Bleeding time, 338
Blood pressure, 251, 252, 257
Blood pressure regulation, 254
Blood velocity, 230, 231
Bone reabsorption, 13
Bovine hemoglobin, 541
Bowel injury, 442
BQ123, 367
Bradykinin, 2
Brain constitutive NOS (bNOS), 195, 352
Brain injury, 523
Brain, 21
Bronchodilation, 486
BRPIF, 9

C reactive protein, 414
C57BL/6 mice, 212
Ca^{2+}, 9
Caesarean section, 356
Caged nitric oxide, 459
Calcification, 173
Calcitonin gene related-peptide (CGRP), 431
Calcium channel blockers, 268
Calcium ionophore (A23187), 2
Calcium phosphate coprecipitation, 573
Calcium pumps, 70
Calmodulin, 7, 25, 28, 198, 235
Calmodulin antagonists, 506
Calpain inhibitor, 326
cAMP, 161, 165, 187, 382, 399
cAMP phosphodiesterase, 239
cAMP responsive element, 187
Capillary migration and proliferation, 188
Capillary permeability, 351
Carbon monoxide, 134

Cardiac hypertrophy, 258
Cardiac myocytes, 54
Cardiac necrosis, 308
Cardiac output, 330
Cardiac transplantation, 479, 581
Cardioprotection, 316, 317
Cardiopulmonary bypass, 475
Cardiovascular disease, 173
Carotid artery disease, 285
Cartilage, 397, 413, 415
Cartilage metabolism, 418
CAT1, 26
Catalytic domain, 507
Cationic amino acid transporter 1 (CAT1), 26
Cationic liposomes, 557
Caveolae, 20
Caveolin, 25
Caveolin-1, 54
Caveolin-3, 54
CCAAT-enhancer-binding protein β, 131
CD 11/CD 18, 101, 428, 438
CD 12, 428
CD 36, 144
CD 39, 231
CD4$^+$T cells, 376
Cell death, 318
Cell proliferation:
    inhibition by diazeniumdiolates, 466
Cellular antioxidant systems, 524
Cellular thiols, 62
Central nervous system, 10, 375
Ceramide, 134
Cerebral ischemia:
    differing roles of nNOS and eNOS, 218
Cerebrospinal fluid, 376
Cerebrovascular blood flow, 219
Cerebrovasculature, 218
Cervical ripening, 350, 358
c-fos, 386
cGMP pathways, 293
cGMP, 3, 101, 112, 578
Chemically modified tetracycline (CMT), 14
Chemoattractant, 283
Chemokines, 283, 287, 426, 571
Chemotaxis, 238

Cholecystokinin, 431
Chondrocyte, 398
Chondrocytes, 399, 413, 415, 441
Chondrosarcoma, 12
Chorioallantoic membrane (CAM), 189
Chronic allograft rejection, 581
Chronic exercise, 175
Chronic obstructive pulmonary disease, 485
Chronic pulmonary disease, 140
Chronic rejection, 581
Chronic thromboembolic, 268
Cicaprost, 367
Circulatory failure, 323, 324, 326
Circulatory shock, 78, 310, 324, 328, 329, 432, 440
Circulatory shock:
    hypodynamic, 439
Cirrhotic rats, 175
Citrulline, 8
c-jun, 134
Clot retraction, 233
cNOS mRNA, 134
CNS trauma, 524
CNS, 375
Coagulation cascade, 232
Coagulation factors, 229
Coagulopathy, 492
Collagen synthesis, 418
Collagenase IV (MMP-9), 121
Collagen-vascular disease, 268
Complex lesions, 293
Cone-plate viscometer, 176
Congenital diaphragmatic hernia, 482
Congenital heart disease, 268, 475
Consumption coagulopathy, 241
Copper chelating activity, 531
Coronary arteries, 196
Coronary artery bypass grafting, 479
Coronary artery, 292
Coronary microcirculation, 304
Coronary vasospasm, 307
Corticotropin-releasing hormone (CRH), 350
COX, 155
COX-1 activation:
    NO-mediated, 160
COX-1, 155, 156, 361

COX-2 activation:
  NO-mediated, 162
COX-2, 11, 155, 156, 162, 182, 361,
  399, 401, 435
CRE, 187, 188
c-Rel, 103
c-Src, 128, 134
Cu,Zn superoxide dismutase, 60
Cultured hepatocytes, 204
Cupferron, 456, 457
Cyanamide, 463
Cyanide, 133
Cyclic amidines, 512
Cyclic AMP, 44
Cyclic guanosine monophosphate (cGMP),
  3, 189, 256, 286
Cyclic nucleotide phosphodiesterases, 236
Cyclic strain, 171, 179
Cyclic strain:
  endothelin, 181
  iNOS gene expression, 184
  PDGF, 181
  proliferation and migration, 179
  morphological changes, 179
  tPA, 181
Cycloheximide, 430
Cyclooxygenase, 155, 231, 405, 434
Cyclosporine A, 342
Cysteine, 523
Cysteine:
  enzyme cofactor in NO production, 461
Cystic fibrosis, 558
Cytochrome oxidase, 435
Cytochrome P480, 132
Cytokines, 39, 241, 269, 283, 286, 287, 323,
  329, 349, 375
Cytomegalovirus, 285
Cytotoxicity due to PARS activation, 73

DEA/NO, 465
Decidua, 350, 352, 354
Delayed type hypersensitivity, 385
Demyelination, 378, 383
Deoxy-hemoglobin-NO complexes, 539
DETA/NO, 465
Dexamethasone, 43, 202, 326
Diabetes mellitus, 74, 84, 240, 283, 341

Diaminobenzimidazole, 514
Diaminohyroxypyrimidine, 506
Diapirin cross-linked hemoglobin, 541
Diazeniumdiolates, 456, 464, 466
Diazonium salts, 455, 460
Dietary arginine, 290, 292, 293
Dihydrodiazete dioxides, 456, 461
Dihydropteridine reductase, 508
Dimethylarginine dimethylaminohydrolase
  (DDAH), 296
Dimethylarginines, 339
Diphenyleneiodonium, 506
Direct enzyme inhibitors:
  based on the substrate, 507
  based on ligation of the catalytic
    heme center, 507
Disseminated intravascular coagulation, 241
Ditutyryl-cAMP, 165
DLD-1 cells, 428
Dlg, 52
DNA, 70
D-NMMA, 7
DOCA salt-sensitive hypertensive rats, 260
DOPA synthesis, 60
Dopastin, 456, 457
Duchenne muscular dystrophy, 52
Dynamic exercise, 173
Dystrophin-glycoprotein complex, 20

Ebselen, 525
ecto-ADPase, 231
Edema, 163
EDHF, 7
EDRF, 1, 99, 216, 255
EGF, 43, 131
EGF-receptor, 131
Egr-1, 100, 104
Eisenmenger complex, 475
Eisenmenger physiology, 268
Eisenmenger's syndrome, 275
Elastic membrane, 176
Electron nuclear double resonance (ENDOR)
  spectroscopic techniques, 21
Electron paramagnetic resonance (EPR), 377
Electroporation, 572
Embolic stroke, 284
Embryonic stem (ES) cells, 210

Encephalitis, 377
Endothelial cell injury, 199
Endothelial cells, 54, 60, 74
Endothelial damage, 351
Endothelial denudation, 286
Endothelial dysfunction, 75, 100, 199, 290,
    301, 325, 363
Endothelial dysfunction:
    in circulatory shock, 79
Endothelial heparan sulfate, 234
Endothelial nitric oxide synthase, 22, 125
Endothelial NOS knockout mice, 209
Endothelial NOS mutant mice:
    cardiac contractility and relaxation, 217
    hypertension and blood pressure response,
        216
    response to vascular injury and
        atherosclerosis, 218
Endothelial ulceration, 284
Endothelial-derived NO, 112
Endothelial-leukocyte interaction, 101
Endothelin, 268, 367
Endothelin-1 receptor subtypes, 125
Endothelin-1, 121, 122, 124, 271
Endothelium, 558
Endothelium-dependent modulation of
    vascular tone, 195
Endothelium-dependent relaxation, 173, 259
Endothelium-derived hyperpolarizing
    factor, 7, 255
Endothelium-derived relaxing factor, 1
Endotoxemia, 323, 328
Endotoxic shock, 324, 329
Endotoxin, 43, 165, 323, 329, 425, 544
Energy metabolism, 132
eNOS activating protein (ENAP), 26
eNOS expression:
    upregulation of, 41
eNOS gene:
    expression and regulation of, 41
eNOS gene transfer, 555
eNOS gene transfer:
    and atherosclerosis, 561
    and cerebral vasospasm, 560
    and pulmonary hypertension, 563
    and restenosis, 560
    and vein bypass graft failure, 563

eNOS mRNA, 41
eNOS promoter function, 41
eNOS, 22, 40, 121, 182, 187, 209, 352, 556
eNOS:
    oligomeric status, 23
    protein-protein interactions, 25
    tissue distribution, 53
    subcellular targeting, 54
    mice deficient in, 209
Enterocytes, 425
Epidermal growth factor (EGF), 429, 284
Epithelial cells, 74
EPR spin-trapping, 21, 22
ERα, 199
ERβ, 199
ERKO mice, 199
Erythrocytes, 486
Erythropoietin, 134
ES cells, 210
E-selectin, 100, 312
Essential hypertension, 251, 261
Estrogen, 290
Estrogen receptors, 199
Estrogen:
    vasculoprotective effect, 200
ET-1, 130
Ethinyl estradiol, 198
Excision repair, 72
Exercise, 173
Exhaled NO, 270
Experimental allergic encephalomyelitis
    (EAE), 376
Extracellular matrix proteins, 129
Extracellular matrix, 269, 284, 293, 359
Extracorporeal membrane oxygenation,
    476
Extrinsic pathways, 232

FAD, 18
Female reproduction, 349
Female sexual steroid hormones:
    cardioprotection by, 195
Fenton-reaction, 531
FGF, 43
Fibrinogen, 231
Fibrin deposition, 231, 269
Fibrinogen receptor, 231

Fibrinolysis, 229, 233, 241, 269
Fibroblast mitogenesis, 122
Fibroblasts, 112
Fibrochondrocytes, 415
Fibronectin, 129, 231, 419
Fibrotic lung diseases, 485
Fibrous cap, 284
Fibrous plaque, 284
Fissures, 285
Five-electron oxidation, 17
FK-409, 462
Flavoprotein oxidoreductases, 132
Flavoprotein reductase inhibitors, 506
Flexercell strain unit, 179
Flk-1, 124
Flt-1, 124
fMLP, 308
FMN, 18
Foam cells, 100, 283
Foetal depression, 364
Foetal stress, 350
Formyl-methionyl-leucyl-phenylalanine
  (fMLP), 305
Free radical, 3
Fuoxans, 456, 461

$G\alpha_{13}$, 44
GAF, 113
Gallbladder, 433
GAS, 44
Gastrin, 431
Gastrointestinal dysfunction, 440
Gastrointestinal function, 431
Gastrointestinal steal phenomenon, 439
Gastrointestinal tract, 425
Gastroparesis, 432
G-cyclase, 3
Gelatinase, 13
Gender difference, 196
Gene delivery methods, 573
Gene knockout approaches:
    developmental effects, 211
    genetic background, 212
    limitations to, 211
    tissue specificity, 211
Gene knockout technology, 209
Gene therapy, 276

Gene transfer:
    to the cerebral vasculature, 558
Genetic factors, 283
Genetic hypertension, 251
Genetic injury, 439
Gestation, 349, 354
Gliosis, 379
Global ischemia, 304
Glomerulonephritis, 339
Glucocorticoids, 106
Glutathione peroxidase, 240
Glutathione:
    enzyme cofactor in NO production, 461
    S-nitroso, 464
Glycolysis, 132, 436
Glycoproteins, 231
Gp 120 glycoprotein, 164
Growth factors, 283
Growth factors:
    regulation by oxygen tension and nitric
      oxide, 131
GTP cyclohydrolase I, 198, 328, 577, 508
GTP/GDP exchange, 109
Guanosine 3',5'-monophosphate (cGMP),
  236
Guanosine 5'-triphosphate, 236
Guanylate cyclase, 3, 155, 236, 431
Gut necrosis, 432

HDL oxidation, 199
Heart failure, 276
Heart-lung transplantation, 475
Heme, 539
Heme binders, 506
Heme domain dimer, 28
Heme proteins, 132, 159
Heme-binding protein, 133
Heme-containing enzymes, 514
Heme-containing sensor, 121
Hemodynamic factor, 230
Hemodynamic forces, 173, 285
Hemoglobin, 4
Hemoglobin:
    clinical studies, 549
    its role as a nitric oxide scavenger, 539
    NO donor, 539
    release of nitrosothiols, 539

Hemoglobin (effects on):
  endotoxin and bacterial growth, 548
  isolated vessel preparations, 541
  oxygenation, 545
  regional blood flow, 546
  renal function, 547
Hemoglobin solutions, 539
Hemorrhage, 229, 286
Hemorrhagic diathesis, 229
Hemostasis, 229
Hemostatic defects, 242
Hemostatic disorders, 229
Hemostatic response, 229
Heparan sulfates, 129
Heparin, 234
Hepatocytes, 415
Herpes simplex virus, 43, 285
Herpes simplex virus type-1(HSV-1), 377
Heterocyclic inhibitors, 514
HIF-1, 134
HIF-1 binding site, 128
High altitude pulmonary edema, 485
High cholesterol diet, 283, 290
High-density lipoproteins (HDL), 241
Histamine, 2
Histocompatibility antigen class II (MHCII), 376
Homocysteine, 283
Hormone replacement therapy, 196
Hormones, 349
Host-defense, 323, 325, 434
Human atherosclerotic plaques, 64
Human cartilage, 12
Human hypertension, 260
Human neutrophils, 60
Human pulmonary hypertension, 274
Human recombinant hemoglobin, 541
Human saphenous vein endothelial cells, 313
Human umbilical vein EC, 41, 182
Hydrocortisone, 12
Hydrogen peroxide (H_2O_2), 59, 104, 384
Hydronephrotic kidney, 162
Hydroxyguanidines, 462
Hydroxyl radical, 434, 531
Hydroxylamines:
  formation of NO from, 461
Hydroxy-L-arginine:
  intermediate in arginine metabolism, 462
  as NO donor, 463
Hypercholesterolemia, 64, 240, 283, 289, 290, 294, 296
Hypercholesterolemic animals, 292
Hypercholesterolemic patients, 198
Hyperpermeability, 438
Hypertension, 173, 240, 251, 283, 285, 289, 351, 555
Hypertension:
  genetic, 251
  renovascular, 251
Hypertrophy, 257
Hypotension, 241, 323, 324
Hypoxanthine, 312
Hypoxia, 121, 268, 272, 273, 336, 435, 437
Hypoxia:
  vascular response to, 123
  regulation of vasoactive genes by, 123
  transcription factors regulated by, 135
  acute responses to, 139
  chronic responses to, 140
Hypoxia-inducible transcription factor-1 (HIF-1), 135
Hypoxia-reoxygenation, 316, 429
Hypoxic pulmonary vasoconstriction, 272, 544
Hypoxic vasoconstriction, 269, 272

ICAM-1, 41, 100, 306, 312
ICI 182780, 199
IFN-γ, 113, 140, 376, 428
IGF, 129
I_{KB} kinase, 104, 430
$I_{KB\alpha}$, 104, 289
I_{KB}-α:
  induction of, 110
I_{KB}-protease, 326
Il-1α, 113
IL-1β, 140, 158, 163, 204
IL-13, 382
IL-17, 404
IL-2, 416
IL-4, 43, 383
IL-6, 101, 131
IL-8, 101, 134, 426
Ileus, 432

I/8 *Index*

Iloprost, 367
Imidazole, 506, 514
Immune response, 456
Immunoglobulin superfamily, 312
Immunomodulator, 413, 416
Immunosuppressive, 413
Impaired vascular NO production, 555
In situ hybridization, 415
Inactivated hemagglutinating virus of
 Japan, 576
Indazoles, 506, 515
Indomethacin, 165, 350, 414
Inducible nitric oxide synthase, 26, 113, 195
Inducible nitric oxide synthase:
 tissue distribution, 51
 regulation by oxygen tension, 126
Infant mortality, 349
Infarct size, 59
Infarction, 230
Inflammation, 328
Inhaled NO, 276, 478, 572
Inhaled NO:
 delivery, 487
 dosage, 487
 monitoring, 487
 toxicity, 487
Inherited dyslipidemia, 561
Inhibitors of phospodiesterase, 276
iNOS, 26, 40, 69, 107, 140, 162, 165, 209,
 241, 292, 352
iNOS:
 catalytic activities, 30
 gene therapy approaches with, 569
 NO production, 31
 oligomeric status, 28
 subcellular targeting, 53
iNOS cDNA, 576
iNOS enhancer, 430
iNOS gene:
 transcriptional activation of, 44
 promoter structure, 44
iNOS gene expression:
 induction of, 43
iNOS gene transfer:
 efficacy of, 576
 other applications of, 581
iNOS holo-enzyme, 28

iNOS inhibitor, 378
iNOS knock-out mice, 324, 325
iNOS protein, 576
Insulin-like growth factor-I, 418
Integrin receptor:
 expression, 231
Intercellular adhesion molecules, 283
Interferon-γ, 40
Interleukin (IL) 1β, 403
Interleukin-1, 40, 415
Interleukin-4 (IL-4), 376, 416
Interleukin-6, 414
Interleukin-8, 384, 417
Interleukin-10, 376
Interleukin-12, 376
Interleukin-13, 376
Intestinal hyperpermeability, 75
Intima, 562
Intimal hyperplasia, 173
Intimal hyperplasia:
 pathogenesis of, 569
Intimal thickening, 286
Intracellular signal molecules, 44
Intrapulmonary shunt fraction, 471
Intrauterine infection, 350
Intrinsic pathways, 232
IRE, 429
IRF-1, 45, 113, 429
Iron chelators, 134
Iron ion, 539
Ischemia, 230, 284
Ischemia/reperfusion, 287, 301, 302, 307,
 309, 310, 312, 313, 434,
Ischemic heart disease, 173
Ischemic myocardium, 308
Isoamyl nitrite, 457
Isobutyl-methylxanthine (IBMX), 383
Isolated rat hearts, 304
Isosorbide dinitrate, 456, 457
Isothioureas, 328, 506, 510
ISRE, 45

Janus kinase, 430
JNK, 100

Kallikrein, 232
Ketosis, 364

Kidney, 51
Krebs cycle, 435
Kupffer cells, 165

Labour, 351, 354
LAD coronary artery, 302
Laminar flow, 176, 230
Laminar shear stress, 181
Laminin, 129
L-arginine analogs, 506
L-arginine binding site, 40
L-arginine paradox, 294
L-arginine transport, 275
L-arginine, 7, 30, 198, 507
L-arginine:
    NO pathway, 10
    supplementing dietary, 571
Lazaroids, 532
Lazaroid class of synthetic antioxidants, 523
L-citrulline, 17, 507
LDL cholesterol, 283
LDL oxidation, 41
LDL, 64
Left ventricular failure, 257, 258
Leishmania major, 325
Lentivirus, 556
Leukocyte adherence, 287, 305, 311
Leukocyte adhesion molecules, 134
Leukocyte infiltration, 318
Leukocyte-endothelial interaction, 310
Leukocytes, 285, 571
Lewis rat, 378
LHRH, 161
Lipid peroxidation inhibitors:
    mechanism, 532
    assays, 532
Lipid peroxidation, 288, 523
Lipopolysaccharide (LPS), 39, 241, 428
Lipoprotein (a), 283
Liposome-mediated uptake, 573
Liposomes, 576
Lipoxygenase, 231
L-NAME, 102, 128, 186, 188, 327, 364, 419, 542
L-NIL, 159, 162, 419
L-NMMA, 7, 159, 188, 327, 403, 419

L-NNA, 20
Long term potentiation, 54, 221
Low-density lipoprotein (LDL), 241, 296
LPS, 39, 53, 140, 156, 202
L-selectin, 312
L-thiocitrulline, 509
Lung capillary bed, 267
Lung circulation, 267
Lung disease, 268
Lung, 51, 471
Lung transplantation, 484
Lysophosphatidylcholine, 40, 41

Macrophages, 74, 113, 164, 339, 456, 465
MAHMA/NO, 465
Malaria, 350
MAP kinases, 44, 100, 108
Mass spectrometry, 7
Mast cells, 438
Maternal hypertension, 364
Matrix metalloprotease (MMPs), 13, 358, 418
MCP-1 expression, 292, 294
MCP-1, 287
M-CSF, 101
MDF, 44
Mdx mice, 52
Mechanism-based NOS inhibitors, 507
Media, 562
Medullary interstitial space, 254
Megakaryocyte, 237
MEKK1, 108
Mercaptoethylguanidine (MEG), 440
Mercaptopropionyl glycine (MPG), 308
Mesenteric circulation, 305
Mesenteric microcirculation, 312
Mesenteric microvasculature, 305
Mesenteric perfusion, 440
Metal complexes of NO, 456, 459
Metalloproteinases, 284, 293, 350, 359, 398
Methemoglobin, 490, 539
Methotrexate, 420, 506
Methylated proteins, 296
Methylene Blue, 4
MHC class II antigens, 113
Microcirculation, 308, 310
Microglia, 376, 379, 384, 387

Microvascular endothelial dysfunction, 305
Microvascular endothelium, 311, 312
Microvasculature, 285
Mitochondria, 62, 73
Mitochondrial dysfunction, 387
Mitochondrial respiration, 325
Mitogenic factors, 571
Mitral valve surgery, 479
Molecular conjugates, 557
Molsidomine, 242, 459
Monocrotaline, 273
Monocyte:
    adherence, 287, 290
    infiltration, 290
Monocytes, 189
MRL/lpr mice, 419
MSP, 43
Mucosa, 426, 438
Mucosal barrier, 433
Multiple organ dysfunction syndrome
    (MODS), 323, 326
Multiple sclerosis, 375
Mutagenesis, 431
Myelin, 375, 381
Myelin basic protein, 376
Myocardial contractility, 257
Myocardial damage, 257
Myocardial depressant factor (MDF), 316
Myocardial function, 545
Myocardial infarction, 242, 283, 284
Myocardial ischemia/reperfusion, 59,
    308, 315, 316
Myointimal hyperplasia, 286, 289
Myometrial contractility, 350
Myometrium, 351, 354
Myristoylation, 54

nNOS:
    spectral characterization, 20
N-(1-iminoethyl)-L-ornithine, 509
Na^+/K^+ ATP-ase, 70
N-acetylcysteine (NAC), 100, 105, 113, 240,
    384
NAD, 69
NADH oxidase, 101
NADPH oxidase, 132
NADPH, 7

NADPH-cytochrome P-450 reductase, 18
N-allyl-L-arginine, 509
NANC nerves, 9
Necrotic core, 284, 293
Neointimal lesion formation, 241
Neointimal SMC, 570
Neonatal and infant respiratory failure, 480
Neonatal mortality, 349
Neopterin, 328, 379
Neovascularization, 188
Neurodegeneration, 524
Neuronal injury, 60, 74
Neuronal nitric oxide synthase:
    bidomain structure, 18
    flavoprotein domain, 18
    heme domain, 18
    oligomeric status of, 19
Neuronal NOS, 258
Neuronal NOS knockout mice, 209
Neuronal NOS mutant mice:
    pyloric stenosis, 213
    airway responsiveness, 214
    role in renin release, 215
    aggressive behavior, 215
Neuronal postsynaptic density protein, 21
Neuronal tyrosine hydroxylase, 43
Neurotoxicity, 523
Neutrophil amplification, 301
Neutrophil infiltration, 315, 316
Neutrophils, 53, 238, 285, 287, 306,
    405, 438
NF_{-KB} activation, 107, 294
NF_{-KB}, 44, 100, 103, 138, 289, 306, 326,
    383, 384, 385, 398, 405, 429, 430
NF_{-KB}:
    enhanced nuclear translocation of, 109
N^G-monomethyl-L-arginine, 7
N^G-nitro-L-arginine, 20
N-hydroxy-L-arginine, 507
Nicotinamide, 384
Nitrate/nitrite, 539
Nitrates:
    NO formation from, 458, 461
Nitration of tyrosine residues, 62
Nitric oxide, 3, 17, 69, 130, 155, 229
Nitric oxide:
    activation of the COX enzymes by, 158

and the regulation of vasoactive genes, 121
  cyclic strain upregulation of, 182
  gene transcription, 99
  inhibition of PG synthesis, 164
  in the treatment of intimal hyperplasia, 571
  toxicity of, 523
Nitric oxide and tissue injury, 59
Nitric oxide chemistry:
  conversion to nitrite, 460
  diazonium salt formation, 456
  nitrosating agents, 456
  nylong form, 455
  reaction with secondary amines, 456, 464
  reaction with superoxide, 459, 461
Nitric oxide donors, 305, 455
Nitric oxide sensor, 132
Nitric oxide synthase expression:
  regulation by hemodynamic forces, 171
Nitric oxide synthase inhibitors, 505
Nitric oxide synthase, 129, 234
Nitric oxide synthase:
  isoforms of, 17
  gene regulation, 39
  modulation by PGs, 165
  regulation by 17β-estradiol, 195
  oxidation of hydroxyarginine by, 463
  gene therapy of vascular diseases, 556
  genetic approach to, 209
Nitric oxide-derived oxidants, 59
Nitrites:
  formation from NO, 460
  inorganic, 460
  NO formatin from, 458
Nitro-cysteine, 526
Nitro-glutathione, 526
Nitrogen dioxide, 490
Nitrogen oxides, 17
Nitroglycerin, 242, 456, 457, 458
Nitrosamines:
  formation from nitrites/nitrates, 458
  potential for NO-donation, 459
Nitrosation, 455, 456, 460, 461, 463
Nitrosonium ion, 59
Nitrosothiols, 100, 109
Nitrosylation, 288

Nitrotyrosine, 64, 241, 293, 379
Nitro-tyrosine residues, 524
Nitrous acid, 460
Nitrous oxide:
  formation from nitroxy, 459, 463
Nitrovasodilators, 3
Nitrovasodilators:
  amyl nitrite, 456, 457, 458
  isosorbide dinitrate, 456, 457
  nitroglycerin, 456, 457, 458
  sodium nitroprusside, 456, 457, 459
Nitroxyl:
  formation from Angeli's salt, 463
  formation from cyanarmide, 463
  formation from furoxans, 461
  formation from hydroxylamines, 456, 463
  formation from oximes, 462
  nitrous oxide from, 459
  oxidation to NO, 463
NMDA, 10, 164
NMDA receptor, 52, 60
N-methyl-L-arginine, 506
N-nitro-L-arginine, 506
nNOS, 40, 209, 556
nNOS:
  alternatively spliced, truncated form of, 22
  mice deficient in, 209
  subcellular targeting, 52
  superoxide production, 22
  tissue localization, 21
NO, 3, 140, 187, 195, 479, 577
NO:
  as an anti-inflammatory mediator, 113
  comparative toxicity of, 71
  delivered by inhalation, 471
  role in the regulation of vascular tone, 555
NO bioavailability, 562
NO deficiency, 239
NO donor, 72, 101, 158, 188, 242, 243, 276, 287, 306, 310, 313, 315, 316, 357, 441, 572
NO gas, 139
NO synthase, 7
NO_2^-, 8
NO_3^-, 9

NO$_2$Arg, 159
NOCs, *see* diazeniumdiolates
NO-mediated cytotoxicity, 580
Non-cGMP-dependent effects, 112
Nonenzymatic glycation, 341
NONOATE, 439
NONOates, *see* diazeniumdiolates
Non-selective NOS inhibitors, 505
Nonsterodial anti-inflammatory drugs
   (NSAIDs), 405
Non-viral methods, 573
Non-viral vectors, 557
Norepinephrine, 161
Normocholesterolemic animals, 292
NOS, 173
NOS:
   active site models, 516
   electron transfer, 507
   expression of, 40
   shear stress induced expression of, 181
   transcriptional regulation of, 186
NOS activity:
   autoregulation, 127
NOS autoinactivation, 507
NOS catalytic domain, 507
NOS dimers, 507
NOS inhibitor, 327
NOS inhibitors:
   approaches to, 505
NOS-I (nNOS), 17, 195, 505
NOS-II (iNOS), 17, 195, 202, 505
NOS-II:
   protein structure, 26
   inhibited by 17β-estradiol, 201
NOS-III (eNOS), 17, 22, 195, 505
NOS III:
   mRNA, 273
   protein, 273
NOS III-deficient mouse, 251, 271
NSAID, 159, 163

Obstructive pulmonary disease, 272
Occlusion/reperfusion, 305
Oct-2, 43
Oestrogen, 349
Oligodendrocyte, 375, 379, 386, 387
Organ injury, 323

Organic nitrates, 242
Organic nitrites, 456, 458
Osteoarthritis, 397, 414
Osteoarthrosis, 397
Osteopontin, 284
Ovariectomized rats, 202
Ovariectomy, 196
Oxatriazoles, 459
Oxidant injury, 484
Oxidant stress, 429
Oxidant-sensitive transcription factors, 100
Oxidative burst, 307
Oxidative modification of LDL, 199
Oxidative processes, 70
Oxidative stress, 59, 256, 261, 285, 287,
   289
Oxidized forms of nitric oxide, 59
Oxidized lipoprotein, 284, 287, 290
Oxidized low density lipoproteins, 306
Oximes:
   NO release from, 462
   review of chemistry, 456
   vasoactivity, 462
Oxygen, 121, 486
Oxygen delivery, 325
Oxygen free radical, 429
Oxygen sensor, 132
Oxygenation, 471
Oxyradicals, 73

P selectin, 237
p21 ras, 100, 109, 134
p50 (NF-$_{KB1}$) 103
p52 (NF-$_{KB2}$) 103
p53, 292
p65 (Rel A), 103
PAF, 134, 432
PAI-1, 233
Pancreatic islet cells, 51
Parallel-plate flow chamber, 176
Parkinson's disease, 60
PARS, 69
PARS:
   potential regulation of iNOS expression,
   85
PARS activation, 74
PARS cleavage, 78

PARS knockout mouse, 88, 436
Partial reduction of oxygen, 59
Parturition, 352
Paw inflammation, 162
PDGF, 569
PDGF-B, 121, 124
PDTC, 100, 105, 113, 399
PDZ, 21
PDZ motif, 52
PECAM-1, 312
Penicillamine methyl ester, 523
Pentagastrin, 431
Pentoxifylline, 383
Perfusion-bioassay, 5
Peripheral nitregic nerves, 51
Peripheral vascular disease, 173, 242
Peristalsis, 440
Permeability, 283
Peroxidase, 156
Peroxynitrate anion, 287, 293
Peroxynitrite Reactivity:
    molecular mechanisms, 54
Peroxynitrite scavengers, 523
Peroxynitrite, 22, 59, 69, 104, 155, 160,
    239, 241, 313, 316, 325, 379, 434, 523
Peroxynitrite:
    DNA injury, 72
    single strand breakage, 72
    formation by superoxide, 459, 461
    reaction with sugars, 461
    oxidative biochemistry of, 60
Peroxynitrite-mediated nitration of
    tyrosine, 63
Peroxynitrous acid, 61, 436, 438, 525
Persistent pulmonary hypertension of the
    newborn, 480
PGE_1, 189
PGE_2, 156, 359, 401
$PGF2\alpha$, 156
PGG_2, 156
PGH_2, 156
PGI_2, 134
Phenylimidazoles, 506
Phosphatases, 386
Phosphodiesterase, 439
Phosphoinositol-specific phospholipase
    C (PLC), 238

Physical forces, 171
Pial vessel, 220
Piloty's acid, 456, 463
PIN, 22
PKC activation, 313
PKC inhibitors, 134
Placenta, 350, 355
Plaque instability, 293
Plaque rupture, 284, 286
Plasmin, 233
Plasminogen activator inhibitor, 233
Plasminogen activator:
    tissue-type, 233
    urokinase-type, 233
Plasminogen, 233
Platelet aggregation, 122
Platelet deposition, 231
Platelet factor 4, 231, 240
Platelet inhibitors, 238
Platelet recruitment, 237
Platelet:
    activation, 229
    adhesion, 229, 231, 285
    aggregation, 229, 231, 285
Platelet-derived growth factor, 231, 284
Platelets, 8, 112, 284, 438
PLGF, 131
PMA, 43
PMNs, 308
Pneumonia, 350, 482
Poiseuille's law, 231
Poly (ADP) ribosyltransferase (PARS),
    325, 383
Poly (ADP-ribose) synthetase, 69
Polyamines:
    NO donors from, 465
Polymorphonuclear (PMN) leukocytes, 240,
    301, 309, 417
Postmenopausal women, 196
Postoperative pulmonary hypertension, 479
Post-prandial hyperemia, 439
Post-translational processing, 236
Pre-eclampsia, 344, 349, 362
Pregnancy, 344, 349, 352, 363, 368
Prekallikrein, 232
Premature atherosclerosis, 283
Pressure-induced natriuresis, 252, 254

Preterm labour, 349
Primary pulmonary hypertension:
  animal model, 269
Progesterone, 349, 359
Progressive obliterative pulmonary vascular
  disease, 475
Pro-inflammatory genes, 102
PROLI/NO, 465
Prostacyclin, 129, 231, 239, 268, 479
Prostaglandin El, 479
Prostaglandin, 417
Prostaglandins biosynthesis, 156
Prostaglandins, 155, 359
Prostanoid contractile factors, 255
Prostanoids, 129
Protein C, 233
Protein kinase C, 382
Protein tyrosine phosphatases, 100
Proteinuria, 351
Proteoglycan, 398, 419
Prothrombin, 232
Prothrombinase, 232
PSD-93, 21
PSD-95, 21, 52
P-selectin, 306, 310, 312
Pulmonary arterial pressure, 267, 471
Pulmonary arteries, 274
Pulmonary circulation, 267
Pulmonary conduit arteries, 273
Pulmonary disease, 480
Pulmonary fibrosis, 268
Pulmonary hemodynamics, 271, 274
Pulmonary hypertension, 121, 267, 471, 474,
  539, 543
Pulmonary hypertension:
  experimental, 272
  postcapillary, 268
  precapillary, 268
Pulmonary hypertensive crises, 471
Pulmonary microvessels, 268
Pulmonary parenchymal disease, 471
Pulmonary remodeling, 271
Pulmonary transvascular flux, 328
Pulmonary vascular disease:
  primary pulmonary hypertension, 473
  pulmonary thromboembolism, 474

Pulmonary vascular resistance (PVR),
  267, 268, 274
Pulmonary vascular tone, 270, 274
Pulmonary vasoconstriction, 477
Pulmonary vasodilation:
  selective, 471
Pulmonary vessel, 128
Pulsatile flow, 172
Pulsatile stretch 257
Pyelonephritis, 350
Pyloric stenosis, 425
Pyrrolidine dithiocarbamate (PDTC), 105,
  429, 430

Quantitative reverse transcriptase PCR, 182

Radiation sensitizers, 459
Radical chain propagation, 70
Radiocontrast nephropathy, 338
Raf, 44, 128
Rap 1β, 237
Ras, 44
Rat islet cells, 163
RAW 264.7 cells, 14
Reactive oxygen intermediate, 133
Reactive oxygen species, 104
Rebound pulmonary hypertension, 491
Recombinant endothelial nitric oxide
  synthase gene expression:
  therapeutic implications, 555
Recombinant human superoxide dismutase
  (hSOD), 302
Recombinant viruses, 556
Redox state, 238
Redox-regulated transcription factors, 386
Redox-responsive transcriptional pathways,
  287
Reduced thiols, 240
Reductase domain, 507
Refractoriness, 365
Regional blood flow, 173
Regression, 292
Rel B, 103
Relaxin, 359
Rel-related proteins, 103
Remyelination, 386

Renal function, 252
Renovascular hypertension, 258
Reoxygenation, 305
Reperfusion injury, 60, 81, 301
Reproductive cycle, 349
Respiratory distress syndrome, 481
Restenosis, 289, 569
Restenosis:
    inhibition by diazeniumdiolates, 466
Retinal circulation, 242
Retinoid, 376
Retrovirus, 556, 574
Retroviruses, 577
Rheologic properties, 230
Rheumatoid arthritis, 397, 413
Rheumatoid synovium, 401
Right ventricular dysfunction, 479
Rolipram, 383
Roussin's salts, 459
Rupture, 284

Sabra hypertensive-prone rat, 260
Saliva, 426
Salmonella, 428
Salt-sensitive hypertension, 251, 259
Sandwich mount procedure, 2
Scavengers of peroxynitrite, 70
Schwann cell, 387
Scleroderma, 141
Secondary pulmonary hypertension, 269
Selectins, 312
Sepsis, 323, 328, 426, 539
Septic shock, 140, 162, 241, 323, 324
Serine protease inhibitors, 429
Serine protease, 233
Serine-threonine protein kinase, 133
Serotonin, 2, 285
Serum cholesterol, 283
Severe left ventricular dysfunction, 490
Sexual steroid hormone, 195, 349
Shear rate, 230
Shear stress, 5, 122, 171, 257, 272
Shear stress:
    morphological changes, 179
Shock, 425
Sickle cell disease, 121, 141, 485

Sickle hemoglobin, 471, 485
SIN-1, 113, 460, 459
Site-specific gene targeting, 555
Skeletal muscle, 51
SMC migration, 284, 570
SMC proliferation, 569
Smooth muscle cells, 74
Snake venom-like protease, 403
SNAP, *see* S-nitroso-N-acetyl-penicillamine
S-nitrosoalbumin, 464
S-nitrosocaptopril, 464
S-nitrosoglutathione, 113, 416
S-nitrosohemoglobin, 539
S-nitroso-N-acetylcysteine, 238
S-nitroso-N-acetyl-penicillamine, 138, 527
S-nitroso-N-acetyl-penicillamine:
    peptide derivatives, 464
    structure, 457
S-nitrosothiols, 238, 456, 463, 466, 539
S-nitrosylation, 100, 109, 134
SNP, *see* sodium nitroprusside
SOD, 5, 524
Sodium channels, 70
Sodium nitrite, 460
Sodium nitroprusside, 138, 293, 456, 457,
    459, 479
Sodium retention, 254
Sodium salicylate, 11
Sp-1, 188
SPER/NO, 465
Sphincter of Oddi, 433
Sphincteric tone, 433
Sphingomyelin, 134
Spinal cord, 51
Splanchnic ischemia, 307
Splanchnic ischemia/reperfusion, 301, 315,
    316
SPM-5185, 572
Spontaneous hypertension, 254
Spontaneously hypertensive (SH) rats, 175
S-substituted isothioureas, 328
Stabilization of NF-$_{KB}$, 107
Stable homodimer, 20
STAT, 100, 113, 430
Steroid, 441
Stomach, 51

Streptozotocin, 459
Stress ulcer, 426
Stroke, 285
Stromelysin, 13
Subarachnoid hemorrhage, 523, 533, 555
Subendothelial matrix, 231
Subendothelial space, 284
Substance P, 2, 188, 237
Sulfated proteoglycans, 144
Sulfhydryl groups, 70
Sulfhydryl oxidation, 523
Sulfhydryl scavengers:
    mechanism, 525
Sulfonated, 456
Superoxide anion, 5, 239, 285, 287,
    289, 293, 325, 384, 434, 439
Superoxide dismustase, 5, 70, 239, 308, 524
Superoxide production, 30
Superoxide radical, 59, 199, 294, 301, 306,
    307, 312, 313, 318, 343, 523
Superoxide radical release, 304
Surgical bypass grafting, 569
SV129 mice, 212
Sympathetic ganglia, 51
Synovial fibroblasts, 415
Synovial fluid, 414
Synoviocytes, 415
Synoviocytes:
    type A, 415
    type B, 415
Synovitis, 417
Synovium, 397, 415
Systemic blood pressure, 539
Systemic inflammatory response syndrome
    (SIRS), 323
Systemic sclerosis (scleroderma), 121, 141
Systemic vascular tone, 254

Tetracyclines:
    doxycycline, 13
    minocycline, 13
Tetrahydrobioproterin, 316, 379, 579
TGFβ, 44, 122, 131, 399, 569
TGF-α, 131, 382
Th1, 416
Th2, 416

Thiols:
    amyl nitrite and, 458
    furoxans and, 461
    nitroglycerin metabolism and, 458
Thionitrites, *see* S-nitrosothiols
THP-1, 296
Thrombin, 2, 232
Thrombocytopenia, 241
Thrombomodulin, 233
Thrombosis, 229, 269, 555
Thrombosis:
    inhibition by diazeniumdiolates, 466
Thrombospondin, 121, 129, 231
Thrombospondin-1:
    regulation by oxygen tension and nitric
        oxide, 128
Thromboxane A2, 285
Thrombus, 229, 340
Tirilazad, 523, 533
Tissue acidosis, 437
Tissue factor, 232
Tissue factor, 284
Tissue injury, 312, 434
Tissue necrosis, 306, 318
Tissue plasminogen activator:
    S-nitrosyl, 464
T-lymphocytes, 106
TNF-α, 113, 118, 325, 403
TNFβ, 382
TNF-RE, 45
Tocolytics, 357
Tolerance:
    to nitrates, 458
Topoisomerase-mediated repair, 72
Total anomalous pulmonary venous
    connection, 477
Transfection techniques, 574
Transforming growth factor beta (TGFβ),
    376
Transgene expression efficacy, 555
Transient ischemic attacks, 285
Transplant coronary artery disease, 285
Transplantation arteriosclerosis, 581
Transporter system, 237
Trifluoroperazine, 506
Tumor necrosis factor-alpha, 40, 414

TxA$_2$, 156
Tyrosine hydroxylase, 70
Tyrosine kinase, 429
Tyrosine kinases, 134, 386

U-46619, 271
Ubiquitin-proteasome pathway, 105
Ulceration, 284
Unstable angina, 284
Uremia, 242, 338
Urinary nitrate excretion, 295
Urinary nitrate, 261
Urinary tract infections, 53
Uterine arteries, 368
Uterine contractility, 351
Uterine smooth muscle, 352
Uterotonins, 350
Uterus, 51

Vascular cell targeting:
    cell type, 557
    specific promoters, 557
Vascular contractile failure:
    in circulatory shock, 79
Vascular dysfunction, 562
Vascular gene therapy, 555, 572
Vascular gene transfer, 555, 556
Vascular hyporeactivity, 325
Vascular hyporesponsiveness, 324, 545
Vascular inflammation, 100
Vascular injury, 229, 231
Vascular O$_2$ sensing, 132
Vascular remodeling, 140, 268, 269, 286
Vascular smooth muscle cells, 112, 140, 162
Vascular smooth cells:
    migration, 286
    proliferation, 290
Vascular tone, 216

Vascular tone:
    local regulation of, 122
Vasculoprotection, 200
Vasoactive intestinal peptide (VIP), 425
Vasoconstrictor substance, 261
Vasodilators, 471
Vasodilators:
    furoxans as, 461
    oximes as, 462
    streptozotocin as, 459
Vasomotor tone, 541
Vasopressin, 2
VCAM-1, 41, 100, 105, 306, 312
Vector technology, 555
VEGF receptor, 124, 128
VEGF, 121, 128, 129
VEGF:
    regulation by oxygen tension and nitric
        oxide, 127
Venular endothelium, 305, 310
Vessel bifurcation, 172
Viral DNA, 557
Viral methods:
    of gene delivery, 574
Viral vectors, 556
Viscosity, 176
Vitronectin, 231
Voltage dependent K$^+$ channel, 52
vWF, 231

W-7, 506
Weibel-Palade bodies, 237, 313
Wound healing, 439

Xanthine oxidase, 312
Xanthine-oxidase system, 306

ZO-1, 52